INTRODUCTION TO THEORETICAL PHYSICS

INTRODUCTION TO THEORETICAL PHYSICS

VOLUME II

BY

ARTHUR HAAS, Ph.D.

PROFESSOR OF PHYSICS IN THE UNIVERSITY OF VIENNA

TRANSLATED FROM THE THIRD AND FOURTH EDITIONS BY

T. VERSCHOYLE, M.C., B.Sc., A.R.C.S.

LONDON

CONSTABLE & COMPANY LTD

10–12 ORANGE STREET W.C. 2

1925

Printed in Great Britain by The Whitefriars Press, Ltd.,
London and Tonbridge.

TRANSLATOR'S NOTE

OWING to the rapid advances in Atomic Theory and in the experimental researches upon which it is based, Professor Haas has kindly supplied a brief summary of the more important work carried out between the publication of the German edition and of the present English edition of Vol. II. This summary has been placed under the heading of " Addenda," and is divided into sections numbered in accordance with the sections in Part III to which they refer. To draw the reader's attention to this new matter, the sections concerned in Part III have been marked with an asterisk.

As was the case with the first volume, the translator is very deeply indebted to Dr. R. W. Lawson, of the University of Sheffield, for the trouble taken in his revision of the manuscript, and for permission to use Dr. Lawson's translation of the author's " The New Physics " in the preparation of the present translation. Professor Haas's help and advice have been invaluable throughout. Finally, great thanks are due to Messrs. Constable & Co., Ltd., for undertaking the publication of this work; and the translator is especially grateful to both publishers and printers for the readiness with which they have met all his wishes, and for the attention that they have given to every detail.

T. VERSCHOYLE.

CONTENTS

PART III

ATOMIC THEORY

CHAPTER XI

THE ELEMENTARY QUANTA

CHAPTER XII

THEORY OF THE HYDROGEN ATOM

CHAPTER XIII

RÖNTGEN RAYS

CHAPTER XIV

Theory of the Elements

CHAPTER XV

General Theory of Spectra and of Atomic Structure

PART IV

THEORY OF HEAT

CHAPTER XVI

Statistics

CHAPTER XVII

Thermodynamics

PART V

THEORY OF RELATIVITY

CHAPTER XVIII

Special Theory of Relativity

CHAPTER XIX

Theory of Gravitation

APPENDIX

INTRODUCTION TO THEORETICAL PHYSICS

PART III

ATOMIC THEORY

CHAPTER XI

THE ELEMENTARY QUANTA

§ 87. The Elementary Quantum of Electricity.

IN the whole of theoretical physics there is no older conception than that of the *individual structure of matter*, according to which matter consists of minute, invisible, *individual particles*. This idea originated long ago with the Greek philosopher *Democritus*, who gave the name *atoms* to the ultimate invisible particles into which matter could be resolved. This term has also been retained in modern science for the *smallest particles of the chemical elements*, and it is to the building up of atoms to form so-called molecules that the newer chemistry, founded by *Dalton*, ascribes the multiplicity of chemical compounds.

Numerous physical phenomena, however, have clearly shown that *atoms* must undoubtedly *contain electrical charges*. Above all, the fact of *electrolysis* demonstrates that atoms are influenced by electrical forces. Moreover, after the establishment of the electromagnetic theory of light, the long known phenomenon of the *dispersion of light* effected by matter could not be explained without the assumption of a mutual action between electromagnetic waves and electrical charges contained in the atoms of matter. The hypothesis of an atomistic structure of matter thus necessarily led to the hypothesis of an *atomistic structure of electricity*, and hence to the conception of an *elementary quantum of electrical charge* (or elementary charge). Not only theoretical considerations, however, have given rise to this fundamental assumption of modern atomic theory, for a *direct empirical confirmation* is to be found in some *experiments of Millikan*, which have, moreover, rendered possible a direct and extremely accurate determination of the elementary quantum of electricity.

Millikan's experiments consisted in the observation with a microscope of *minute, individual particles of matter* carrying *feeble electric charges*, and moving under the two-fold influence of gravity and of a field directed vertically upwards.

This method of so-called "*individual observation*"[1] was devised in 1909 by *Ehrenhaft* and a little later also by Millikan, and was further developed and perfected in the following years.[2]

Millikan worked with *oil droplets* with a diameter of about 10^{-4} cm. The advantage of oil droplets, as compared with particles of other substances, is that they do not change in size owing to evaporation. By means of a spray-diffuser a fine mist of oil is blown into a chamber, and through a small hole in the bottom of the latter a droplet is from time to time admitted into a condenser placed beneath the chamber, the condenser consisting of two parallel horizontal plates. In consequence of the friction due to the spraying, these droplets are almost always electrically charged. The plates of the condenser can be connected through a commutator to a battery of accumulators giving some 10,000 volts. The particle in the interior of the condenser is illuminated by a strong source of light placed at one side, and it is at the same time observed by a microscope through a window in the condenser; it then appears as a bright point on a black background, although its true shape cannot be recognized. When the electric field is cut out, a particle of this kind slowly falls with a constant velocity, in consequence of its own weight and the strong frictional resistance. When, however, the battery is switched on, the particle slowly rises, again with a constant velocity, provided that the electric force is not much greater than that of gravity. (With the given size of droplet and potential difference the conditions are just right.) By always switching on or off at the correct moment, it is possible to keep a droplet moving up and down between the two plates of the condenser as long as is required.

On account of the large friction, the *velocity* of the particle in both the upward and the downward movement is *proportional to the effective force*, as follows both from mechanical considerations and by direct experimental confirmation. Denoting, therefore, the velocity of the downward motion by v_1 and that of the upward motion by v_2, the mass of the particle by m and its charge by Q, the acceleration due to

[1] The essential novelty of this method lay in the fact that it dealt with *individual* particles, whereas earlier methods had always dealt with the *mean value* of the charges on a *large number* of particles.

[2] A collected report of Millikan's researches is to be found in his book, "*The Electron, its Isolation and Measurement*" (Chicago, 1917). His first paper was published in August, 1909 (in the reports of the British Association Meeting at Winnipeg); Ehrenhaft's first communication appeared in March, 1909 (in the *Anzeiger der Wiener Akademie*).

gravity by g, and the electric strength of field (*i.e.*, the quotient of the potential difference and the distance between the plates) by E, we must have the following relation :—

$$(1) \qquad \frac{v_2}{v_1} = \frac{QE - mg}{mg}.$$

The velocity is easily determined by measuring the time taken by the particle in traversing the interval between two cross-threads in the observing microscope. Table I gives the values in seconds found in one experiment by Millikan for the times (t_1) required for the downward motion, and for the times (t_2) required for the upward motion between the cross-threads ; in this case the distance between the threads corresponded to an actual interval of rise or fall of 0·5222 cm.

TABLE I

t_1	t_2	t_1	t_2
13·6	12·5	13·8	34·6
13·8	12·4	13·7	34·8
13·4	21·8	13·8	16·0
13·4	34·8	13·6	34·8
13·6	84·5	13·5	34·6
13·6	85·5	13·4	21·9
13·7		13·8	
13·5		13·4	
13·5		Mean value 13·595	

As the table shows, the velocity of the downward motion of the particle is always the same. On the other hand, with a given potential difference, the values of t_2 (and hence of v_2 also) vary intermittently, whilst earlier values recur later on during the up-and-down motion.

The jumps in the values of t_2 must obviously be due to *collisions* between the positively charged oil droplet and *air ions* resulting in an alteration in the charge of the moving particle. When a negative ion is taken up, the upward motion is retarded and hence the value of t_2 increased, whereas a decrease in the value of t_2 is due to the addition of a positive ion or the loss of a negative ion.

Now denoting by Q' the charge on the ion, and by v_2' the velocity of the upward motion after addition of this ion, we have in accordance with eqn. (1)

$$(2) \qquad \frac{v_2'}{v_1} = \frac{QE + Q'E - mg}{mg}.$$

Hence, on subtracting eqn. (1) from eqn. (2) we find

$$(3) \qquad \frac{v_2' - v_2}{v_1} = \frac{Q'E}{mg}.$$

In Table II the first column contains the values found for the velocity of the upward motion by dividing the interval of rise or fall, 0·5222 cm., by the values of t_2 given in Table I.[3] In the second column we find, in the first place, the differences between the successive numbers in the first column.

TABLE II

Velocity of the upward motion.	Difference.
$\frac{0·5222}{12·45} = 0·04196$	
	$\frac{0·01806}{2} = 0·00903$
$\frac{0·5222}{21·8} = 0·02390$	
	$\frac{0·00885}{1} = 0·00885$
$\frac{0·5222}{34·7} = 0·01505$	
	$\frac{0·00891}{1} = 0·00891$
$\frac{0·5222}{85·0} = 0·00614$	
	$\frac{0·00891}{1} = 0·00891$
$\frac{0·5222}{34·7} = 0·01505$	
	$\frac{0·01759}{2} = 0·00880$
$\frac{0·5222}{16·0} = 0·03264$	
	$\frac{0·01759}{2} = 0·00880$
$\frac{0·5222}{34·7} = 0·01505$	
	$\frac{0·00891}{1} = 0·00891$
$\frac{0·5222}{21·85} = 0·02396$	

Table II shows us, however, that the above differences always have one of two values, of which the one is exactly twice the other (in other experiments we find also triple and quadruple values). In the case of a given droplet the

[3] When two approximately equal values of t_2 follow successively in Table I, the mean value has been taken for Table II.

quantities m, v_1, E, and g are to be regarded as constants and hence it follows, in accordance with eqn. (3), that the *charges on the air ions* must always represent *integral multiples of an elementary charge.* Thus the data of Table II afford a *direct experimental proof of the atomistic structure of electricity.*

Moreover, the actual charge on the droplet (Q) may readily be compared with the ionic charge (Q'), and hence with the elementary quantum. We have only to start from eqn. (1), which gives the relation

$$(4) \qquad \frac{v_2 + v_1}{v_1} = \frac{Q\,E}{m\,g};$$

whence, taking eqn. (3) into account, we find

$$(5) \qquad \frac{v_2 + v_1}{v_2' - v_2} = \frac{Q}{Q'}.$$

Since the velocities v_1, v_2, and v_2' can be measured, the charge Q can therefore be compared with the smallest charge, as integral multiples of which the ionic charges manifest themselves. Millikan found that the *droplet's charge* also always proves to be an *exact integral multiple of the same elementary charge.* With the greatest accuracy he was able to detect all integral multiples of the elementary charge up to about 50, and he showed that the number of elementary quanta associated with an oil droplet could even amount to 150. As the multiplying factor increases, the accuracy of the count of course decreases, for the limits of error in this method are about a half *per cent.*, so that, with a charge of more than 200 elementary quanta, the error is already more than one electrical quantum.

Now it is evident from eqn. (1) that the absolute magnitude of the *elementary quantum of electricity* itself can be calculated, as soon as the *weight of a droplet* is known. Moreover, this weight can be determined with the aid of a relation connecting the constant velocity of fall with the radius of the particle. As already found by *Stokes* in 1845, the velocity of fall of a sphere of radius a and density ϱ in a medium of density ϱ', and with a coefficient of viscosity η, is given by the formula

$$(6) \qquad v = \frac{2}{9}\,\frac{g\,a^2\,(\varrho - \varrho')}{\eta}.$$

Millikan and his collaborators now arranged the experiments which served to determine the weight of a droplet in such a way that the assumptions, upon which the validity of Stokes's law is based, were satisfied with the greatest possible exactness.[4] In so far, however, as these assumptions could not be completely satisfied, the errors thereby caused were again accurately investigated. By means of the law of fall thus modified,[5] the radius of the droplets could be calculated with the greatest accuracy, and therefore, the density being known, their exact weight also followed.

Measurements extending over many years enabled Millikan in this way to determine the value of the *elementary quantum of electricity* (now generally denoted by e) with an *accuracy of one in a thousand*; as a mean of his experimental results he found

$$(7) \quad e = 4{\cdot}774 \times 10^{-10} \text{ electrostatic units } (\pm\, 0{\cdot}004 \times 10^{-10}).$$

It does not here seem necessary to enter any further into the details of Millikan's measurements, for in later sections various other methods will be described, which, quite independently of his researches, lead indirectly to exactly the same value for e.[6]

Historically, it may be noted that the first direct determinations of the elementary quantum of electricity were made in 1897 and 1898 by *Townsend* and *J. J. Thomson.* These researches, which gave for e an approximate value of 3×10^{-10} electrostatic units, were not concerned, however, with individual particles of matter, but with slowly falling *clouds of mist* in which *air ions* formed condensation nuclei for the drops of water. With the aid of Stokes's law of fall, the radius of the water drops could be derived from the

[4] It was particularly important that the oil droplets should be perfectly spherical—a result obtained by special artifices.

[5] For further details, *vide* Millikan's book quoted above.

[6] It also follows from this that objections which have been raised against Millikan's method can in no way call in question the existence of an elementary quantum of electricity, which has been definitely proved by numerous other methods.—Objections have been made to Millikan's work by *Ehrenhaft* and his pupils, who have experimented with spherical particles of various substances (mercury, silver, gold, oil, *etc.*) in various chemically inert gases (nitrogen, argon); the radii of the particles lay between 5×10^{-5} cm. and about 5×10^{-6} cm. (the limit of observability). From these researches, some of which were published before Millikan's first paper, Ehrenhaft concluded that as the radius of the particle (and therefore its electrostatic capacity) decreased, the charge also decreased to such an extent that it might be considerably less than the elementary quantum. On the other hand, Ehrenhaft's results have been ascribed to conceivable *variations in density*, which might be due to a spongy or flaky structure of extremely minute particles (smaller than Millikan's oil droplets) or to an adsorption of gas. *Cf.* Ehrenhaft's comprehensive papers in the *Ann. der Phys.*, **56** (1918), pp. 1–80, and **63** (1920), pp. 773–815; also Th. Sexl, *Zeitschr. f. Phys.*, **16** (1923), pp. 34–41. For the opposing point of view, *cf.* the summaries in *Die Naturwissenschaften* by R. Bär [**10** (1922), pp. 322 and 344], and by E. Regener [**11** (1923), p. 17].

velocity of fall of the mist, and hence, for a known weight of mist, the number of drops could be ascertained. Since, furthermore, the total electric charge on the cloud could be measured, the charge on an individual ion could be calculated.[7] By arbitrarily identifying this charge with an elementary quantum, a value of the right order of magnitude was obtained for the latter.

§ 88. Electrolysis, and the Mass of the Hydrogen Atom.

Long before the absolute values of atomic masses were known to physicists, chemists had been in a position to determine the *relative weights of the atoms, i.e.*, the *ratios* existing between their masses, with the help of the so-called *Law of Multiple Proportions*, which was established by *Dalton* in 1805. As Dalton discovered, and as he formulated in this well-known law, a definite *number*, characteristic of the element, can be allotted to every chemical element in such a way that the quantities of the elements contained in a chemical compound are related to each other as *integral multiples* of the numbers characteristic of the elements concerned.

If, following convention, we make the characteristic number for *oxygen* equal to 16 *exactly*, we then obtain the so-called *atomic weights of the elements*. The smallest atomic weight is that of hydrogen—1·008 (or, more accurately, 1·0077)—the largest that of uranium (238·2). We need not enter here further into the methods by which chemists determine atomic weights ; we shall have something to say about these methods later (in connection with Avogadro's Law).

The number by which the figures representing the atomic weights must be divided, in order to obtain the absolute masses of the atoms in grams, is nowadays generally known as the *Loschmidt number*. It is so called in honour of the physicist who, in 1865, was the first to determine approximately the absolute magnitude of atoms, by means of a theoretical method based on the gas laws (which we shall discuss later). Hence, denoting the mass of the hydrogen atom in grams by m_{H} and the Loschmidt number by L, we have

$$(1) \qquad m_{\mathrm{H}} = \frac{1{\cdot}0077}{L} \text{ grams.}$$

In like manner, we obtain the mass of any given atom by dividing the atomic weight of the element in question by the

[7] For further details, *vide, e.g.*, Millikan's book, ch. iii.

Loschmidt number. The number of grams of a substance equal to the number which represents the atomic or the molecular weight of the substance is known as a *gram-atom* or as a *mol* (gram-molecule) respectively. Thus, for example, a gram-atom of oxygen is 16 grams, a mol of water about 18 grams, and so on. The Loschmidt number therefore gives the number of atoms in the gram-atom, or of molecules in the mol of any given substance. If, now, we know the value of the elementary quantum of electricity, we are then able to calculate the Loschmidt number with great accuracy, on the basis of the results of *electrochemical measurements.*

In 1833 *Faraday* derived the *fundamental law of electrochemistry* from his observations on *electrolysis.* This law states that the *amount* of a particular substance *produced* electrolytically in a given time is completely determined by the *current* and directly *proportional* to it. Furthermore, the amounts of the constituents liberated from different electrolytes by a current of given strength are "*chemically equivalent,*" *i.e.*, they are related to each other as the quotients of the atomic weight and the chemical valency. Thus a current which liberates 1·008 grams of univalent hydrogen in a unit of time will liberate 8 grams of bivalent oxygen of atomic weight 16 in the same interval, *etc.*

The fundamental law of electrochemistry can be very simply explained on the assumption that, in an electrolyte, the molecules are entirely or partially split up, or *dissociated,* into oppositely charged components—the so-called *ions*—and that the conduction current in the electrolyte consists in a *transport* of such ions. The positive ions thus wander in the direction of the current to where it leaves the liquid, at the so-called *cathode,* whereas the negative ions wander in the opposite direction to where the current enters the liquid, at the so-called *anode.* Faraday's Law then follows at once on the hypothesis that the charge on all univalent ions is one elementary quantum of electricity, that on bivalent ions two, and that on trivalent ions three elementary quanta, *etc.*

Denoting the number of atoms liberated in a unit of time at the cathode by N and the valency by z, the current, which is the quantity of electricity passing the cathode in unit time, is given by

$$(2) \qquad I = N\,z\,e.$$

On the other hand, the mass of a single atom is given by the quotient of the atomic weight and the Loschmidt number, *i.e.*, by the quantity A/L, where A is the atomic weight. The amount of the substance in question liberated in a unit of time is therefore determined by the relation

$$(3) \qquad M = \frac{NA}{L},$$

or, in view of eqn. (2), by

$$(4) \qquad M = \frac{A}{z}\frac{I}{L\,e}.$$

Now this equation is an actual expression of the fundamental law of electrochemistry. For, since L and e are universal constants, it follows on the one hand that, with a given substance, and therefore with given values of A and z, the mass (M) liberated in unit time is proportional to the current (I) ; and on the other hand it follows that, with a given current, the value of M is proportional to the quotient of atomic weight and valency.

Since, in accordance with what we have said above, the quantity M/A represents the number of gram-atoms, we also recognize from eqn. (4) that the *product of the Loschmidt number and the elementary quantum of electricity* simply represents the *current*, measured in electrostatic units, which *liberates one gram-atom of a univalent element* in a unit of time.

Very accurate measurements, in which principally silver has been used, have demonstrated that a gram-atom of a univalent substance would be deposited in a unit of time by a current of 96,494 ampères. In order to express this current in electrostatic measure, we have to multiply it by one-tenth of the velocity of light [according to § 64 eqn. (9)]. Since, by § 76 eqn. (1), the velocity of light is equal to $2{\cdot}999 \times 10^{-10}$ cm. sec.$^{-1}$, we thus find

$$(5) \qquad L\,e = 2{\cdot}894 \times 10^{14} \text{ E.S.U.}$$

Hence, introducing the value already given for the elementary quantum of electricity, we obtain for the *Loschmidt number*

$$(6) \qquad L = 6{\cdot}062 \times 10^{23},$$

with an error which can scarcely be more than 1 in 1,000.

It follows from eqn. (1) that the *mass of the hydrogen atom* is

$$(7) \qquad m_H = 1{\cdot}662 \times 10^{-24} \text{ grams.}$$

For the ratio of charge to mass, *i.e.*, for the so-called *specific charge of the ionized hydrogen atom*, we have, in accordance with eqn. (1),

$$(8) \qquad \frac{e}{m_H} = \frac{L\,e}{1{\cdot}0077},$$

or, by eqn. (5),

$$(9) \qquad \frac{e}{m_H} = 2{\cdot}871 \times 10^{14} \text{ E.S.U./gram.}$$

§ 89. Negative Corpuscular Radiation and the Electron.

In 1858 *Plücker* discovered the remarkable rays which are emitted by the cathode in highly evacuated Geissler tubes, and which are therefore known in general as *cathode rays*. Their properties were more closely investigated, particularly by *Hittorf* (1869) and *Crookes* (1879), both of whom made the surprising discovery that these rays, which ordinarily travel in *straight lines*, are very easily *deflected* by a *magnet*. Moreover, *Goldstein* (1876) showed that they may also be *deflected* by an *electric field*.[1]

As was first pointed out by *J. J. Thomson* and *Kaufmann* (1897), all these properties of the cathode rays can be simply explained on the assumption that they consist of quickly moving *electrically charged particles*, which at the same time possess *inertial mass*. Their rectilinear propagation follows from their inertia, and the observed deflections both in a magnetic and in an electric field necessarily follow from their electrical charge; the sense of the deflection in both cases shows that the *charge* must be *negative*. Unlike rays of light or heat, which are forms of electromagnetic waves, the cathode rays therefore represent a so-called *corpuscular radiation* consisting of a stream of inertial particles.

Now (as was shown in § 71) it follows from the theory of electricity, without the aid of any kind of atomistic hypothesis, that, in a *transverse homogeneous magnetic field*, a moving electrically charged mass must describe a *circular path* with constant velocity in a plane at right angles to the magnetic lines of force. In accordance with § 71 eqn. (11), the radius of the circle is given by the simple relation

$$(1) \qquad r = \frac{c\,v}{\gamma H},$$

where H is the magnitude of the magnetic strength of

[1] The study of the cathode rays was greatly aided by an artifice of *Lenard's*, which consisted in letting the rays escape from the tube through a piece of thin metal foil inserted in the wall of the tube.

field, γ the specific charge of the moving mass, and v its velocity.

It follows likewise from the theory of electricity (according to § 71), irrespective of any kind of atomistic considerations, that, in a *transverse homogeneous electric field*, a moving charged body describes a *parabola*, the curvature of which depends only on the specific charge and the velocity, in addition to the electric field-strength. If we take the original direction of the path as x-axis, the direction of the electric force as z-axis, and the position of the moving body at the time $t = 0$ as origin, we have then [by § 71 eqn. (3)] the relation

$$\frac{z}{x^2} = \frac{1}{2}\frac{\gamma}{v^2}E, \tag{2}$$

where E is the magnitude of the electric field-strength.

As a matter of fact, observations show that the particles in the cathode rays do move in a circular path in a transverse magnetic field, and in a parabola in a transverse electric field. Hence, if we apply the two equations to cathode rays, and regard γ and v as unknown, we may calculate the velocity and the specific charge of the particles from the known measured values of the magnetic and electric strengths of field.

Exceedingly large values are obtained for the *velocity*, reaching to as much as *one-third of the velocity of light.* Naturally, the velocity depends on the potential applied to the cathode tube (*viz.*, on the potential difference through which the particles travel). If V be the tube potential and Q the charge on a cathode particle, the product $Q\ V$ must, of course, be equal to the kinetic energy of the particle as it leaves the tube, *i.e.*, to $\frac{1}{2}mv^2$. Hence

$$v = \sqrt{2\gamma V}. \tag{3}$$

Thus the velocity varies as the square root of the tube potential.

The value of the *specific charge*, as obtained from measurements according to eqns. (1) and (2), is

$$\gamma = 5{\cdot}3 \times 10^{17} \text{ E.S.U./gram.} \tag{4}$$

On comparing this value with that obtained for the specific charge of the ionized hydrogen atom, we see that it is about 1,800 times as large. If we make the very likely assumption that the *charge on a cathode ray particle* amounts to *one elementary quantum of electricity*, we must then ascribe to the particle a *mass* 1/1800th of that of the *hydrogen atom*.

When this assumption was first suggested, it appeared to physicists an extremely bold one, because, up to about the end of the 19th century, the mass of the hydrogen atom had been regarded as the smallest mass possible. In 1896, however, the discovery of the *Zeeman effect* (which we shall discuss later), which consists in the resolution of spectral lines in a magnetic field, led to the same conclusion. Moreover, the subsequent progress of theoretical physics has appeared entirely to confirm all deductions made on the hypothesis of smallest particles with a mass 1/1800th of that of the hydrogen atom. These particles with a negative charge of one elementary quantum of electricity are now generally known as *negative electrons*, or simply as *electrons*.[2] It follows from eqn. (4) and the known value of the elementary quantum of electricity that the *mass* of an electron, which is generally denoted by m in atomic theory, is given by

$$m = 9{\cdot}0 \times 10^{-28} \text{ grams.} \tag{5}$$

As was mentioned above, the velocity of the cathode rays depends on the tube potential; and since the latter cannot be indefinitely increased, an upper limit is thus set for the velocity of the rays, this limit depending on the state of experimental technique. Besides the artificially produced negative corpuscular rays, however, physics is also acquainted with spontaneously occurring rays, likewise consisting of electrons, but pssesosing a much higher velocity. These rays form part of the radiation emitted by so-called *radioactive substances*.

In 1896 *Becquerel* made the surprising discovery that *uranium ores* constantly emit rays without the intervention of any outside agency. These rays are able to blacken a photographic plate through an opaque covering, and also to ionize the air, thus rendering it conducting, so that a charged electroscope set up close at hand is discharged. Two years later (1898), *Pierre* and *Marya Curie* found that the so-called Becquerel rays (as they were originally termed) are, in point of fact, principally emitted by a metal contained in the pitchblende. At that time this metal was still unknown, but, after tedious researches, M. and Mme. Curie isolated it from pitchblende and named it *radium*.[3] A closer investigation

[2] The term " electron " is due to Stoney (1881).

[3] *Metallic* radium was not obtained until 1910, by Mme. Curie; up till then only salts of radium were known.

of the " radioactive radiation " emitted by radium and other substances showed that it consists of *three* entirely different kinds of rays, which are distinguished as *alpha-*, *beta-*, and *gamma-rays ;* a separation of the three kinds of rays is rendered possible by their *different behaviour in a magnetic field.* Whereas the γ-rays are not deflected at all, and are thus to be regarded as a radiation of electromagnetic waves, the α- and the β-rays are found to be corpuscular rays that are deflected by a magnet. Furthermore, it appears that the α-rays (which will be discussed later) carry positive charges, whereas the β-rays carry negative charges.

The *velocity* and *specific charge* of the *beta-rays* could be determined by the same method as in the case of cathode rays. The value obtained for the specific charge was the same as for cathode rays, which apparently proved the β-rays to consist likewise of *electrons.* They are thus essentially identical with cathode rays, but they are distinguished from them by a *greater velocity*, for which values ranging from 30 *to* 99·8% *of the velocity of light* have been obtained.

The motion of β-particles is the *most rapid material motion* known in physics, and for this enormous velocity so closely approaching that of light[4] are clearly revealed the *deviations from classical mechanics* which, according to the *theory of relativity*, are to be expected in the case of such rapid motions. It follows from this theory—the subject will be fully discussed later—that *mass* depends on *velocity* in accordance with the relation

$$m = \frac{m_0}{\sqrt{1 - \frac{v^2}{c^2}}}, \tag{6}$$

where m_0 denotes the value of the mass for velocities small compared with the velocity of light (c). This formula shows that, with a velocity of $v = c/2$, the mass is some 1·15 times as large as m_0, whilst, with a velocity of 99·8% of the velocity of light, the mass is as much as 16 times as large. Measurements by *Kaufmann* and other investigators have actually established the increase in mass with increasing velocity, as demanded by the relativity theory, and have also confirmed eqn. (6) quantitatively.[5]

Essentially the same as cathode and β-rays are the negative corpuscular rays which are a consequence of the so-called

[4] It appears from the theory of relativity that velocities greater than the velocity of light are impossible, as is also shown by eqn. (6).

[5] An increase in mass connotes a decrease in γ.

photo-electric effect. In continuation of an earlier research by *Hertz*, *Hallwachs* in 1888 made the important observation that a *metal plate* becomes *positively charged* when illuminated with *ultra-violet light. Lenard* recognized the cause of this phenomenon to be a negative corpuscular radiation emitted by the illuminated metal. This radiation was found to have the same specific charge as cathode rays, so that obviously it must also consist of *electrons.* Their velocity, however, proved to be much smaller than that of cathode rays, only amounting to some thousandths of the velocity of light.

Every corpuscular radiation of definite velocity and definite specific charge may be characterized by a definite value of the electric potential, which may be called the *equivalent potential* and, when multiplied by the charge of the radiated particle, gives its kinetic energy. Thus, in accordance with eqn. (3), we have the relation

$$V = \frac{v^2}{2\gamma}, \tag{7}$$

or, by § 64 eqn. (10), when the potential is expressed in volts,

$$V = 150\frac{v^2}{\gamma} \quad (V \text{ in volts, } \gamma \text{ in electrostatic measure}). \tag{8}$$

In the case of cathode rays we find, of course, as equivalent potential the tube potential, or about 30,000 volts for velocities amounting to one-third of the velocity of light. The potential corresponding to the fastest β-rays would be about 4,000,000 volts, for v is then almost equal to c, and γ is 16 times smaller than in the case of cathode rays (in consequence of the increase in mass). On the other hand, the rays produced by ultra-violet light have an equivalent potential of only a few volts.

§ 90. Positive Corpuscular Radiation.

The negative rays which we have hitherto been discussing find their counterpart in the *positive corpuscular radiation.* This was discovered by *Goldstein* in 1886, in the form of the so-called *canal rays.* Like the cathode rays, the canal rays emanate from the cathode in a highly evacuated Geissler tube, but travel in the opposite direction to the cathode rays.[1] It was, moreover, later observed that the anode itself emits positive rays, known as *anode rays.*[2]

[1] In order that the rays may escape from the cathode, the latter must be perforated by small canals ; hence the name "canal rays."

[2] The anode rays were discovered in 1906 by *Gehrcke* and *Reichenheim.*

There is, however, otherwise no essential difference between canal rays and anode rays. Their velocity and specific charge can be determined by the same method as in the case of negative corpuscular rays, the sense of deflection clearly demonstrating that the particles in canal and anode rays are *positively charged.* Their *velocity* is found to be a few thousandths of the velocity of light.

Experiment shows the value of the *specific charge* to be much smaller than in the case of the negative rays; furthermore, this value depends on the particular gas of which traces remain in the evacuated tube. The largest value of the specific charge is obtained when the tube is filled with hydrogen before evacuation; in that case the value is identical with that obtained from electrolytic measurements for the specific charge of the hydrogen ion. If the tube be filled with mercury vapour before evacuation, the specific charge is found to have a value 1/200th of the above, corresponding to the fact that the atomic weight of mercury is 200. The particles in positive corpuscular rays must therefore be regarded as *electrically charged atoms*, the charge on each amounting to one, or sometimes to several elementary quanta.[3]

Now, in transverse electric and magnetic fields the particles in a beam of corpuscular rays must be deflected, and, if the electric field be arranged at right angles to the magnetic field, the deflections will be such that all particles with the *same specific charge* will strike a plate, placed at right angles to the beam, along a *parabolic curve.* Hence, if a positive radiation be composed of atoms with different specific charges, these must form parabolas of different curvatures, and upon this fact rests the possibility of *positive ray analysis*, as was first recognized and applied by *J. J. Thomson.*

The method of positive ray analysis has been developed by *Aston* (1919) into one of the most important methods of experimental research.[4] Owing to the focussing effect

[3] In addition to charged atoms, we may also meet with charged atomic complexes, *e.g.*, molecules of hydrochloric acid which possess a positive charge of one elementary quantum, and so on.

[4] In Aston's method the electric and magnetic fields are applied so that the deflections are at 180° instead of at 90° to each other. A very thin ribbon of rays is spread out by the electric field into an "electric spectrum," and a group of the deflected particles is allowed to pass through the magnetic field, which is so arranged that the resulting deflection is opposite in direction to that produced by the electric field. It may be shown that all particles with the same specific charge come to a real focus, and that the positions of the foci so generated lie along a straight line, which is sharply defined on a suitably placed photographic plate.

obtained, it is possible to produce much greater resolution with his form of apparatus than with the original method. In this way the masses of arbitrary particles can be exactly compared with each other; and in addition to some other very important experimental results which we shall discuss fully later, Aston's researches established the most important fact that the *masses of arbitrary positive ray particles* are always integral multiples of the mass of the hydrogen atom (or, more accurately speaking, *almost always whole numbers in terms of the mass of the oxygen atom, taken as* 16).

Like cathode rays, canal and anode rays are artificial products of experimental physics. A *natural* counterpart to the negative β-rays is represented by the *positive alpha-rays* which, as we have already mentioned, are emitted similarly to the β-rays by radioactive substances. Their positive electrical nature is proved by the fact that they are deflected by a magnet, but in the opposite sense to β-rays.

The *velocity* of α-rays generally lies between *five* and *seven per cent. of the velocity of light,* varying with the chemical nature of the radioactive substance emitting them. Measurements of the *specific charge* yield a value of

$$(1) \qquad \gamma = 1{\cdot}44 \times 10^{14} \text{ E.S.U./gram.}$$

Thus the specific charge of the α-particle is *half as large as that of the ionized hydrogen atom.*

Now it is a fact of great importance that the α-particles emitted by a radioactive substance can be *counted,* for the *charge on an individual α-particle* can then be directly determined. This is rendered possible by a remarkable phenomenon known as *scintillation.* If we bring a radioactive preparation into the neighbourhood of a fluorescent screen (of zinc blende or a very thin section of diamond), we observe a continual flashing of discrete points of light. This suggests that each flash of light is due to the impact of a *single* α-particle.[5] We can place the radioactive preparation behind a screen of some substance which absorbs the α-radiation, and in which there is a small hole allowing an exactly calculable fraction of the total α-radiation emitted by the preparation to pass through. This fraction is allowed to strike a small portion of the scintillation screen, which is observed under a microscope. In this way we can readily

[5] This assumption is justified by the fact that the scintillation method gives the same value for the disintegration constant of radium as do other methods (which we shall discuss subsequently).

count the number of scintillations occurring in a definite time, and, knowing the size of the hole, we can calculate how many α-particles are emitted every second by a given preparation. Determinations of this kind were first made by *Rutherford* and *Geiger* in 1908.

Furthermore, by cutting out the β-rays emitted simultaneously with the α-rays by means of a magnetic field, a metal plate can be charged positively by the radioactive radiation. In order to find the electric charge on a single α-particle, we have then only to divide the total charge conveyed to the plate in a given time by the known number of α-particles which strike the plate in that time. Rutherford and Geiger's measurements showed that the value of the charge on an α-particle is exactly *double* that of the *elementary quantum of electricity*. Since, on the other hand, the specific charge is half that of the ionized hydrogen atom, we must ascribe to the α-particle a *mass equal to that of four hydrogen atoms*, and a *positive charge of two elementary quanta*.

On *passing through ordinary substances*, *α-rays* are far more strongly *absorbed* than β-rays, and, moreover, the nature of the absorption is entirely different. For, whereas β-rays are absorbed by matter in accordance with an *exponential law*, the action of α-rays (such as the blackening of a photographic plate or the ionization of air) ceases quite suddenly, as soon as the rays have penetrated a certain distance into a given medium. This distance is known as the *range*, and it depends on the chemical character of the radiating substance as well as upon the nature of the absorbing medium. The greater the initial velocity of the α-rays on their emission from the radioactive substance, the larger is the range, which, as *Geiger* has empirically discovered, is *proportional to the cube of the initial velocity*. In ordinary air the ranges of known α-rays lie between 2·7 and 11·3 cm.; in hydrogen the α-rays of RaC have a range of 28 cm. (at room temperature and ordinary pressure). An aluminium plate 0·06 mm. thick suffices to absorb completely all but the fastest α-rays, whereas β-rays can penetrate through aluminium plates some millimetres in thickness.

The passage of α-rays through matter has been investigated with particular success by *C. T. R. Wilson* (1912). His experiments were based upon the well-known fact that, in *supersaturated water vapour*, electrically charged particles serve as *condensation nuclei* for *water drops*. Thus, on bringing a radioactive preparation into a chamber containing supersaturated water vapour, Wilson was able to *photograph* the collection of water drops created by the passage of an α-particle. A photograph of the streaks of mist thus produced gives also a picture of the *actual paths of the α-particles.* In general the photographs (Figs. 59 and 60) show *straight lines*, many of which, however, appear *bent at the end* through fairly large angles. This fact, which we shall discuss in detail later, is of fundamental importance for modern atomic theory.

Besides canal, anode, and α-rays, a fourth kind of positive

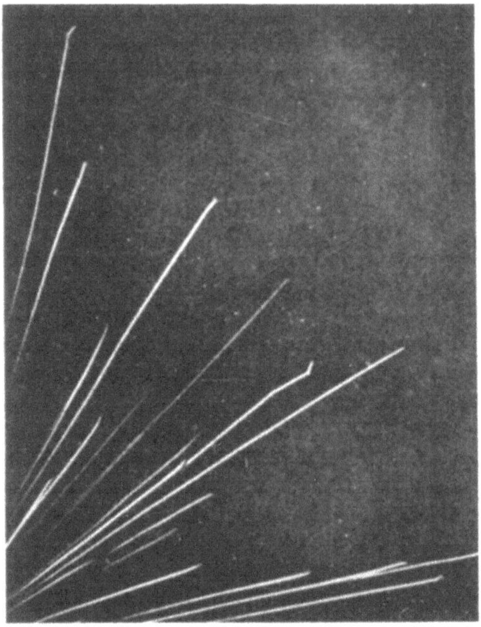

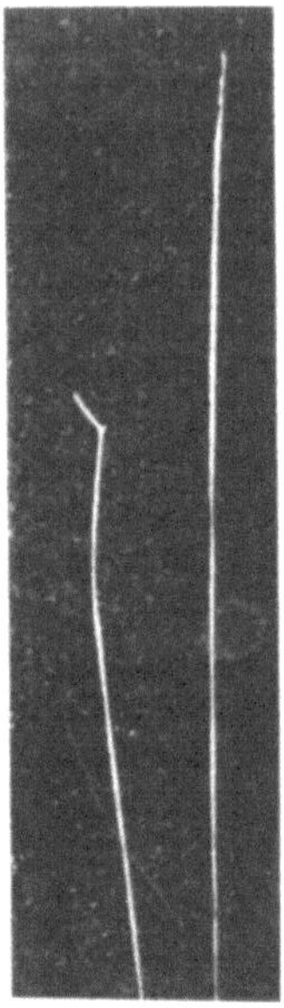

FIG. 59. FIG. 60.
Paths of α-rays.

corpuscular rays is known in physics, *viz.*, the so-called *H-rays*, discovered by *Marsden* in 1914. When he passed the α-particles emitted by radium C through *hydrogen*, in which their range is 28 cm., Marsden was able to detect scintillations on a screen even at a distance of 80 cm. These scintillations could obviously not be due to the α-rays, but must have been caused by other rays which, in their turn, were produced by the α-particles. A determination of the specific charge gave the same value for these newly discovered rays as is obtained from electrolytic measurements for hydrogen ions. From the chemical symbol for hydrogen, these rays are therefore called H-rays.

Equivalent potentials can easily be calculated for positive corpuscular radiation. From eqn. (1) and § 89 eqn. (8) we find for α-rays, whose velocity, as already mentioned, amounts to 5 to 7% of the velocity of light, a value of about two to four million volts. If we wish to express in ergs the kinetic energy of a rapid α-particle possessing a velocity of 2×10^9 cm./sec., we have to multiply half its mass (*i.e.*, twice the mass of a hydrogen atom) by the square of its velocity. We thus find

$$(2) \qquad E_{kin} = 1{\cdot}3 \times 10^{-5} \text{ ergs.}$$

The velocity of the fastest β-particles is, it is true, some 15 times as large, and therefore the square of the velocity about 200 times as large. On the other hand, their mass is some 7,000 times as small, and even when we take into account the 16-fold increase in mass for the enormous velocity of 99·8% of the velocity of light, the mass is still some hundreds of times as small as that of an α-particle. The kinetic energy of β-particles is therefore considerably less than that of α-particles.

§ 91. The Elementary Quantum of Action.

During an investigation of the *photo-electric effect* in 1902, *Lenard* had already observed the remarkable fact that the *velocity* of the electrons liberated by ultra-violet light is entirely *independent of the intensity of the light used.* The velocity cannot be increased by raising the intensity, but only by raising the frequency of the light. Increasing the intensity augments only the *number* of liberated electrons, but not their velocity.

Quantitative measurements of the photo-electric effect are based on the fact that the emission of the electrons from the metal plate subjected to the radiation can be prevented by giving the plate a *positive charge.* With a given frequency of the incident light, we can thus determine for a plate of a given metal a quite definite *counterpotential,* on reaching which the emission of electrons just comes to a complete stop. If V be the counterpotential, experimental observation shows that *V increases according to an exact linear relation with the frequency of the light.* We may write

$$(1) \qquad V = C\nu - P,$$

where C and P are positive constants. We find from experiment that only the constant P depends on the nature of the metal,[1] whereas the constant C is independent thereof and must therefore represent a *universal constant*. Now the product $e\,V$ denotes the kinetic energy of the most rapid electrons liberated, which suggests a multiplication of eqn. (1) by e. If we write h for the product $C\,e$, eqn. (1) becomes

$$(2) \qquad e\,(V + P) = h\,\nu.$$

Since the product $e\,V$ represents work or energy, whilst frequency has the reciprocal dimensions of time, the dimensions of the constant h must be those of a *product of energy and time*. Such a product in mechanics is termed *action*, and the constant h is therefore called the *elementary quantum of action*. The constant h thus forms a *fourth* elementary quantum, in addition to the elementary quantum of electricity and the masses of the electron and the hydrogen atom. Its value is found from measurements of the photo-electric effect to be

$$(3) \qquad h = 6{\cdot}55 \times 10^{-27} \text{ erg sec.}$$

The hypothesis of the elementary quantum of action was introduced into theoretical physics by *Planck* in 1900. By means of this hypothesis Planck succeeded in solving a difficult fundamental problem in the theory of thermal radiation, which we shall discuss at length later. For this purpose he assumed that the emission of wave radiation occurs *discontinuously* in such a way that *elements of energy* of magnitude $h\,\nu$ play a part in the process.

In 1905, *Einstein* extended Planck's conception by introducing the idea of *light-quanta* whose energy is likewise to be determined by the product $h\,\nu$, where ν is the frequency of the light.[2] Einstein formulated the *law* that every *mutual transformation between wave radiation and corpuscular radiation* takes place in such a way that either a *single light-quantum* is utilized in the liberation of a single electron, or, in the converse case, a single light-quantum is produced by the energy of a single electron.

The first case occurs in the photo-electric effect, where wave radiation is transformed into corpuscular radiation, and, as a matter of fact, later exact measurements by

[1] It is true that the presence of adsorbed or occluded gases has a great influence.

[2] Planck had already been able to determine the value of h from his theory of thermal radiation.

Millikan have demonstrated the accurate fulfilment of Einstein's law for the photo-electric effect. For, in the empirically obtained eqn. (2), a light-quantum of the exciting light has been equated to the sum of the maximum kinetic energy ($e V$) with which a liberated electron leaves the metal subjected to radiation, and the work ($e P$) which the electron must perform in escaping from the metal. The *constant* P, which likewise represents a potential, when multiplied by the elementary quantum of electricity is thus a measure of the *work expended in freeing an electron*, and it thus becomes intelligible why the value of P should agree in order of magnitude with the difference of potential between two metals not too close together in Volta's well-known potential series. The values of P, which is termed the "*contact potential*," amount to only a few volts for the various metals. The smallest value is for *potassium*, in which, as is well known, the photo-electric effect is the greatest. In accordance with eqn. (2), the expression $e P/h$ determines the *limiting frequency* below which the metal cannot be photo-electrically excited.[3]

We shall not discuss until later the converse phenomenon to the photo-electric effect, the transformation of corpuscular radiation into wave radiation; this phenomenon also confirms Einstein's law of light-quanta. Moreover, Einstein's hypothesis also affords a simple explanation of a relation holding for the transformation of primary into *secondary wave radiation*. According to Einstein's law, *one secondary light-quantum* must arise *from each primary quantum*. If this be the case, however, the *principle of energy* teaches that the *secondary frequency cannot be greater than the primary*; it must be *smaller* when energy is also expended for other purposes during the transformation. Secondary wave radiation is termed *fluorescent light*. As a matter of fact, *Stokes* in 1852 established the *rule*, known by his name, that the *wave-length* of the fluorescent light is *longer* than that of the exciting light.[4]

§ 92. Application of the Quantum Principle to Periodic Motions.

Besides oscillatory processes, a second important field for the application of the principle of the elementary quan-

[3] This limit has been determined for many substances. In the case of copper, for example, it lies at about 3,000 Å.U.; in the case of sodium, at about 6,000 Å.U. The limiting wave-length for potassium is at any rate greater than 10,000 Å.U.

[4] For apparent exceptions, *cf.* § 123.

tum of action or, in short, of the *quantum principle*, is afforded by *periodic motions*. Quite generally, motions may be described in terms of coordinates and velocities, or, in the form of Lagrange's generalized equations of motion, in terms of *coordinates* and *momenta*. We can at once recognize a possible connection between this second method of description and the elementary quantum of action. For, denoting the quite arbitrary coordinates by q_i and the momenta by p_i, we have, in accordance with the definition of momentum (*vide* § 22),

$$p_i = \frac{\partial E_{kin}}{\partial\left(\frac{dq_i}{dt}\right)}, \tag{1}$$

where E_{kin} is the kinetic energy. Hence the product of coordinate and momentum must always have the same dimensions as the product of energy and time, *i.e.*, it must have the *dimensions of action*.

This suggests the assumption that, in those motions in which the coordinate q_i varies periodically, the *integral*

$$\int p_i \, dq_i = n\,h \tag{2}$$

when taken over a period, *i.e.*, that it is equal to an integral multiple of the elementary quantum of action.

Particularly simple results follow from the suggested hypothesis in the case of a *circular motion*. If the circle be given as such, the single positional coordinate (q) is then represented by the *angle* included between the radius-vector drawn from the centre to the revolving particle and a definite direction drawn through the centre (the so-called zero direction). Thus, if a be the radius of the circular orbit and v the linear velocity, we have

$$v = a\,\frac{dq}{dt}\,. \tag{3}$$

The kinetic energy is given by

$$E_{kin} = \frac{m\,v^2}{2} = \frac{m\,a^2}{2}\left(\frac{dq}{dt}\right)^2; \tag{4}$$

hence, by eqns. (1) and (3), we find for the momentum

$$p = m\,a\,v. \tag{5}$$

Therefore, in accordance with the terminology of mechanics, the quantity p represents nothing else than the so-called *angular momentum* of the motion (*vide* § 7). Since the motion takes place with constant velocity, p is a constant in the case of a circular motion. Hence, denoting the angular momentum by U,

$$\int p\,dq = p\int_0^{2\pi} dq = 2\pi U, \tag{6}$$

the integral being taken over a period. Since, however, in accordance with the quantum principle [eqn. (2)], this integral is to be a whole-number multiple of the elementary quantum of action, it follows from that principle that those *circular orbits are privileged* for which the *angular momentum multiplied by* 2π represents an *integral multiple of the elementary quantum of action.* For such orbits we therefore have the relation

$$2\pi U = n\,h. \tag{7}$$

CHAPTER XII

THEORY OF THE HYDROGEN ATOM

§ 93. The Nuclear Atom.

As we have seen, the phenomena of corpuscular radiation render manifest the existence, on the one hand, of particles with a *negative charge* of one elementary quantum of electricity and with a mass invariably equal to 1/1,800th of that of the hydrogen atom, and, on the other hand, of corpuscles with a *positive charge* of one or of a few elementary quanta and possessing a *mass* which is always approximately an *integral multiple* of that of the hydrogen atom.

It would appear that *atoms* must be *composed* of these negative and positive corpuscles, and important clues as to the nature of this structure are afforded by the previously described observations on the *passage of alpha-rays through matter.* Now these observations show that an α-particle can describe *rectilinear paths* of a few centimetres in air. A cubic centimetre of air, however, contains approximately 5×10^{19} atoms, as follows from the known density and from the known value of the Loschmidt number.[1] On the other hand, it has long been known from the kinetic theory of gases that the diameter of an atom must be of the order of magnitude of 10^{-8} cm., so that, in air, about 1/100,000th part of the total volume is occupied by the atoms, and a line 1 cm. long therefore passes through thousands of atoms. Hence an α-particle is capable of *penetrating through many thousands of atoms* without its direction being appreciably affected. Occasionally, however, as Wilson's photographs show, *a single atom deflects the α-particle through a very large angle.*

This remarkable fact can be readily explained, on the basis of *Coulomb's Law,* by the assumption that the deflection is caused by an *electrical force* exerted on the electrically positive α-particle by an *atomic nucleus,* the *volume* of which constitutes *only a very small part of the total volume of the atom.* We might, at first, equally well suppose the electrical

[1] The number of molecules is half as large.

force to be one of attraction as of repulsion, and hence the atomic nucleus to be either negative or positive.

Now it follows from the principle of action and reaction that the extraordinarily large deflections of the α-rays actually observed are only possible if the *mass of the nucleus* is considerably *larger* than the mass of the deflected α-particle. Observations show, in fact, that the magnitude of these large deflections increases with the atomic weight of the deflecting substance. We must therefore assume that the *mass of the atom* is principally *concentrated in the nucleus.*

Again, as the mass of the electrons is vanishingly small in comparison with that of the positive electrical particles, we must suppose that the nuclei of the atoms are electrically *positive.* Since, in general, the atoms as a whole are electrically *neutral*, this positive charge must be *compensated* by negative charges ; and, because negative particles are found *only* as electrons in cathode and β-rays, the positive nuclear charge must obviously be compensated by *electrons* surrounding the nucleus. Provided that the *motion* of these electrons *around the nucleus* be sufficiently rapid, the electrical force of attraction which would otherwise cause them to unite with the nucleus will be counterbalanced by the *centrifugal force.*

These considerations lead to a *model of the atom* which was first introduced into theoretical physics by *Rutherford* in 1911. According to this model, the atoms consist of positively charged nuclei around which, to borrow a frequently used simile, electrons revolve like planets around the sun.[2]

The *simplest* imaginable atomic model is obviously one in which a *single electron* revolves about an *atomic nucleus* charged with *one positive elementary quantum.* The idea at once suggests itself to associate this model with the *lightest* of all atoms, the *hydrogen atom*, and it is actually found that all conclusions which may be theoretically derived from this assumption are completely confirmed by experiment. Hence,

[2] Rutherford based his atomic model, which he had developed a year before Wilson's researches, on his own observations on the passage of α-rays through metal foils.—To-day only a historical significance is attached to *J. J. Thomson's model of the atom* which, at that time, was a subject of much discussion. Thomson supposed the atom to consist of a uniformly dense spherical volume charge of positive electricity, within which electrons described circular orbits about the centre of the sphere. The force acting on an electron is then directly proportional to its distance from the centre (because only a portion of the spherical charge is effective). Although the assumption of such forces proved most fruitful for the explanation of dispersion and magneto-optical phenomena, Thomson's atomic model had to be relinquished when observations on the passage of α-rays through matter rendered it untenable.

we may also speak of *hydrogen nuclei*, instead of *positive corpuscles or electrons*, but the word "*protons*" is now generally adopted.[3]

§ 94. Bohr's Model of the Hydrogen Atom.

In Rutherford's model of the hydrogen atom just described, there at first appeared to be really no way of determining the radius of the atom, or, more accurately, the radius of the orbit described by the electron.[1] In 1913 *Bohr* succeeded in solving this riddle, by *applying the hypothesis of the elementary quantum of action to Rutherford's theory of the atom.*

Bohr's fundamental assumption states that (in the sense of § 92) the *angular momentum of the normal hydrogen atom, multiplied by 2π, shall be equal to the elementary quantum of action;* this relation determines the normal radius of the hydrogen atom.

In accordance with § 92 eqn. (7), however, *in addition to the normal state, other quite definite states* are considered possible, *viz.*, those for which 2π times the *angular momentum* is *equal to an integral multiple of the elementary quantum of action.* These *abnormal* states are described as *multiple-quantum* states, and the corresponding whole number is called the *quantum number* of the state. Thus we speak of a two-quantum, a three-quantum, a four-quantum state, and so on; in the hydrogen atom the *fundamental* or *normal orbit* of the electron is a *one-quantum* orbit.

We will now calculate the values of the *radius of the orbit* and of the *energy* for a state of arbitrary quantum number in Bohr's atomic model. In view of later applications we shall consider, besides the hydrogen atom itself, the more general case in which the nucleus, around which a *single* electron is revolving, has a positive charge of z elementary electrical quanta; such an atom may be called a *hydrogen-analogue.* Since z is greater than unity, it obviously cannot be neutral, but must behave as a *positive ion.*

The forces acting on the revolving electron are, on the one hand, the *electrical attraction of the nucleus*, and, on the other hand, the *centrifugal force.* In accordance with Newton's Third Law of Motion, the *two forces* must be *equal.* By

[3] Rutherford suggested the term "*protons*" for positive electrons. This terminology renders the use of the single word "electron" preferable to the term "negative electron" for the negative constituent.

[1] In Thomson's model of the atom, which was based on a sphere of given radius, this difficulty, of course, did not arise.

Coulomb's Law, the force of attraction is equal to $z\,e^2/a^2$; the centrifugal force is $m\,v^2/a$, where m is the mass of the electron.

Equating the two expressions, we find

$$(1) \qquad m\,v^2\,a = z\,e^2 .$$

Again, in a permissible orbit of the quantum theory, the following relation must be obeyed [according to § 92 eqns. (5) and (7)] :—

$$(2) \qquad m\,a\,v = \frac{n\,h}{2\pi} .$$

If the *quantum number* (n) be given, eqns. (1) and (2) contain as unknowns only the orbital radius (a) and the linear velocity (v), so that these two magnitudes can be determined from the two equations. We may distinguish by the subscript n the values assumed by these magnitudes for the n-quantum state. The value of a_n is obtained by squaring eqn. (2) and dividing by eqn. (1) ; we thus find

$$(3) \qquad a_n = \frac{n^2\,h^2}{4\pi^2\,z\,e^2\,m} .$$

The *radii* of the permissible orbits therefore vary as the *squares of the quantum numbers.*

We find the *velocity* in the n-quantum orbit on dividing eqn. (1) by eqn. (2) ; we have then

$$(4) \qquad v_n = \frac{2\pi\,z\,e^2}{n\,h} .$$

The linear velocities therefore vary inversely as the quantum numbers. In the case of the normal hydrogen atom, for which both z and n are unity, we find, on introducing the known values of e, h, and m into eqns. (3) and (4), that the radius of the orbit is about 5×10^{-9} cm., and the velocity about 2×10^8 cm./sec., or somewhat less than one-hundredth of the velocity of light.

Finally, let us calculate the *energy* possessed by the revolving electron in the n-quantum orbit. This energy is made up of kinetic and potential energy. The *kinetic* energy is found by multiplying half the square of the velocity by the mass (m). The *potential* energy follows from Coulomb's Law :—

$$(5) \qquad E_{pot} = -\frac{z\,e^2}{a} .$$

A *negative sign* has to be prefixed because, on account of the attraction between nucleus and electron, an *increase in the distance* a requires an *addition of energy*, and the energy can only increase with a (to which it is inversely proportional) if it be *opposite in sign*. In spite of the negative sign, the energy itself is, of course, not negative, for in eqn. (5) there has still to be added a positive constant which has the same value for all states. Besides, the potential energy represents the *work* which would be required to remove the electron, were it at rest, to an infinite distance.

On comparing eqns. (1) and (5) we have

$$(6) \qquad E_{pot} = - m v^2 .$$

The potential energy is therefore twice as large as the kinetic energy, and opposite in sign ; hence the *total energy* (the sum of kinetic and potential) is given by

$$(7) \qquad E_n = - \frac{m v^2_n}{2} ,$$

or, according to eqn. (4), by

$$(8) \qquad E_n = - \frac{2\pi^2 z^2 e^4 m}{n^2 h^2} .$$

The *energy values* belonging to the different quantum orbits thus vary *inversely as the squares of the quantum numbers.*

§ 95. The Spectral Formula of Bohr.

In Bohr's model of the atom, every *transition between two orbits with different quantum numbers* is associated with a *change in the energy of the atom.* On account of the negative sign in § 94 eqn. (8), larger quantum numbers correspond to larger values of the energy. Hence, if the *quantum number be lowered* in the transition, *energy* is *liberated.* On the other hand, only an *addition of energy* will permit of a transition in which the *quantum number is raised.*

Now it is a fundamental assumption in Bohr's theory, that the liberation of energy by a transition of this kind must always produce a *single light-quantum,* and that, in the reverse process, the addition of energy will always be effected by the taking up of a single light-quantum. If, therefore, we consider two orbits whose quantum numbers are n and s, where $n > s$, radiation of a definite frequency (ν) will be *emitted* for a transition from the n-quantum to the s-quantum orbit, whilst, in the reverse transition, radiation of the same frequency ν will be *absorbed.* The *frequency,* however, is determined by the so-called *frequency condition of Bohr* just quoted, in that, when the frequency is *multiplied by the elementary quantum of action*, the product is *equal to the difference in the energy values.* We thus have

(1) $$h\,\nu = E_n - E_s.$$

On introducing the values given for the energy in § 94 eqn. (8), the frequency condition assumes the form of the *spectral formula* derived by Bohr in 1913 :—

(2) $$\nu = \frac{2\pi^2 z^2 e^4 m}{h^3}\left(\frac{1}{s^2} - \frac{1}{n^2}\right).$$

Since both s and n can run through the whole series of integers, there results a *double manifold of frequency values.* This double manifold is defined as the *line spectrum* of the atom. It may occur either as an *emission spectrum* or as an *absorption spectrum.* It is composed of *series* which, in their turn, represent simple manifolds and are obtained by giving the one quantum number (s) a fixed value, whilst the other quantum number (n) passes through the series of whole numbers.

Bohr's frequency condition leads at once to the *representation of spectral lines by spectral terms.* For, by eqn. (1), the frequency of every line can be written as the *difference between two terms*, each of which represents one of the possible energy values divided by the elementary quantum of action. The whole of the lines in the spectrum can thus be *referred to a simple manifold of terms.* Each series can hence be represented as a difference between a *fixed term* and a *variable term*, the *variable number* of which passes through the series of integers. The fixed term thus determines the *limiting frequency of the series*, which is approached by the lines as the variable number increases.

If, now, n, r, and s be three arbitrary whole numbers, the identity

(3) $$(E_n - E_s) = (E_n - E_r) + (E_r - E_s)$$

necessarily holds good. This identity may also be written in the form

(4) $$(E_r - E_s) = (E_n - E_s) - (E_n - E_r)\,.$$

On replacing the energy differences in these equations by the equivalent frequency values, we recognize at once the possibility of deriving new spectral lines from ones already known, by means of *additive or subtractive combinations.* For example, denoting the spectral line emitted on a transition from the five-quantum to the three-quantum orbit by

$\nu_{3,5}$, *etc.*, we can conclude from the given lines $\nu_{3,5}$ and $\nu_{2,3}$ the existence of a line

$$\nu_{3,5} + \nu_{2,3} \, (= \nu_{2,5}) \, .$$

Similarly, given the lines $\nu_{2,3}$ and $\nu_{2,4}$, we can deduce the existence of a line

$$\nu_{2,4} - \nu_{2,3} \, (= \nu_{3,4}) \, ,$$

etc. This so-called *combination principle* was established by *Ritz* in 1908, before Bohr's theory had been developed.

As regards the *intensity of a spectral line*, this obviously depends on how frequently the transition to which it is due occurs.

§ 96. Line Spectrum of Hydrogen.

If, in Bohr's spectral formula, the so-called nuclear charge number (z) be made equal to unity, the formula thus specialized should represent the *line spectrum of hydrogen*. As a matter of fact, complete agreement is found to exist between this purely theoretical formula and the results derived from experimental spectroscopy.

Of the hydrogen lines, those which have been longest

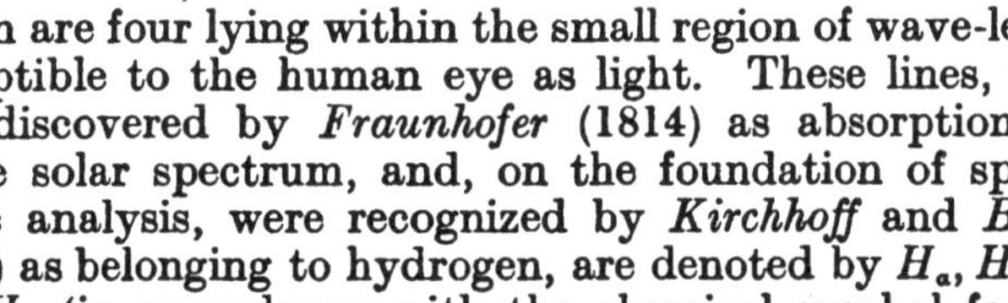

FIG. 61.—The optical hydrogen series.

known are four lying within the small region of wave-lengths perceptible to the human eye as light. These lines, which were discovered by *Fraunhofer* (1814) as absorption lines in the solar spectrum, and, on the foundation of spectroscopic analysis, were recognized by *Kirchhoff* and *Bun en* (1860) as belonging to hydrogen, are denoted by H_α, H_β, H_γ, and H_δ (in accordance with the chemical symbol for that element). The first and strongest line lies in the red, the second in the blue-green, and the third and fourth in the violet (Fig. 61). They form the first members of a series which continues with decreasing intensity into the ultra-violet, with a limiting wave-length of about 3,600 Å.U. (As was explained in § 86, an Ångström unit is equal to 10^{-8} cm.)

As long ago as 1885, *Balmer* made the extremely important discovery that very simple numerical relations exist between the wave-lengths of the individual lines in this series. The frequencies of the individual lines can, in fact, be represented by the so-called *Balmer formula*,

$$\nu = R\left(\frac{1}{4} - \frac{1}{n^2}\right), \tag{1}$$

where n denotes any one of the series of whole numbers from three upwards, whilst R is a constant frequency whose value, as found from the observed frequencies of the lines, is given by

$$R = 3{\cdot}291 \times 10^{15} \text{ sec.}^{-1}. \tag{2}$$

Since wave-length and frequency are connected by the simple relation

$$\nu = \frac{c}{\lambda}, \tag{3}$$

eqn. (1) may also be written in the form [1]

$$\frac{1}{\lambda} = \frac{R}{c}\left(\frac{1}{4} - \frac{1}{n^2}\right). \tag{4}$$

Balmer's formula is obeyed with wonderful exactness, and moreover, spectroscopic measurements can be carried out to an extraordinary degree of accuracy, amounting to about a millionth of the measured value. Hence, it has been possible to determine the constant R very exactly from spectroscopic measurements, and the value thus found from the hydrogen spectrum (which fact may be expressed by adding the subscript H) is [2]

$$\frac{R_H}{c} = 109{,}677{\cdot}69 \text{ cm.}^{-1} (\pm\ 0{\cdot}06)\,. \tag{5}$$

Of the so-called *optical series* of hydrogen, as represented by Balmer's formula, the first twenty lines [3] have been observed in Geissler tubes, and a further thirteen in the spectra of nebulous stars, in which matter is obviously far more attenuated than in even the most highly evacuated Geissler tubes. In the absorption spectrum, however, as many as fifty lines have been detected in laboratory experiments.

It had already occurred to *Ritz*, on the basis of his *com-*

[1] Balmer himself gave his formula the form

$$\lambda = C\,\frac{n^2}{n^2 - 4},$$

where C is a constant equal to $4c/R$.

[2] The value given in eqn. (5) is based on Paschen's measurements.

[3] Thus up to $n = 22$.

bination principle, that, besides the optical series of hydrogen previously described, at least one other series must exist in the *infra-red*, the frequencies of which would be representable by the formula

$$(6)\qquad \nu = R\left(\frac{1}{9} - \frac{1}{n^2}\right), \ (n = 4, 5 \ldots).$$

The first two lines in Ritz's predicted series were actually discovered by *Paschen* in 1909.

It follows from Bohr's theory that the optical series in the emission spectrum arises from a transition into the two-quantum end state, whereas the infra-red series is due to a transition into the three-quantum end state. According to Bohr's theory, we may thus at least expect the existence of a series in the emission spectrum arising from a transition into the one-quantum normal state. This series would be represented by the formula

$$(7)\qquad \nu = R\left(1 - \frac{1}{n^2}\right), \ (n = 2, 3 \ldots).$$

In 1914 this series was actually discovered by *Lyman*. It lies in a somewhat remote region in the *ultra-violet*, and the constant R forms the series limit. Its first line is that due to a transition from the two-quantum state to the one-quantum normal state; and, inasmuch as it arises from the simplest imaginable transition in the simplest of all atoms, this line occupies quite a special position among all the spectral lines of all the elements. Its frequency is equal to three-quarters of R, so that its wave-length is 1215·7 Å.U. Actually, the line gives a very sharp image in all photographic exposures made in the extreme ultra-violet.

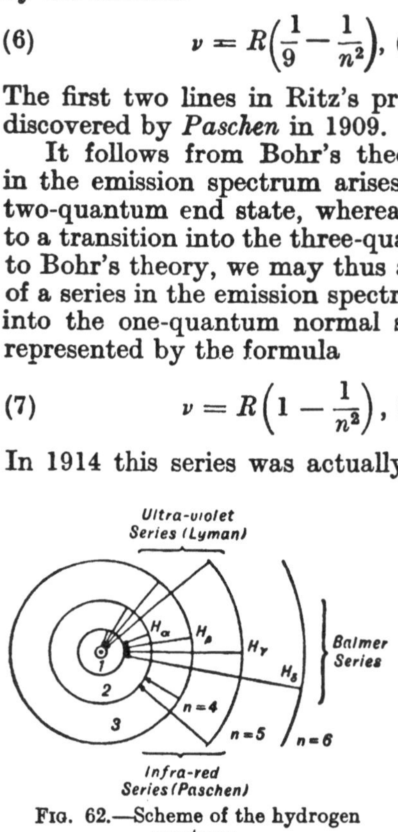

FIG. 62.—Scheme of the hydrogen spectrum.

Finally, in 1922, *Brackett* discovered in the infra-red the first two lines of the series due to a transition to a *four-quantum end state*.[4] (On the same occasion, he also discovered

[4] F. S. Brackett, "*A New Series of Spectrum Lines*," *Nature*, **109** (1922), p. 209. The wave-lengths of the first two lines of the new series are 4·05 μ and 2·63 μ.

the third, fourth, and fifth lines of the Paschen series, of which only the first two lines had up till then been known.)

The origin of the Lyman, Balmer, and Paschen series is diagrammatically represented in Fig. 62, in which the various circular quantum orbits and the transitions from one to another orbit are shown. Another very clear illustration of the origin of the spectral lines is given by the graphical representation of the *energy-levels in the atom* (Fig. 63). By

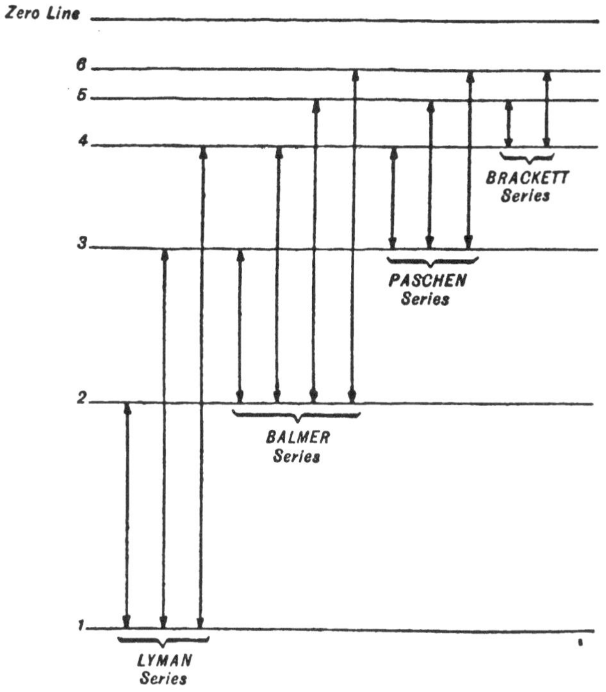

FIG. 63.—Energy-levels in the hydrogen atom.

means of a so-called *zero-line*, we represent the "zero" energy possessed by an electron revolving at an infinite distance from the nucleus. The energy in the one-quantum (normal) state is represented by a second line drawn parallel to the zero-line and below it (it is drawn below because the sign of the energy is negative). Since the energy in the n-quantum state is equal to the normal energy divided by the square of the quantum number, the line representing the two-quantum state lies *between* the one-quantum line and the zero-line. The line for the three-quantum state lies between that for the two-quantum state and the zero-line, and so on. If we wished our drawing to be to scale, the distances between the individual lines and the zero-line would have to be equal to 1/4th, 1/9th, 1/16th, 1/25th, 1/36th, *etc.*, of the distance between the one-quantum or normal line and the zero-line. In Fig. 63, however, the scale has been intentionally distorted, in order to obtain a clearer diagram.

The *level nearer to the zero-line* always represents the *higher level*, and energy is liberated in every transition which increases the distance from the zero-line, whereas, conversely, energy must be supplied for a transition which diminishes that distance. The method of representation in Fig. 63 is of great advantage, especially in the structurally more complicated atoms. If we denote each level by its corresponding *term* (the energy value divided by h and also, for convenience, by c), the *wave-length* of a spectral line due to a transition between two levels is given very simply by the *reciprocal value of the difference between the terms;* the reciprocal values of the terms themselves represent the *limiting wave-lengths of the various series.* The terms are proportional to the work which must be expended in removing an electron from the orbits to which the individual levels correspond, to beyond the limits of the atom. In the case of the hydrogen atom, the values of the successive terms are simply the quantity R_H/c, as determined by eqn. (5), and 1/4th, 1/9th, 1/16th, 1/25th, 1/36th, *etc.*, of this quantity.

The constant frequency denoted by R is known as *Rydberg's constant*, because, in 1895, Rydberg discovered that this constant has real significance not only for the hydrogen spectrum, but also for the spectra of other elements (especially the alkali metals) : thus it represents a *universal physical constant.* Comparing Bohr's theoretically derived formula with the empirical formula established by Balmer, we obtain the relation

$$(8) \qquad R = \frac{2\pi^2 e^4 m}{h^3} .$$

This relation, which connects the fundamental constant of spectroscopy with the fundamental quantities of the electron theory and the elementary quantum of action, was first derived by the author of this book in 1910,[5] and more

[5] *Sitz. Ber. der Wiener Akad., Abt. IIa*, 1910, pp. 119–144 [eqn. (67)]; also *Physikal. Zeitschr.*, **11** (1910), p. 537. This derivation, in which the quantum principle was applied for the first time to the theory of the atom, was, however,

accurately later by Bohr from his theory. Eqn. (8), moreover, permits of an exact *determination of the elementary quantum of electricity upon a spectroscopic basis.* For we may write eqn. (8) in the form

(9) $$R = \frac{2\pi^2 e^2}{\left(\frac{e}{m}\right)\left(\frac{h}{e}\right)^3},$$

and the values of e/m and h/e are known from measurements on the cathode rays and the photo-electric effect respectively. Introducing the values in question, we find for e the same value as follows from the direct determinations of Millikan, although not with the same degree of accuracy.[6]

The product of the Rydberg constant and the elementary quantum of action represents the *energy* (taken with opposite sign) *of the hydrogen atom in the one-quantum (normal) state,* as follows at once from the representation in terms, and as may also be seen by comparing eqn. (8) with § 94 eqn. (8). Denoting this normal energy by E*, we therefore have

(10) $$E^* = -hR,$$

and hence

(11) $$E^* = -2{\cdot}15 \times 10^{-11} \text{ ergs.}$$

We can also readily evaluate the other constants characteristic of the normal state of the hydrogen atom. Giving n and z the value unity in § 94 eqn. (4), we find for the *orbital velocity* of the revolving electron

(12) $$v^* = \frac{2\pi e^2}{h} = 2{\cdot}188 \times 10^8 \text{ cm. sec.}^{-1}.$$

For the ratio of this velocity to that of light, which represents an important constant, we obtain the value

(13) $$\alpha = \frac{v^*}{c} = 7{\cdot}29 \times 10^{-3}.$$

We shall denote the *radius of the orbit* in the normal state by a^*. Now the total energy, which (by § 94) represents half the potential energy, is equal to $(-e^2/2a)$, and thus

(14) $$a^* = -\frac{e^2}{2E^*}.$$

based on Thomson's model of the atom (*vide* § 93), and therefore led to an incorrect numerical factor (instead of the correct factor 2).

[6] This in itself seems to prove that objections to Millikan's determinations are not necessarily at the same time objections to the existence of an elementary quantum of electricity.

Hence it follows from eqn. (11) that

$$a^* = 5{\cdot}30 \times 10^{-8} \text{ cm.} \tag{15}$$

Finally, we obtain the *revolution frequency* in the normal state on dividing the velocity v^* by the circumference of the orbit. We have

$$\omega^* = \frac{v^*}{2\pi a^*}, \tag{16}$$

or, by eqns. (12), (14), and (10),

$$\omega^* = 2R; \tag{17}$$

whence we find

$$\omega^* = 6{\cdot}58 \times 10^{15} \text{ sec.}^{-1}. \tag{18}$$

§ 97. Line Spectrum of Ionized Helium.

Whereas the charge on the hydrogen nucleus is a single positive elementary quantum of electricity, in the case of *α-particles* we have become acquainted with positive structures with twice as large a charge and four times as large a mass. Since, on the other hand, the atomic weight of the second lightest element, helium, is four, the idea suggests itself of identifying the α-particle with the *helium nucleus*. A direct confirmation of this idea is found in the fact, which was first established in 1903 by *Ramsay* and *Soddy*, that helium is produced by radioactive substances emitting α-rays.[1]

Therefore, if we put the nuclear charge number z equal to two in Bohr's spectral formula, we might well expect the formula thus specialized to represent the *spectrum of the ionized helium atom*, in which only a single electron revolves round the doubly charged nucleus, the second electron, which is present in the neutral condition, having been removed.

For a transition from an n-quantum to an s-quantum orbit, the frequency produced in ionized helium must be four times as large as that produced in hydrogen, seeing that ν is proportional to z^2. The spectrum of ionized helium would therefore be represented by the formula

$$\nu = 4R\left(\frac{1}{s^2} - \frac{1}{n^2}\right), \tag{1}$$

which may also be written

$$\nu = R\left[\frac{1}{\left(\frac{s}{2}\right)^2} - \frac{1}{\left(\frac{n}{2}\right)^2}\right]. \tag{2}$$

[1] A strict proof of this was established somewhat later by Rutherford. For further details, *vide* § 112.

In 1897, *Pickering* actually discovered in the spectrum of the star ζ *Puppis* a series corresponding to eqn. (2) with $s = 4$; moreover, in 1912, *Fowler* discovered in a Geissler tube spectrum a series described by eqn. (2) when s is put equal to 3.[2]

Originally, both the Pickering and the Fowler series were *erroneously ascribed to hydrogen.* Bohr's theory indicates that they are *helium* series, and it was a triumph for that theory when, in 1914, *Paschen* and *Bartels* succeeded in observing the Fowler series in Geissler tubes filled with pure helium and completely uncontaminated by any traces of hydrogen or compounds of hydrogen.

According to the formulæ derived for the line spectra of hydrogen and helium, every second line of the Pickering series should really coincide with a line of the Balmer series,

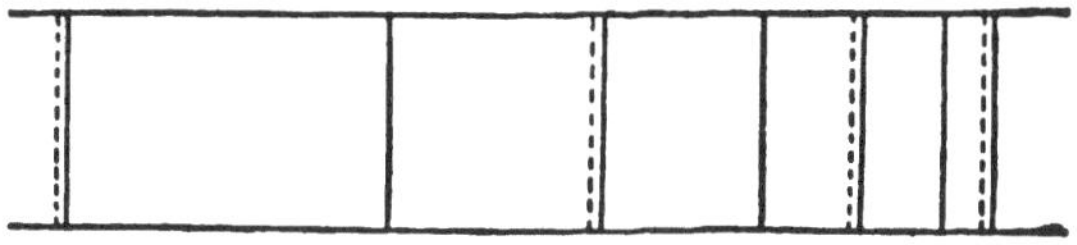

FIG. 64.—Displacement towards the violet of the helium lines.

for the Pickering series is obtained by putting $s = 4$ in eqn. (2) so that, for even values of n, the equation becomes identical with § 96 eqn. (1). Actually, however, the corresponding lines of the Pickering series are found to be slightly displaced in the direction of increasing frequency, and thus towards the violet in the visible region of the spectrum (*cf.* Fig. 64, where the broken lines belong to hydrogen, and the full lines to helium). Although this so-called *displacement towards the violet of the helium lines* might at first appear as a defect in Bohr's theory, it really proved to be a brilliant confirmation of it, when Bohr perfected his original theory by taking the motion of the atomic nucleus into account.

§ 98. The Motion of the Atomic Nucleus.

The previous considerations have been based upon the simple assumption that, in the hydrogen atom and the analogous ionized helium atom, the electron revolves round the nucleus. Some important results, however, followed from a *development of the theory by Bohr* in 1913, when he

[2] Formerly, two Fowler series were distinguished, according as n is odd or even. This distinction, however, appears entirely superfluous from the point of view of Bohr's theory.

took into account the *simultaneous motion of the atomic nucleus.* In reality, the electron does not revolve round the nucleus, but both revolve *about their common centre of gravity.*

Denoting by x and y the respective distances separating the electron and the nucleus from the centre of gravity and by M the mass of the nucleus, we have the ratio

$$x : y = M : m, \quad (x + y = a)\,; \tag{1}$$

whence it follows that

$$x = a\,\frac{M}{M + m}, \quad y = a\,\frac{m}{M + m}. \tag{2}$$

Let the velocity of the electron be v', and that of the nucleus v''. Then, since both electron and nucleus have the same angular velocity, which may be denoted by w,

$$v' = x\,w, \quad v'' = y\,w\,. \tag{3}$$

From this we find the value of the kinetic energy to be given by

$$E_{kin} = \frac{w^2}{2}\,(m\,x^2 + M\,y^2)\,,$$

or, by eqn. (2),

$$E_{kin} = \frac{a^2\,w^2\,M\,m\,(M + m)}{2\,(M + m)^2}\,.$$

In order to simplify the subsequent calculations, let us put

$$\frac{M\,m}{M + m} = \mu\,; \tag{4}$$

then

$$E_{kin} = \frac{a^2\,w^2}{2}\,\mu\,. \tag{5}$$

Again, the centrifugal force must be equal to the force of attraction, so that

$$m\,x\,w^2 = \frac{z\,e^2}{a^2}. \tag{6}$$

Now since, according to eqns. (2) and (4), $m\,x$ is equal to $a\,\mu$, we have

$$\mu\,a^2\,w^2 = \frac{z\,e^2}{a}, \tag{7}$$

and hence, by eqn. (5),

$$E_{kin} = \frac{z\,e^2}{2a}. \tag{8}$$

The potential energy is given by

$$E_{pot} = -\,\frac{z\,e^2}{a}, \tag{9}$$

and hence the total energy in the n-quantum orbit is

$$E_n = -\frac{z e^2}{2a_n} \tag{10}$$

(in agreement with § 94).

For the *angular momentum* we find the value

$$U = m x v' + M y v'' = w (m x^2 + M y^2) , \tag{11}$$

so that

$$U = \frac{2E_{kin}}{w} ,$$

or, by eqn. (5),

$$U = a^2 w \mu. \tag{12}$$

According to the fundamental quantum relation, the angular momentum multiplied by 2π must be equal to the product of the quantum number and the elementary quantum of action. We therefore have

$$2\pi a^2 w \mu = n h . \tag{13}$$

On squaring this equation and dividing by eqn. (7), we obtain

$$a_n = \frac{n^2 h^2}{4\pi^2 z e^2 \mu} . \tag{14}$$

Hence, in accordance with eqn. (10), the energy is given by

$$E_n = - \frac{2\pi^2 z^2 e^4 \mu}{n^2 h^2} . \tag{15}$$

Comparing this formula with § 94 eqn. (8), we recognize a complete agreement, except that m is now replaced by the quantity μ. If we define the frequency R by § 96 eqn. (8), *Bohr's improved spectral formula* becomes

$$\nu = z^2 R \frac{\mu}{m} \left(\frac{1}{s^2} - \frac{1}{n^2}\right),$$

or,

$$\nu = \frac{R}{1 + \frac{m}{M}} \left[\frac{1}{\left(\frac{s}{z}\right)^2} - \frac{1}{\left(\frac{n}{z}\right)^2} \right] . \tag{16}$$

Thus, neither in the case of hydrogen nor in that of helium can the constant of the line spectrum be exactly equal to the *Rydberg constant*, as defined by § 96 eqn. (8), in the narrower sense of the term. Rydberg's constant may be denoted by R_∞, because eqn. (16) reduces to the earlier derived spectral formula when the atomic nucleus is assumed to be infinitely heavy in comparison with the revolving electron.

We shall denote the constants of the hydrogen and the helium spectra by R_H and R_{He}, and the nuclear masses of the two atoms by M_H and M_{He} respectively. By eqn. (16) we then have the relations

$$(17) \qquad \begin{cases} R_H = \dfrac{R_\infty}{1 + \dfrac{m}{M_H}}, \\ R_{He} = \dfrac{R_\infty}{1 + \dfrac{m}{M_{He}}}. \end{cases}$$

But the mass of the helium nucleus is approximately four times that of the hydrogen nucleus, so that

$$(18) \qquad R_{He} > R_H ;$$

and the explanation of the *displacement towards the violet of the helium lines* is at once evident.

Exact measurements by *Paschen* have given as values for the two quantities R_H and R_{He}

$$(19) \qquad \begin{cases} \dfrac{R_H}{c} = 109{,}677{\cdot}691 \ (\pm 0{\cdot}06) \\ \dfrac{R_{He}}{c} = 109{,}722{\cdot}144 \ (\pm 0{\cdot}04), \end{cases}$$

(the value for R_H has already been given in § 96).
Again, the *atomic weights* of hydrogen and helium are

$$(20) \qquad H = 1{\cdot}0077, \quad He = 4{\cdot}00.$$

We may therefore put

$$(21) \qquad \frac{M_{He}}{M_H} = 3{\cdot}97,$$

although, in a more exact calculation, we should also have to take into account the amount contributed toward the total atomic weight by the electron which revolves around the nucleus. Let us put, for brevity,

$$(22) \qquad \frac{m}{M_H} = x.$$

Then, by eqn. (17),

$$(23) \qquad \frac{R_H}{R_{He}} = \frac{1 + \dfrac{x}{3{\cdot}97}}{1 + x}.$$

From this it follows that

$$x = \frac{R_{He} - R_H}{R_H - \frac{1}{3{\cdot}97} R_{He}}, \tag{24}$$

and hence

$$\frac{1}{x} = \frac{M_H}{m} = \frac{109{,}678 - 27{,}638}{44{\cdot}453} = 1846\,. \tag{25}$$

From the *displacement towards the violet of the helium lines,* Bohr's theory thus enables us to *calculate* with great accuracy a fundamental constant of the electron theory, *viz.*, the *ratio between the mass of the hydrogen nucleus and the mass of the electron.*

Eqn. (25) may also be written in the form

$$1846 = (e/m) : (e/M_H)\,. \tag{26}$$

Now the quotient e/M_H is known from electrolytic measurements (*cf.* § 88), and hence *from the displacement toward the violet of the helium lines* we can also *calculate the specific charge of the electron.* The value thus obtained ($5{\cdot}30 \times 10^{17}$ E.S.U./gram) agrees exactly with that derived from direct measurements on the electric and the magnetic deflection of cathode rays; and this complete agreement naturally constitutes a great triumph for Bohr's theory.

On introducing the value given by eqn. (25) into the first of eqns. (17), we find

$$\frac{R_\infty}{c} = \frac{R_H}{c} \times 1{\cdot}000542 = 109{,}737{\cdot}1\,; \tag{27}$$

and hence, by § 96 eqn. (8),

$$109{,}737{\cdot}1 \times c = \frac{2\pi^2 e^4 m}{h^3}. \tag{28}$$

Eqns. (26) and (28) attain a still higher degree of accuracy when we also take into account the share contributed towards the atomic mass by the revolving electron. In that case, eqns. (24) and (26) are replaced by more complicated formulæ. In accordance with §§ 87 and 88, we can regard the values of e and of M_H as given with very great accuracy by Millikan's determinations and electrolytic measurements. If we then solve the improved eqns. (26) and (28) with respect to m and h, we obtain for these two quantities also more accurate values than would be possible from measurements on cathode rays, or on the photo-electric effect. As *Flamm* has shown,[1] we find

(29) $h = 6{\cdot}545 \times 10^{-27}$ erg. sec. ($\pm\ 0{\cdot}012 \times 10^{-27}$),

(30) $\frac{e}{m} = 5{\cdot}304 \times 10^{17}$ E.S.U./gram. ($\pm\ 0{\cdot}009 \times 10^{17}$),

and finally,

(31) $m = 9{\cdot}00 \times 10^{-28}$ gram. ($\pm\ 0{\cdot}02 \times 10^{-28}$).

§ 99. Sommerfeld's Theory of Elliptical Orbits.

Two years after Bohr had so successfully evolved his theory of spectra, *i.e.*, in 1915, this theory was the subject of an important extension by *Sommerfeld*, who replaced the circular orbits of the Bohr atomic model by more general *elliptical orbits*, in one *focus* of which the *atomic nucleus* is situated. We know, indeed, from mechanics that, in a periodic motion executed by a body about a centre of attraction under the influence of a force which is inversely proportional to the square of the distance, a *circular orbit* only represents a *special case* of the more general elliptical orbit.

When the considerations, which, based on the quantum theory, enabled Bohr to derive his formulæ for the various quantum states of the hydrogen atom, are now correspondingly modified and generalized, they lead to the following result (the process of reasoning is too complicated to be repeated here) :—

For the *semi-major axis* of the orbital ellipse, the same manifold and the same values are obtained as were found in Bohr's original theory for the radii of the circular orbits. Denoting the semi-major axis by a and the radius of Bohr's circular orbit by r, we have the relation

(1) $$(a_n)_{\text{SOMMERFELD}} = (r_n)_{\text{BOHR}},$$

where for r_n is to be substituted the right-hand side of § 94 eqn. (3). With every value of the major axis as thus determined, however, there may now be associated a *series* of possible values for the semi-minor axis (b), according to the simple formula

(2) $$b_{n,k} = a_n \frac{k}{n}, \quad (k = 1, 2 \ldots n).$$

[1] *Physikal. Zeitschr.*, **18** (1917), p. 515.

In Sommerfeld's theory, therefore, each orbit requires two *quantum numbers* for its definition: a "*principal quantum number*" (n), and a "*subordinate quantum number*" (k), which, however, can never be larger than the principal number. We shall speak symbolically of a state n_k, or of an n_k-orbit, thereby writing the subordinate number as subscript to the principal number.[1]

Contrasted with Bohr's theory, Sommerfeld's theory thus points to a *greater variety* of possible states. In Bohr's theory, there is only a simple series of states distinguished by the numbers

1, 2, 3, 4, 5, *etc.*

In Sommerfeld's theory there exists a greater manifold, characterized by the numerical symbols

1_1; 2_1, 2_2; 3_1, 3_2, 3_3; 4_1, 4_2, 4_3, 4_4; 5_1, 5_2, 5_3, 5_4, 5_5; *etc.*

The *principal quantum number* determines the *length* of the major axis; the *subordinate number* the *eccentricity*, and therefore the *shape* of the orbit. The smaller the subordinate number, for a given principal number, the more eccentric

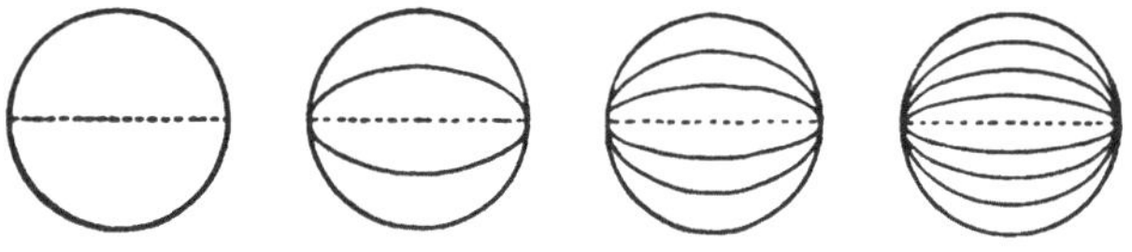

FIG. 65.—Orbital ellipses in accord with the quantum theory.

and elongated is the ellipse. If the *subordinate number is equal to the principal number*, the orbit is *circular*, and we have the special case on which the original Bohr theory was based.

In Fig. 65 the possible forms of the ellipses for a given principal quantum number are represented; for the sake of greater distinctness the ellipses are drawn concentrically,

[1] If the elliptical motion be described in polar coordinates, r and ϕ (r being the distance between the revolving electron and the nucleus situated at a focus), then, denoting the momenta with respect to r and ϕ by p_r and p_ϕ, Sommerfelds' theory postulates that § 92 eqn. (2) must be satisfied for each of the two momenta individually and independently. We must therefore have

$$\int p_\phi d\phi = n'h; \quad \int p_r dr = n''h,$$

where n' and n'' are two whole numbers, and both integrals are taken over a period; n' is called the *azimuthal*, and n'' the *radial quantum number*. The azimuthal quantum number is identical with the subordinate number, whilst the principal quantum number is equal to the sum of the azimuthal and radial quantum numbers.

whereas the actual arrangement is confocal. In Fig. 66 this confocal arrangement is reproduced, but on different scales for different values of the principal number : for the sake

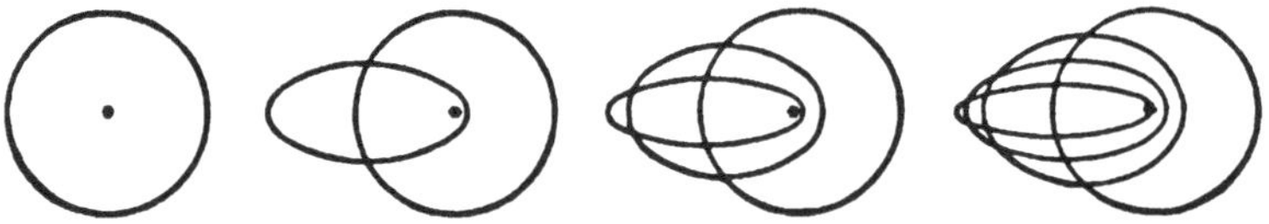

FIG. 66.—Orbital ellipses in accord with the quantum theory.

of clearness, the major axis has been given the same length throughout, although it is really proportional to the square of the principal quantum number. This fact is taken into account in Fig. 67, where we see reproduced on the proper scale the elliptical orbits for $n = 1$ to 4.

If n and s are two principal numbers, it follows from Sommerfeld's theory that a total of $n \times s$ *transitions* between the n- and s-*quantum* states are conceivable. This number, however, is considerably *reduced* by the so-called *selection principle*, which was introduced by *Rubinowicz* in 1918, and was also derived by *Bohr* as a special case of a more general principle (the "correspondence principle"). The selection principle states that *only those transitions are permissible* in which the *subordinate number varies by unity.*[2]

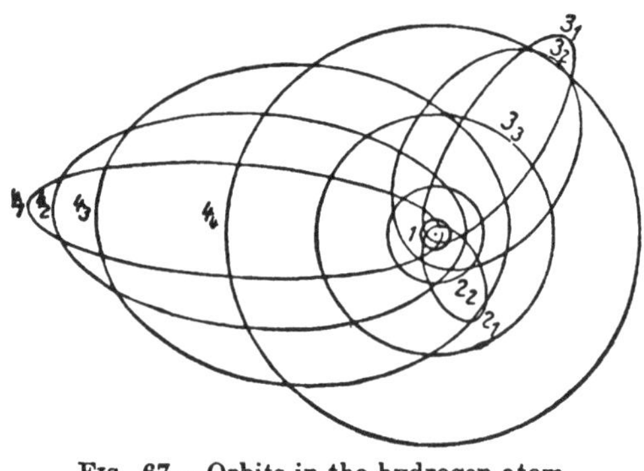

FIG. 67.—Orbits in the hydrogen atom.

Thus, for example, if we consider the states whose principal number is s, we can at once see that a total of $(2s - 1)$ regular transitions from states with an arbitrary principal number n will lead to these states, where $n > s$. For s_1 can be combined only with n_2, s_2 only with n_1 or n_3, s_3 only with n_2 or n_4, *etc.* Hence, only a single transition leads from any

[2] For further details of the selection principle and apparent exceptions thereto, as well as of the correspondence principle, *vide* Sommerfeld's "*Atombau und Spektrallinien*," ch. v. (English translation by H. L. Brose : London, Methuen), and Bohr, "*Three Essays on Spectra and Atomic Constitution*," 2nd essay (Cambridge University Press).

higher quantum state to the one-quantum state; three transitions lead from any higher to the two-quantum state; five from a higher to the three-quantum state, and so on; whereby the one-quantum state consists of a single modification, the two-quantum of two, and the three-quantum state of three modifications, *etc.* Each line in the Lyman series can therefore be due only to one definite transition, whereas each line in the optical Balmer series can be produced by three, and each line in the infra-red Paschen series by five transitions. According to the selection principle, a transition between two circular orbits is only possible when the principal numbers differ by unity; thus, only the first lines in all series can arise from a transition between two circular orbits.

As regards the *energy* possessed by an electron in an elliptical orbit, calculation (too complicated to be repeated here) leads, in the first place, to the result that it *depends only on the principal quantum number.* The value given by this calculation for the energy in an n-quantum orbit is identical with that resulting from Bohr's theory of circular orbits [§ 94 eqn. (8)], irrespective of the value of the subordinate number. The correctness of the results obtained in Bohr's theory is therefore in no way affected by Sommerfeld's development.

Thus, although the above calculation leads at first to the result that, for a given value of n, the energy in all the variously shaped orbits is the same, yet the calculation is based on the essential assumption that the elliptical motion of the electron under consideration is not disturbed in any way. On the other hand, if any kind of *perturbation* be present, the influence of the subordinate number must be felt, whether these perturbations be due to other electrons likewise revolving about the nucleus (which is always the case, except with hydrogen and ionized helium), or whether they be due to the establishment of an external electric or magnetic field. In these cases (which will be more fully discussed later), Sommerfeld's hypothesis of a subordinate quantum number proves to be extraordinarily fruitful. For in all these cases, the energy values appertaining to the same principal number must *differ* from each other; hence, in accordance with Bohr's *frequency condition,* a so-called *resolution of the spectral lines* must take place. No longer does a transition from an n-quantum to an s-quantum state give rise to a single line, as in the original theory which did not concern itself with subordinate numbers, but a *group* of more or less closely associated lines is produced.

It was, however, an especial triumph for the hypothesis of the subordinate quantum number that, in consequence of the different eccentricities of the orbits, the *spectral lines* even in the undisturbed *hydrogen atom* must be resolved. This resolution, which was predicted by Sommerfeld and which has also been completely confirmed by quantitative experiments, is simply a result of the minute *correction* to be applied to the above-mentioned calculations, when, in contradistinction to classical mechanics, the *theory of relativity* is taken into account. Whereas classical mechanics leads to the result quoted above, according to which the energy for a given principal number is independent of the subordinate number, a certain degree of dependence is nevertheless found to exist when, in accordance with the theory of relativity, the increase in mass with increasing velocity is taken into account; for, of course, the velocities differ in the variously shaped elliptical orbits.

§ 100. The Fine Structure of Spectral Lines.

When, in accordance with § 89 eqn. (6), the *dependence of the mass upon the velocity* demanded by the theory of relativity is taken into account, a complicated calculation (which cannot be repeated here) leads to the result that, in the first place, a steady *rotation of the perihelion of the ellipse*, slow compared with the period, is superimposed upon the motion in the ellipse (Fig. 68). Furthermore, we find for the *energy* in the state of the hydrogen atom (or hydrogen-analogue) characterized by the principal number n and subordinate number k the expression

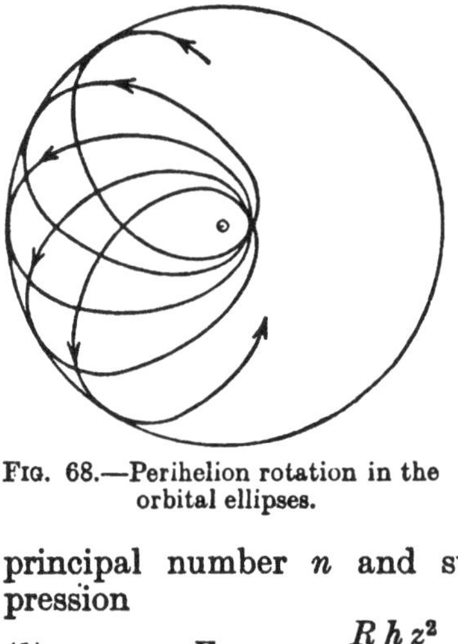

FIG. 68.—Perihelion rotation in the orbital ellipses.

$$(1) \qquad E_{n,k} = -\frac{R h z^2}{n^2} + \frac{3R h z^4 \alpha^2}{4n^4} - \frac{R h z^2 \alpha^2}{n^3 k}.$$

In this expression R, as usual, denotes Rydberg's constant, h the elementary quantum of action, z the nuclear charge number, and α the ratio of the velocity in the normal hydrogen orbit to the velocity of light.

We see from eqn. (1) that the value of the energy is composed of three terms. The first term has the same value as E_n in Bohr's theory. The second and third terms are *relativistic correction terms*, but only the third depends also on the subordinate number k. Thus, by eqn. (1), to every principal quantum number n there correspond n *different energy-levels*. The larger k is, the smaller is the final correction term in eqn. (1), and since the sign of this term is negative as is the sign of the energy, a *larger subordinate number* corresponds to a *higher energy-level* (because of the smaller absolute magnitude of the energy). Fig. 69 gives a schematic representation of the energy-levels, although, to make it clearer, the scale has been distorted.

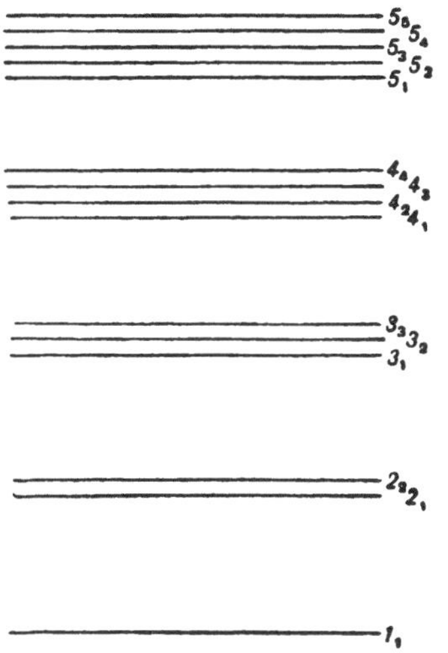

FIG. 69.—Scheme of energy-levels.

The manifold of energy values is, of course, associated with a manifold of *spectral terms*. The frequency difference of the two spectral terms thus corresponding to the two-quantum state of the hydrogen atom is known as the *frequency difference of the hydrogen doublet* ($\Delta\nu_H$). This quantity is therefore equal to the difference between the values of the last term in eqn. (1) divided by h (with $z = 1$ and $n = 2$), when k is put equal to one or to two respectively. Hence

$$\Delta\nu_H = \frac{R\alpha^2}{2^3}\left(\frac{1}{1} - \frac{1}{2}\right), \tag{2}$$

or,

$$\Delta\nu_H = \frac{R\alpha^2}{16}. \tag{3}$$

Substituting for α the value given in § 96 eqn. (13), and for R the value given in § 98 eqn. (27), we find

$$\frac{\Delta\nu_H}{c} = 0{\cdot}365 \text{ cm.}^{-1}. \tag{4}$$

The term differences in the other quantum states stand, of course, in simple numerical relations to the frequency difference of the hydrogen doublet. If, for example, we consider the *three-quantum* state, we find from eqn. (1) for the spectral terms (denoted by T) corresponding to its modifications

$$(5) \qquad T_{3,3} - T_{3,2} = \frac{R\alpha^2}{3^3}\left(\frac{1}{2} - \frac{1}{3}\right) = \frac{8}{81}\Delta\nu_H ,$$

and

$$(6) \qquad T_{3,2} - T_{3,1} = \frac{R\alpha^2}{3^3}\left(\frac{1}{1} - \frac{1}{2}\right) = \frac{8}{27}\Delta\nu_H ,$$

so that

$$(7) \qquad T_{3,3} - T_{3,1} = \frac{32}{81}\Delta\nu_H .$$

Let us now consider the hydrogen line H_α, *i.e.*, the first line of the optical series, which arises by a transition from the three-quantum to the two-quantum state. In accordance with the selection principle, there are three regular possible ways in which this transition may occur, *viz.*,

$$\begin{aligned} &(\mathrm{I}) \quad 3_2 \longrightarrow 2_1 , \\ &(\mathrm{II}) \quad 3_3 \longrightarrow 2_2 , \\ &(\mathrm{III}) \quad 3_1 \longrightarrow 2_2 . \end{aligned}$$

For the frequency differences of the spectral lines we have then the relations (*cf.* Fig. 70)

$$(8) \begin{cases} \mathrm{I} - \mathrm{II} = \dfrac{73}{81}\Delta\nu_H , \\ \mathrm{II} - \mathrm{III} = \dfrac{32}{81}\Delta\nu_H , \\ \mathrm{I} - \mathrm{III} = \dfrac{35}{2}\Delta\nu_H . \end{cases}$$

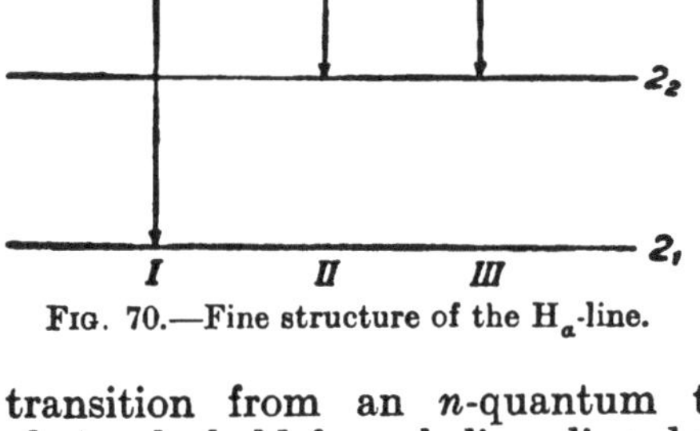

FIG. 70.—Fine structure of the H_α-line.

Considerations exactly similar to those used as an illustration in the case of the H_α-line may, of course also be applied to all the other hydrogen lines. Moreover, all formulæ which hold for a hydrogen line that arises by a transition from an n-quantum to an s-quantum state obviously hold for a helium line due to a similar transition, provided that the quantity $\Delta\nu_H$ be replaced by the quantity $\Delta\nu_{He}$, which is defined in a precisely analogous manner. The last term in eqn. (1) shows us that, on account of the proportionality to the fourth power of the nuclear charge number,

$$\text{(9)} \qquad \Delta\nu_{He} = 16\,\Delta\nu_{H}\,,$$

so that, by eqn. (3),

$$\text{(10)} \qquad \Delta\nu_{He} = R\,\alpha^2\,.$$

Now, in the case of hydrogen, the observation and measurement of the fine structure is extremely difficult, because the intervals between the lines are so small. In the visible region of the spectrum, the difference in wave-length corresponding to the quantity $\Delta\nu_H$ is only about one-twentieth to one-fifth of 1 Å.U.[1] In the case of helium, however, conditions are much more favourable to observation, because, in accordance with eqn. (10), the separation for helium is 16 times as large as for hydrogen (the values of the initial and final quantum numbers being the same), so that the interval between neighbouring lines is considerably larger.

Whereas in the Balmer series the lines only appear double even in spectroscopic apparatus of high resolution (simply because stronger resolution is not available), it has been possible to measure accurately the individual lines as resolved in the spectrum of ionized helium. In this way the value of $\Delta\nu_{He}$ could be experimentally determined, and the value of $\Delta\nu_H$ followed, as it is one-sixteenth thereof. Measurements by *Paschen* gave as a result

$$\text{(11)} \qquad \frac{\Delta\nu_H}{c} = 0{\cdot}3645\ \text{cm.}^{-1}\ (\pm\ 0{\cdot}0045)\,.$$

The complete agreement between the theoretically calculated and the experimentally determined values constitutes a brilliant triumph not only for Bohr's and for Sommerfeld's conceptions, but also for the theory of relativity, the mass formula of which is splendidly confirmed by this agreement.[2]

[1] It follows from the elementary relation

$$c = \lambda\nu$$

that

$$\Delta\lambda = -\frac{c}{\nu^2}\Delta\nu = -\lambda^2\frac{\Delta\nu}{c}.$$

On giving λ a value of between 4×10^{-5} and 8×10^{-5} cm., and $\Delta\nu$ the value in eqn. (4), we find for $\Delta\lambda$ a value of from 5 to 20×10^{-10} cm.

[2] Successful attempts have recently been made to measure the fine structure in the hydrogen spectrum itself; the results confirm Sommerfeld's theory completely. [*Cf.* G. M. Shrum, *Proc. Roy. Soc.*, *A*. **105** (1924), p. 259.] In this connection, attention may be drawn to the fact that it is possible to determine the constants e, h, and m solely from spectroscopic measurements, if Faraday's electrochemical constant be assumed as given. For that purpose, three equations in the theory of spectra are available, *viz.*, § 98 eqns. (26) and (28), and § 100 eqn. (11) [in combination with § 100 eqn. (3).]

§ 101. The Stark Effect.

The wonderful regularities of the hydrogen spectrum show most clearly that all atomic phenomena are governed by *simple integral relations*, but this fact finds even plainer expression in the *changes* produced in the *hydrogen spectrum* by an *external electric field.*

The resolution of spectral lines in an electric field was discovered by *Stark* in 1913, and is therefore generally known as the *Stark effect* (Fig. 71).[1] The *theory* of the effect was developed from the *quantum principle* by *Schwarzschild* and *Epstein* in 1916. Between the results of theory and experi-

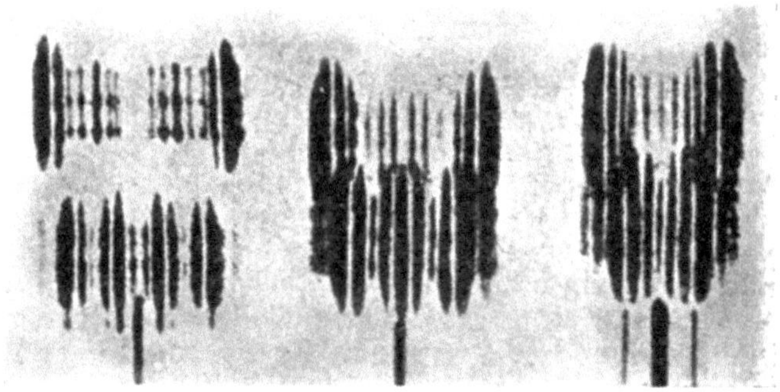

FIG. 71.—Stark effect in the hydrogen lines H_δ, H_γ, and H_β.

ment such an excellent agreement was found to exist that the Stark effect must be regarded as one of the most brilliant empirical confirmations of the quantum principle. The quantum theory of the Stark effect is, however, too difficult to be described in detail here, and we shall only state and discuss the result derived from the theory. According to this, the frequency (ν) of a hydrogen line in an electric field of strength (E) is given by the expression

$$(1) \qquad \nu = \nu_0 \pm z\, E\, \sigma\,,$$

where ν_0 is the frequency of the line in the absence of an external field, and σ is a universal constant of the form

[1] The discovery of the Stark effect only succeeded after the skilful surmounting of great experimental difficulties. The observation of the resolution of the lines is, in fact, only possible in strong fields, which cannot be produced in a Geissler tube, because it is too good a conductor of electricity. Stark therefore used canal ray particles to produce the spectral lines to be resolved, while he introduced a subsidiary electrode into the immediate neighbourhood of the cathode emitting the canal rays, thereby producing a very large potential difference over the short interval of a few millimetres.

$$\sigma = \frac{3h}{8\pi^2 e m}, \tag{2}$$

h denoting the elementary quantum of action, e and m the charge and mass of an electron. The "*ordinal number*" z, which is always an *integer*, depends on the principal quantum numbers n and s which characterize the initial and final states on the emission of the spectral line; it is given by the relation

$$z = nn' \pm ss', \quad \begin{cases} n' = 0, 1, 2 \ldots (n-1), \\ s' = 0, 1, 2 \ldots (s-1). \end{cases} \tag{3}$$

The numbers n' and s' can thus assume all integral values between zero (and including it) and n and s (but excluding them) respectively.

The theory also leads to a rule governing the direction of vibration in the light corresponding to the spectral lines. If the expression

$$(n - s) + (n' - s') = l \tag{4}$$

is an *even* number the *electric vibrations* are *parallel* to the direction of the electric field-strength. On the other hand, if l is an *odd* number, the electric vibrations will occur *transversely* to the direction of the field-strength. There are thus *two different classes of components* to be distinguished, which may be called the p- and the t-components from the initial letters of the words "parallel" and "transverse."

Now, as is well known, light waves are always propagated only in directions normal to the direction of vibration. Hence, if we are observing the resolution in the direction of the lines of force—the so-called *longitudinal effect*—the p-components will not be visible, and only the t-components will be seen in an unpolarized state. If, however, we observe the resolution at right angles to the lines of force—the so-called *transverse effect*—all components will be visible, but the p- and the t-components will be readily distinguishable owing to their different polarization.

The results of the theory may be discussed by reference to two simple *examples*, *viz.*, the *two violet hydrogen lines* H_γ and H_δ. As has already been mentioned, the line H_γ is produced as third line of the Balmer series by a transition from a five-quantum to a two-quantum state. Hence, for the line H_γ, $n = 5$ and $s = 2$, so that n' can assume the values 0, 1, 2, 3, 4, but s' can only be 0 or 1. The ordinal number z is therefore equal either to $5n'$ or to $(5n' \pm 2)$. In the first case the number l, as defined by eqn. (4), is equal to $(3 + n')$; in the second case, it is equal to $(2 + n')$. The ordinal numbers of the p components, for which l must be even, would thus be $5n'$ for odd values of n', and $(5n' \pm 2)$ for even values. Hence the p-components of the line H_γ would have the ordinal numbers

$$2, 5, 8, 12, 15, 18, 22.$$

On the other hand, the ordinal numbers of the t-components, for which l must be odd, would be $5n'$ for even values of n', and $(5n' \pm 2)$ for odd values. Hence the t-components of the line H_γ would have the ordinal numbers

$$0, 3, 7, 10, 13, 17, 20.$$

The complete *agreement* between theory and experiment is to be seen in Fig. 72, in which are reproduced the results of Stark's observations, the length of each line representing the intensity of the component in question, as estimated by Stark from the image on the photographic plate.

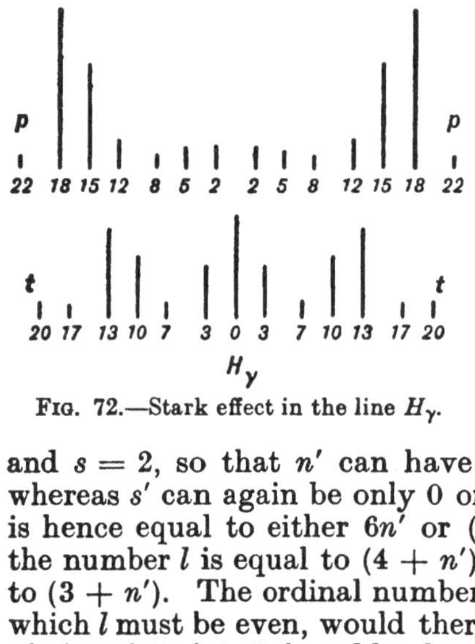

Fig. 72.—Stark effect in the line H_γ.

The fourth line in the Balmer-series, H_δ, arises on a transition from a six-quantum to a two-quantum state. For this line we have thus to put $n = 6$ and $s = 2$, so that n' can have the values 0, 1, 2, 3, 4, 5, whereas s' can again be only 0 or 1. The ordinal number z is hence equal to either $6n'$ or $(6n' \pm 2)$. In the first case the number l is equal to $(4 + n')$; in the second it is equal to $(3 + n')$. The ordinal numbers of the p-components, for which l must be even, would therefore be $6n'$ for even values of n', and $(6n' \pm 2)$ for odd values. Hence the p-components of the line H_δ would have the ordinal numbers

$$0, 4, 8, 12, 16, 20, 24, 28, 32.$$

On the other hand, the ordinal numbers of the t-components, for which l must be odd, would be $6n'$ for odd values of n', and $(6n' \pm 2)$ for even values. The t-components of the line H_δ would therefore have the ordinal numbers

$$2, 6, 10, 14, 18, 22, 26, 30.$$

As may be seen from Fig. 73, the agreement between theory

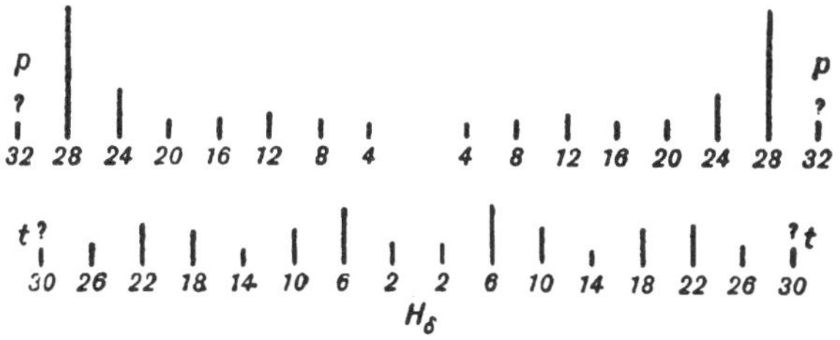

FIG. 73.—Stark effect in the line H_δ.

and the previously known experimental results is excellent also in the case of the line H_δ.[2]

§ 102. The Zeeman Effect.

Just as in an electric field, a *resolution of spectral lines* also occurs in a *magnetic field*. For in accordance with § 71 eqn. (4) the electron, a change in the orbit of which produces a spectral line, is acted on in a magnetic field by a *mechanical force*

$$\text{(1)} \qquad \mathbf{K} = -\frac{e}{c}\left[\mathbf{v}\mathbf{H}\right],$$

where $\mathbf{v}$ is the velocity and $\mathbf{H}$ the magnetic field-strength. Again, if we refer the motion of the electron to a *coordinate system rotating with an angular velocity* $\mathbf{w}$, the theory of relative motion shows the value of the *Coriolis force*, which has then to be introduced as an imaginary accessory force, to be [according to § 14 eqn. (2)]

$$\text{(2)} \qquad \mathbf{K}' = 2m\,(\mathbf{v}\mathbf{w}).$$

Provided that the centrifugal force may be neglected, the actual motion of the electron in a magnetic field therefore appears to be unaffected by the field with respect to a coordinate system rotating with an angular velocity $\mathbf{w}^*$, where $\mathbf{w}^*$ is oppositely directed to $\mathbf{w}$, and has a value given by eqns. (1) and (2) :—

$$\text{(3)} \qquad \mathbf{w}^* = \frac{e}{2mc}\mathbf{H}.$$

Thus, upon the motion executed by the electron in the undisturbed atom there is superimposed a *uniform rotation*

[2] The occurrence of the components with the highest ordinal numbers is, it is true, doubtful in the experimental results, whilst the line of zero ordinal number is absent, possibly because its intensity is too small.

of the orbital ellipse about an axis parallel to the lines of force, since, by eqn. (3), the vectors **w** and **H** are similarly directed. The frequency of rotation is equal to the magnitude of the angular velocity divided by 2π; hence according to eqn. (3), it is given by

$$\omega = \frac{e}{4\pi m c} H. \tag{4}$$

When the considerations of the quantum theory, which hold for the spatial configuration, the size, and the shape of the orbital ellipses in the undisturbed atom, are applied to a coordinate system rotating with the angular velocity determined by eqn. (3), it can be shown that the value of the energy in an arbitrary orbit is given by

$$E = E_0 + k' \omega h. \tag{5}$$

In this expression E_0 denotes the energy in the absence of a magnetic field, h the elementary quantum of action, and k' a whole number which, as follows from the quantum theory, can only change by unity or not at all as a result of a transition. In the latter case, Bohr's frequency condition shows that, for an arbitrary spectral line,

$$\nu = \nu_0, \tag{6}$$

where ν_0 is the frequency in the undisturbed atom. On the other hand, if k' changes by ± 1, it follows from the frequency condition that

$$\nu = \nu_0 \pm \omega. \tag{7}$$

Hence

$$\pm (\nu - \nu_0) = \frac{e}{4\pi m c} H. \tag{8}$$

Just as in the Stark effect, the resolution in a magnetic field is also proportional to the field-strength. There is, however, an essential difference between the two effects, for the magnetic resolution gives rise only to two lines in addition to the original line, and it is completely independent of all quantum numbers, and therefore of the initial and the final state. Even the elementary quantum of action itself is lacking in eqn. (8), for it disappeared from the calculation on the application of the frequency condition.

It also follows on the basis of the quantum theory that the emitted spectral line must be *circularly polarized* when k' changes by unity, but *plane polarized* when k' remains unchanged after a transition. In that event, the linear electric vibrations take place in the direction of the magnetic lines of force, so that, if the resolution be observed in this direction, the component ν_0 will be missing. Hence, in the *longitudinal effect* a *doublet* must be visible, but in the *transverse effect* a *triplet*, *viz.*, all three lines $(\nu_0 + \omega)$, ν_0, and $(\nu_0 - \omega)$. It follows from the theory that the two outer components must be *circularly polarized in opposite senses*, and in such a way that the sense of polarization in the component of smaller wave-length must be the same as the sense of the electric current which would produce, within a solenoid, a magnetic field of the same direction as the field applied.

The resolution of spectral lines in a magnetic field was experimentally discovered by *Zeeman* in 1896, on which account this phenomenon is commonly known as the *Zeeman effect*. In the case of spectral lines which, in the absence of a magnetic field, are simple, complete quantitative agreement is found to exist between the results of experiment and theory. In the case of originally *multiple* lines, however, considerably more complicated resolution phenomena occur, and these are known as the *anomalous Zeeman effect*, which cannot be discussed here.[1]

Immediately after Zeeman's discovery, *Lorentz* developed a *magneto-optical theory* which, since it was evolved four years before the quantum theory, was of course in no way concerned with the elementary quantum of action and quantum numbers. Lorentz's theory was simply based on the assumption that the electrons, to which spectral lines are due, are bound to a definite position of equilibrium by quasi-elastic forces.[2] Although this conception has been superseded by the later developments of atomic physics, Lorentz's theory nevertheless led to eqn. (8) and to the correct rule of polarization. Eqn. (8), as already obtained by Lorentz, thus rendered possible the first determination of the ratio e/m, for measured values of the resolution and of the magnetic field-strength. Moreover, the negative character of the electronic charge was rendered evident in view of the polarization rule,

[1] For a further discussion of the anomalous Zeeman effect *vide*, for example, Sommerfeld's "*Atomic Structure and Spectral Lines*," Ch. vi., § 7; the theory of the normal Zeeman effect is to be found in Ch. v., § 6.—The resolutions observed in the Zeeman effect are considerably smaller than those of the Stark effect. Thus, the resolution in the strongest magnetic fields only amounts to about one Å.U., whereas, in an electric field of 30,000 volt/cm., for example, the resolution of the blue hydrogen line (H_β) extends over 13 Å.U.—A very interesting application of the Zeeman effect is to astrophysical problems, and by its aid *Hale* discovered strong magnetic fields in the *sun*.

[2] Forces are termed quasi-elastic when they are proportional to the distance from a position of equilibrium. *Cf.* § 93, note 2.

by the observed sense of the circularly polarized lines. Since the quotient e/m proved to be more than a thousand times as large as in the case of the ionized hydrogen atom, the Zeeman effect first revealed to physicists the existence of particles considerably lighter than the lightest of the atoms, up till then regarded as indivisible.

CHAPTER XIII

RÖNTGEN RAYS

§ 103. Origin of Röntgen Rays.

WHEN the motion of *rapidly moving electrons* is *checked*, the resulting phenomenon represents a *reversal of the photoelectric effect*. The kinetic energy of corpuscular radiation is then partially transformed into *electromagnetic wave* energy ; and the arrest of a rapid motion of electrons thus produces the remarkable rays discovered by *Röntgen* in 1895, and known as *Röntgen rays* or X-rays.

The most striking property of Röntgen rays is their extraordinary *power of penetration*. They are scarcely absorbed at all by objects which are completely opaque to ordinary light. Among their other properties, particular mention should be made of their strong ionizing power, and the fact that they experience no deflection whatever either in a magnetic or in an electric field ; it thus appeared that Röntgen rays could not be of a corpuscular nature. The penetrating power, or *hardness*, was soon found to be a reliable characteristic for *distinguishing* between various kinds of Röntgen rays ; the greater or smaller the power of penetration, the harder or softer is the relevant radiation said to be.

In Röntgen's original experiments, the newly discovered rays originated in those portions of the glass wall of a cathode ray tube which were bombarded by the cathode rays.[1] The cathode ray tube gradually developed into a *Röntgen ray tube* when, on the one hand, a so-called *anti-cathode, viz.*, a plate of some heat-resistant metal (platinum or tungsten), was introduced into the interior of the tube to serve as emission centre of the Röntgen rays, and, on the other hand, a suitable shape was given to the cathode so as to concentrate the cathode rays at a point on the anti-cathode.[2] It

[1] The discovery was due to Röntgen's observation that a preparation of barium platinocyanide (which fluoresces strongly under the influence of cathode rays) also becomes luminous in the neighbourhood of a cathode tube at places which are out of reach of the rectilinearly propagated cathode rays.

[2] For this purpose the cathode is shaped like a concave mirror. It is necessary to concentrate the cathode rays at a point in order that the Röntgen rays thus generated may produce sharp images.

was soon discovered that the hardness of the Röntgen rays increases with increasing tube potential, and, in tubes of modern construction,[3] it can be regulated at will by means of the potential.

The absence of any electric or magnetic deflection in itself led physicists to suspect that Röntgen rays represent electromagnetic waves. This concept found strong support in the demonstration, afforded by *Barkla* in 1905, that the *secondary Röntgen radiation* emitted by a body upon which primary Röntgen rays impinge is partially *polarized*.[4] The *transverse nature* of Röntgen rays therefore appeared proved.

Moreover, the relationship between Röntgen rays and light must also appear obvious from purely theoretical considerations. Röntgen rays are produced when quickly moving electrons lose their velocity. Now, since an electron is surrounded by an electromagnetic field so long as it is in motion, but only by an electrostatic field from the moment that it comes to rest, it follows at once that the transition to a state of rest, or so-called retardation, must give rise to the *propagation of an electromagnetic disturbance*, a so-called electromagnetic *pulse*. Having recognized this fact, *Stokes* and *Wiechert* in 1896 evolved a theory of the origin of Röntgen rays, in which they regarded the radiation emitted by the anti-cathode as essentially an *impulse radiation*.

If we now make the plausible assumption that the electron is slowed down along the portion of the path traversed *inside an atom*, we find for the " width of the pulse " (corresponding to the wave-length) a value of the order of 10^{-9} cm., or about one ten-thousandth of the wave-length of visible light. Furthermore, Einstein's law, which governs the mutual transformation of corpuscular and undulatory radiation, leads to the same order of magnitude for the wave-length of Röntgen radiation. If, for example, we take the potential (V) of a Röntgen tube as being 100,000 volts (= approx. 300 E.S.U.), and equate the products $e V$ and $h \nu$, we obtain a value of the order of 10^{19} sec.$^{-1}$ for the frequency. This value is again about ten thousand times as large as the frequency of visible light.

[3] In the Röntgen ray tubes devised by *Lilienfeld* and *Coolidge* this is effected essentially by making an incandescent wire the source of the electrons, and the vacuum is so high that a true discharge is no longer possible. The chief difficulty connected with the use of the older tubes, *viz.*, that the hardness of the Röntgen radiation depended on the gas pressure, is thus avoided.

[4] This was proved by the fact that the intensity of the tertiary rays, emitted by a body bombarded by the secondary rays, varied with the position relative to the body.

The assumption of so small a wave-length finds additional support in the failure of all early experimental efforts to demonstrate *interference* or *diffraction* phenomena with Röntgen rays, by means of any ordinary optical method.[5] The usual diffraction gratings could not be applied, for, as follows from the theory of the grating (and as we shall show in detail in the next section), the interval between two lines of the grating must not be much larger than the wave-length of the light investigated. Hence, researches on Röntgen rays would require gratings upon which there were millions of lines to a millimetre. The technical impossibility of preparing such fine gratings seemed to exclude almost every prospect of physicists ever being able to determine the exact wave-lengths of Röntgen rays, until, in 1912, *Laue* conceived the ingenious idea of making use simply of *crystals*, instead of artificial gratings, for diffracting the rays. Laue's idea was based on a theory developed by the mineralogist *Bravais* about 1850, according to which the regular form of crystals must be due to a *regular arrangement of the atoms in the crystal.* It also followed from the known density of crystals, and from the known value of the Loschmidt number, that the mutual distances between the atoms in crystals must be of the order of 10^{-8} cm. Provided that the individual atoms behave as diffraction centres, it was therefore to be expected that *crystals* would actually *behave as gratings when used with Röntgen rays.*

§ 104. Space-Lattices.

In order to understand the behaviour of crystals with respect to Röntgen rays, we shall first consider, as the simplest of all gratings or lattices, a *line-lattice* in which the lattice points are arranged at equal intervals along a straight line. The constant interval between two successive lattice points is called the *lattice constant.*

We now suppose a *wave* to be incident on the lattice, the direction of the ray making with the direction of the lattice an angle whose cosine is α_0.[1] The individual lattice points may be regarded as diffraction centres, each of them consti-

[5] The first diffraction experiments with Röntgen rays were made by Haga and Wind in 1899. Their experiments were repeated by Walter and Pohl in 1908, using an improved method, and Koch's careful photometric measurements, later carried out on their exposures, enabled Sommerfeld in 1912 to calculate from the very slight indication of diffraction a width of about 4×10^{-9} cm. for the predominant pulse. At the present time, however, these experimental results are only of historical interest.

[1] α_0 is therefore the sine of the angle of incidence.

tuting a centre of excitation of new waves. These waves travel in all directions, but it will be immediately evident that they will be *diffracted with especial intensity in those directions*, for which the *difference in path* of the rays proceeding from two successive lattice points amounts to an *integral multiple of a wave-length*. In fact, more accurate theoretical considerations (which we cannot discuss here) show that, in consequence of mutual *interference* between the diffracted rays, an *appreciable* diffraction of the light occurs only in the directions just given.

We shall now denote by α the cosine of the angle included between the direction of the lattice and one of the rays thus diffracted. Then, if a be the lattice constant, the difference

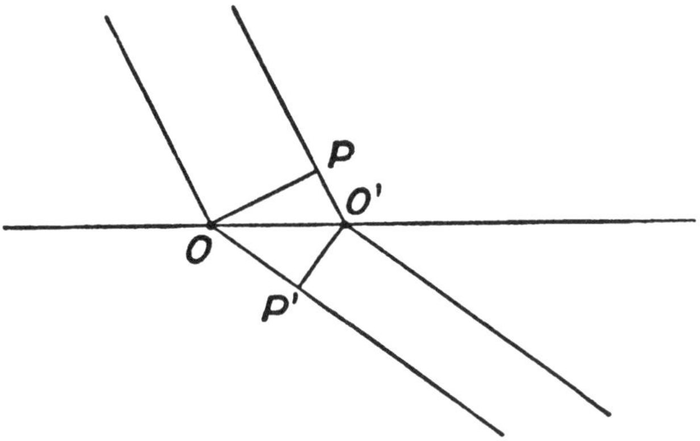

FIG. 74.

in path of the rays diffracted in this direction by two neighbouring lattice points is equal to $a\,(\alpha - \alpha_0)$, as is obvious from Fig. 74. (If O and O' be the two lattice points, and $O\,P$ and $O'\,P'$ the perpendiculars dropped on the incident and diffracted rays respectively, the difference in path is equal to the difference between the intervals $O\,P'$ and $O'\,P$.) Rays of given wave-length λ are therefore diffracted only in those directions (with a cosine α), for which the relation

$$(1) \qquad a\,(\alpha - \alpha_0) = n\,\lambda$$

is satisfied for *integral* values of the so-called *ordinal number* (*order*) n. Conversely, if the incident radiation is composed of rays of all wave-lengths, it is *resolved into spectra* by the diffraction, a different diffraction spectrum corresponding to each value of the ordinal number n. (In passing it may be mentioned that, if we turn the diffracted rays about the direction of the lattice in Fig. 74, we obtain the directions of the light reflected by diffraction, though, of course, it is not a case of reflection in the true sense of the word.)

It follows from eqn. (1) that diffraction can never occur, unless

$$(2) \qquad a > \lambda .$$

Moreover, a must not be *much* larger than λ, because otherwise, in the spectra of low order which are of particular importance, the difference $(\alpha - \alpha_0)$ would be too small and the spectral resolution insufficient.[2]

From the one-dimensional line-lattice, let us now pass to the two-dimensional *plane-lattice*, which will be formed by points with the coordinates

$$(3) \qquad x = k_1 a, \quad y = k_2 b$$

constructed in a plane coordinate system, where a and b are two lattice constants, and k_1 and k_2 can assume every integral value, either positive or negative. The coordinate axes may be inclined at any angle to each other, but we shall limit ourselves to a rectangular system in which the two lattice constants are identical (Fig. 75).

FIG. 75.

Let us suppose rays to impinge on such a plane-lattice, making with the x- and the y-axis angles whose cosines are α_0 and β_0. Eqn. (1) then expands into a system of two equations, *viz.*,

$$(4) \qquad \begin{cases} a(\alpha - \alpha_0) = n_1 \lambda , \\ a(\beta - \beta_0) = n_2 \lambda , \end{cases}$$

where α and β are the cosines of the angles included between the diffracted ray and the x- and the y-axis. Since the two equations are mutually independent, they can be solved with respect to the unknowns α and β. Hence, to every value of the wave-length there corresponds, for a given direction of incidence and given ordinal numbers n_1 and n_2, a definite direction of diffraction determined by the values of α and β [according to eqn. (4)]. Thus a *plane-lattice* also leads to a resolution into spectra of the incident radiation, this resolution forming a *dual manifold of diffraction spectra.*[3]

[2] In the best optical line-gratings (*Rowland* concave gratings), a is always less than ten times the wave-length.

[3] Diffraction phenomena of this kind are to be observed when a street lamp is viewed through a thin umbrella.

We accomplish the transition from the plane-lattice to *space-lattices* by imagining a three-dimensional coordinate system in a space filled with points whose coordinates are

$$(5) \qquad x = k_1 a, \quad y = k_2 b, \quad z = k_3 c\,;$$

where a, b, c are lattice constants, and the numbers k_1, k_2, k_3 assume every integral value, positive and negative. When the coordinate system is rectangular, we term the space-lattice *orthogonal*. In that case, if the three lattice constants a, b, c are all different, we define the space-lattice as being *orthorhombic*; we call it *tetragonal*, when two of the three lattice constants are identical, and *regular* or *cubic* when all three have the same value. If only two of the three coordinate axes are at right angles to each other, the space-lattice is said to be *monoclinic*; and if all three axes are inclined to each other, it is termed *triclinic*.

The following considerations are limited to the special case of a *cubic* space-lattice, in which, therefore, the axes are orthogonal and the lattice constants identical. Let rays meet such a space-lattice in a direction including with the three coordinate axes angles whose cosines are $\alpha_0, \beta_0, \gamma_0$. In place of eqns. (4), we then have the following *triplet of equations*:—

$$(6) \qquad \begin{cases} a(\alpha - \alpha_0) = n_1 \lambda\,, \\ a(\beta - \beta_0) = n_2 \lambda\,, \\ a(\gamma - \gamma_0) = n_3 \lambda\,. \end{cases}$$

There is, however, an essential difference between this triplet of equations and eqns. (4). In eqns. (4) we were able to regard as unknowns two magnitudes which completely determined the direction of the diffracted ray for a given direction of incidence, for given wave-length and order; the number of unknowns was the same as the number of equations. In the present case, however, there are only two unknowns in the three equations (6), if we take again as given the direction of incidence, wave-length, and order. For the three directional cosines α, β, γ determining the direction of the diffracted ray are not mutually independent; two of these cosines are always connected with the third in accordance with the elementary relation [4]

$$(7) \qquad \alpha^2 + \beta^2 + \gamma^2 = 1\,.$$

Since we cannot regard the ordinal numbers as unknowns, inasmuch as they can only assume perfectly definite, per-

[4] This well-known relation also follows from the fact that α, β, γ are the components of a unit-vector.

missible values, *viz.*, integral values (and, indeed, only the lower values require consideration), it appears that, in the three equations, the *wave-length* itself must also be an *unknown* for a given direction of incidence. It is not, indeed, limited to a single value, but it may have any of a *series of possible values* corresponding to the different possible integral values of n_1, n_2, n_3.

These values of the wave-lengths can easily be found from eqns. (6) and (7). We write eqns. (6) in the form

$$(8)\qquad \begin{cases} \alpha = \alpha_0 + n_1 \dfrac{\lambda}{a}, \\ \beta = \beta_0 + n_2 \dfrac{\lambda}{a}, \\ \gamma = \gamma_0 + n_3 \dfrac{\lambda}{a}. \end{cases}$$

We then square these equations and add them together, taking due account of eqn. (7) and the analogous equation connecting the directional cosines $\alpha_0, \beta_0, \gamma_0$. In this way we find

$$1 = 1 + 2\frac{\lambda}{a}(n_1\alpha_0 + n_2\beta_0 + n_3\gamma_0) + \frac{\lambda^2}{a^2}(n_1^2 + n_2^2 + n_3^2)$$

or,

$$(9)\qquad \lambda = -2a\frac{n_1\alpha_0 + n_2\beta_0 + n_3\gamma_0}{n_1^2 + n_2^2 + n_3^2}.$$

In contrast to line- and plane-lattices, a *space-lattice* thus gives *no continuous diffraction spectrum*; its diffractive effect, on the contrary, is to *single out from the incident radiation* those rays whose *wave-lengths* satisfy eqn. (9) for *integral* values of n_1, n_2, n_3.

Eqns. (6) also enable us readily to calculate the *angle* included between the *diffracted* and the *incident ray*. Denoting this angle by 2θ, we have [by § 3 eqn. (14)]

$$(10)\qquad \cos 2\theta = \alpha\alpha_0 + \beta\beta_0 + \gamma\gamma_0 .$$

On squaring eqns. (6) and adding, we find [bearing eqn. (7) in mind]

$$a^2(2 - 2\cos 2\theta) = \lambda^2(n_1^2 + n_2^2 + n_3^2).$$

But, in accordance with a well-known trigonometrical formula,

$$(11)\qquad \sin\theta = \sqrt{\frac{1 - \cos 2\theta}{2}}.$$

Hence,

$$\sin\theta = \frac{\lambda}{2a}\sqrt{n_1^2 + n_2^2 + n_3^2}\,. \tag{12}$$

From the diffracting lattice point O, we cut off an interval OA one unit in length in the direction of the incident ray produced, and an equal interval OB in the direction of the diffracted ray (Fig. 76). We then construct the *median plane* MM' between the incident and the diffracted ray, by bisecting the angle AOB and erecting through the bisector a plane at right angles to the plane AOB. Every point on the median plane must therefore be equidistant from A and B.

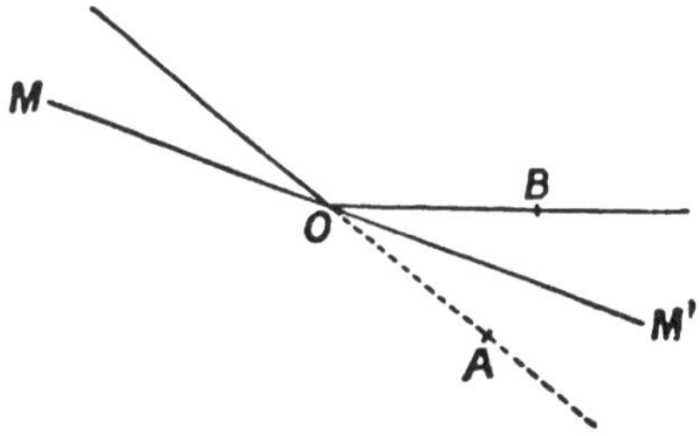

FIG. 76.

With the point O as origin, we next construct a coordinate system with axes parallel to the original axes. With respect to this system, the coordinates of the point A are $\alpha_0, \beta_0, \gamma_0$, while those of the point B are α, β, γ. Since an arbitrary point (x, y, z) on the median plane is equidistant from A and B, it must satisfy the relation

$$(x-\alpha_0)^2 + (y-\beta_0)^2 + (z-\gamma_0)^2 = (x-\alpha)^2 + (y-\beta)^2 + (z-\gamma)^2\,. \tag{13}$$

Bearing in mind eqn. (7) and the analogous one for $\alpha_0, \beta_0, \gamma_0$, we find therefore

$$x(\alpha-\alpha_0) + y(\beta-\beta_0) + z(\gamma-\gamma_0) = 0\,, \tag{14}$$

or, by eqns. (8),

$$x n_1 + y n_2 + z n_3 = 0\,. \tag{15}$$

Referred to a lattice point as origin, this is the *equation of the median plane* between the incident ray and the ray diffracted at that point, the order of which is determined by the whole numbers n_1, n_2, n_3.

§ 105. Lattice Planes.

In a cubic lattice, let us suppose intervals to be cut off along the three coordinate axes from the origin, these intervals being equal to the quotients of the lattice constant and three whole numbers k_1, k_2, k_3, which have no common denominator. The extremities of the three intervals then define a plane, whose equation we wish to determine.

Since the coordinates of any arbitrary plane are interconnected by a *linear* relation, the most general form of the equation of any *plane* is

$$(1) \qquad D_1 x + D_2 y + D_3 z = D_4 ,$$

where the four magnitudes denoted by D are constants. Let

$$(2) \qquad D_4 = D_1 C_1 = D_2 C_2 = D_3 C_3 .$$

Eqn. (1) then becomes

$$(3) \qquad \frac{x}{C_1} + \frac{y}{C_2} + \frac{z}{C_3} = 1 .$$

If now we consider, for instance, the point where the plane cuts the x-axis, both y and z are to be equated to zero for this point of intersection, and we see therefore that the three constants C_1, C_2, C_3 simply represent the intercepts of the plane upon the three coordinate axes. The plane which we considered above cuts off an interval of magnitude a/k, along the x-axis, where a is the lattice constant. Hence the equation of this plane is given by

$$(4) \qquad k_1 x + k_2 y + k_3 z = a .$$

In general, we *define as a lattice plane of the space-lattice* every plane for which the left-hand side of eqn. (4) is equal to an *integral multiple of the lattice constant ;* as the general equation of a lattice plane, we have therefore

$$(5) \qquad k_1 x + k_2 y + k_2 z + s a ,$$

where s denotes a whole number. Since, however, s can assume the values of each one of the whole series of integers, the three so-called *indices of the lattice planes* (k_1, k_2, k_3) determine a *set of parallel planes*, one of which, *viz.*, the one for which $s = 0$, passes through the origin.[1] For the rest, s may of course be either positive or negative.

It is also easy to see that an *infinite number of lattice points* will be distributed in a reticular manner over every lattice plane, which is supposed to be unlimited in extent. If m be any integer, positive or negative, and we put

$$(6) \qquad \begin{cases} x = y = m\, k_3\, a , \\ \qquad z = -\, m\,(k_1 + k_2)\, a , \end{cases}$$

then

$$k_1 x + k_2 y + k_3 z = 0 .$$

[1] For the point ($x = y = z = 0$) satisfies eqn. (5) when $s = 0$.

That is to say, all lattice points represented by eqns. (6) lie on the lattice plane ($s = 0$) which passes through the origin. The same thing obviously holds good for all lattice points which are represented by the equations to be obtained from eqns. (6) by a cyclic interchange, *viz.*,

$$y = z = m\,k_1\,a\,, \quad x = -\,m\,(k_2 + k_3)\,a\,,$$

or,

$$z = x = m\,k_2\,a\,, \quad y = -\,m\,(k_3 + k_1)\,a\,.$$

Let us now consider any arbitrary plane of the set of lattice planes, for which the value of s differs from zero, and let us form, for instance, the *indeterminate equation*

$$k_1 x' + k_2 y' = s\,. \tag{7}$$

Naturally this equation always has integral solutions when both x' and y' may be negative: let x^* and y^* be such a pair of solutions. If then we consider the lattice point

$$x = x^*\,a\,,\ y = y^*\,a\,,\ z = 0\,, \tag{8}$$

we find for this point, according to eqn. (7),

$$k_1 x + k_2 y + k_3 z = s\,a\,;$$

i.e., the lattice point determined by eqn. (8) belongs to the lattice plane in question. We have therefore only to remove the origin to this lattice point in order to be able to apply the proof advanced above to the lattice plane under consideration, and thus to show that it also contains an infinite number of lattice points.

The *distance between two successive lattice planes* in a set determined by a triplet of indices may be denoted by d. This is obviously equal to the *perpendicular* dropped from the origin upon the lattice plane for which $s = 1$. This plane cuts off intercepts a/k_1, a/k_2, a/k_3 along the three coordinate axes. Denoting by φ_1, φ_2, φ_3 the angles included between the perpendicular and the three axes, we have

$$d = \frac{a}{k_1}\cos\varphi_1 = \frac{a}{k_2}\cos\varphi_2 = \frac{a}{k_3}\cos\varphi_3\,. \tag{9}$$

But the sum of the squares of the three cosines is equal to unity. Hence it follows from eqns. (9) that

$$d = \frac{a}{\sqrt{k_1^2 + k_2^2 + k_3^2}}\,. \tag{10}$$

On comparing eqn. (5) with § 104 eqn. (15), we recognize, first, that the *median plane* between the incident ray and the ray diffracted in a space-lattice always represents a *lattice plane*. We see further, however, that the indices of this lattice plane always correspond to the quotients of the ordinal numbers, determining the order of the diffracted ray, and their highest common factor. We have

$$(11) \qquad n_1 = n k_1, \quad n_2 = n k_2, \quad n_3 = n k_3,$$

where n is the highest common factor of the ordinal numbers n_1, n_2, n_3. The diffraction of the incident rays may therefore be regarded as a *reflection at the lattice planes* of the space-lattice, the simple relation expressed by eqn. (11) connecting the ordinal numbers of the diffraction with the indices of the reflecting lattice planes.

The angle made by the incident ray with the lattice plane, *i.e.*, the complement of the angle of incidence, is called the *glancing angle.* In accordance with eqn. (11) and § 104 eqn. (12), it is determined by the relation

$$(12) \qquad \sin\theta = \frac{n\lambda}{2a}\sqrt{k_1^2 + k_2^2 + k_3^2}.$$

In virtue of eqn. (10), however, this may also be written

$$(13) \qquad 2d\sin\theta = n\lambda,$$

where n is, of course, an arbitrary integer.

If now we consider two lattice planes, the distance (AB) between which is equal to d (Fig. 77), the difference in path between two rays reflected at the points A and B will be equal to $(CB + BD)$, where C and D are the feet of the perpendiculars dropped from the point A on the directions of the ray reflected by the second plane. But $(CB + BD)$ is equal to $2d\sin\theta$. Hence, a set of lattice planes *singles out* from the incident radiation those rays for which an *integral multiple of the wave-length* is *equal to the difference in path* between two rays, which are reflected from two successive lattice planes.

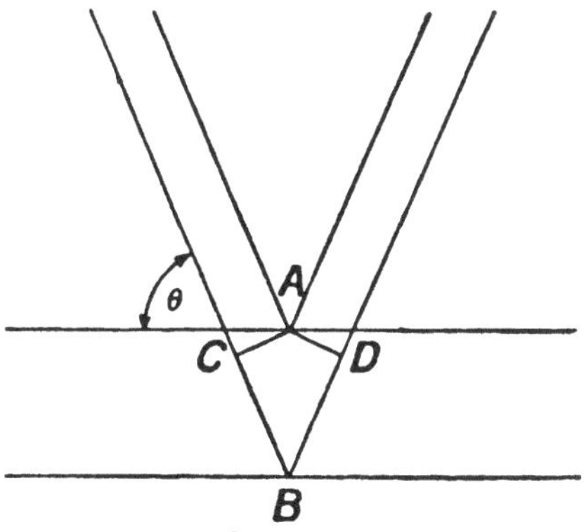

FIG. 77.

As a matter of fact, this law [which is, of course, only another way of expressing § 104 eqn. (9)] is very plausible. For, if the diffraction of the rays takes place in such a way that they appear to be reflected at the lattice planes in accordance with the usual laws of optics, then the reflected rays must strengthen and weaken each other owing to interference. Moreover, an appreciable increase in intensity will only be observed when the difference in path between the rays reflected at two successive planes amounts exactly to an integral multiple of a wave-length.

Furthermore, as we have shown in the previous considerations from two entirely different points of view, a space-lattice only diffracts incident rays of quite definite wave-length from radiation with a given direction of incidence. Thus complete diffraction of a narrow bundle of parallel rays can obviously only be attained if the *space-lattice be rotated*. A complete *resolution into spectra* is then actually attainable, as may be seen from some simple geometrical relations.

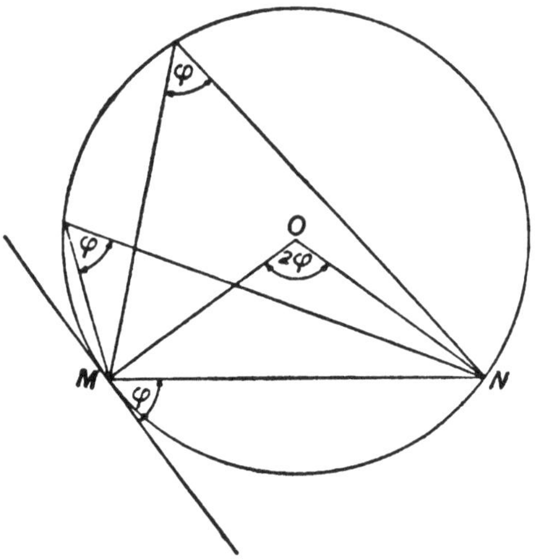

FIG. 78.

In order to derive these relations, we start from a well-known elementary theorem, which states that all the *peripheral angles subtended by the same chord in a circle* are *equal in magnitude*, and also equal to the *tangential angle* included between the chord and the tangent to the circle, drawn at one end of the chord (Fig. 78).[2]

Let us now consider (Fig. 79) a *lattice plane*, and suppose a narrow bundle of rays, passing through a slit R, to meet it at the point O with a glancing angle θ. Denoting the position of the plane by the straight line $A A'$, we therefore have

$$(14) \qquad \angle R O A' = \theta .$$

[2] The second part of the theorem, which is also well known, is most simply proved from the fact that the angle at the centre is twice the angle at the circumference. Hence in Fig. 78 the angle $M O N$ is equal to 2ϕ, so that the angle $O M N$ is equal to half of $(180 - 2\phi)$. The tangential angle, however, is complementary to the angle $O M N$, because the radius is perpendicular to the tangent. Thus the tangential angle is equal to ϕ.

We draw a circle with $R\,O$ as a chord, and the line $A\,A'$ touching it tangentially at O.

Let us now *rotate* the space-lattice about an axis perpendicular to the plane of the figure, so that the lattice plane turns from the position $A\,A'$ into the position $B\,B'$. Let the straight line $B\,B'$ intersect the circle at the point P. Then the angle $R\,P\,O$, a peripheral angle subtended by the chord $R\,O$, is, according to the theorem just quoted, equal to the tangential angle $R\,O\,A'$. Thus we also have

(15) $\angle R\,P\,O = \theta$.

By eqn. (13), however, a definite value of the glancing angle is associated with a definite value of the wave-length, for a given ordinal number n.

When the space-lattice is *rotated*, the point in the lattice plane at which the reflection of a ray of definite wave-length takes place thus *travels along the circumference of the circle* which, passing through the points R and O, touches at O the straight line $A\,A'$ that makes an angle θ with the line $R\,O$. While the lattice plane is being rotated, the ray of definite wave-length thus *glides* along it.

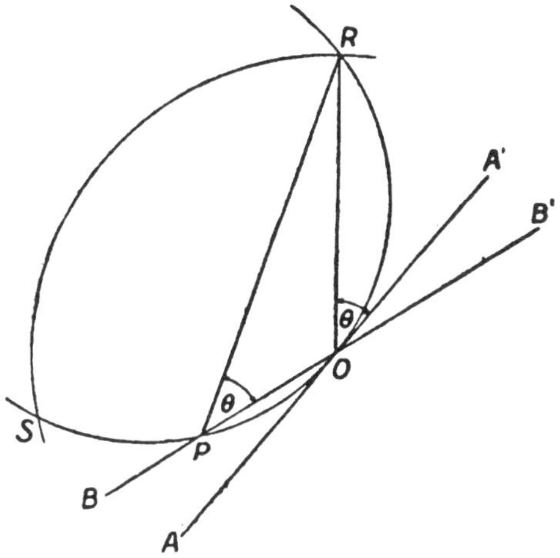

Fig. 79.

We now suppose (Fig. 79) a *second circle* to be described about O as centre, with $O\,R$ as radius. Let this second circle cut the first circle $R\,O\,P$ at a point S. It may be easily shown that every ray which strikes the lattice plane at a glancing angle θ must, after reflection, pass through the point S, whatever may be the position of the lattice plane at the instant of reflection.

We see (Fig. 80) that the angle $O\,S\,R$, a peripheral angle subtended by the chord $O\,R$, is equal to the angle $R\,P\,O$; hence, by eqn. (15),

$$\angle O\,S\,R = \theta. \tag{16}$$

But $S\,O\,R$ is an isosceles triangle, and it follows therefore from eqn. (16) that the angle at the apex of this triangle

(17) $$\angle SOR = 180° - 2\theta .$$

Thus, by eqn. (14),

(18) $$\angle SOA = \theta ;$$

i.e., the ray reflected at the point O, when the lattice plane is in the position $A A'$, passes through the point S after reflection.

Moreover, the angles SOR and SPR are peripheral angles subtended by the same chord SR. Hence by eqn. (17) we have

(19) $$\angle SPR = 180° - 2\theta ,$$

so that, by eqn. (15),

(20) $$\angle SPB = \theta .$$

Therefore, the ray reflected at the point P, when the lattice plane is in the position $B B'$, likewise passes through the point S after reflection. Furthermore, since the position $B B'$ is an entirely *arbitrary* position of the lattice plane, every ray proceeding from the point R, which is reflected at an angle θ from the lattice plane passing through O, passes through the point S after reflection. For a given order of diffraction, however, a quite definite value of the wave-length corresponds to each glancing angle.

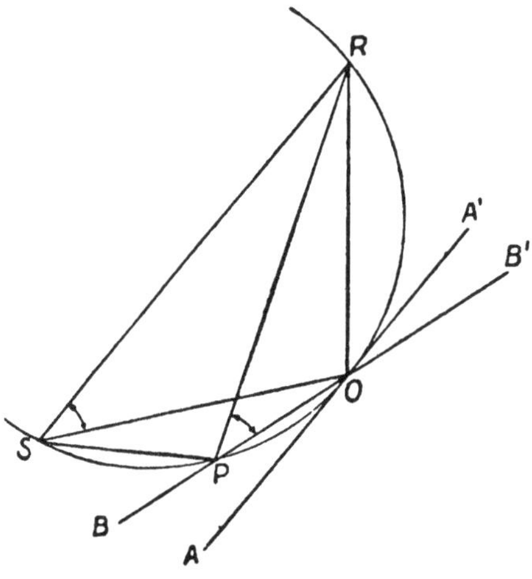

Fig. 80.

Hence, on rotation of the space-lattice, all rays of the same wave-length are, as it were, "*focussed*" at one and the same point.

§ 106. The Interference of Röntgen Rays in Crystals.

If *crystals* possess the structure of *space-lattices*, and their lattice constants, though larger, are not much larger than

PLATE I.

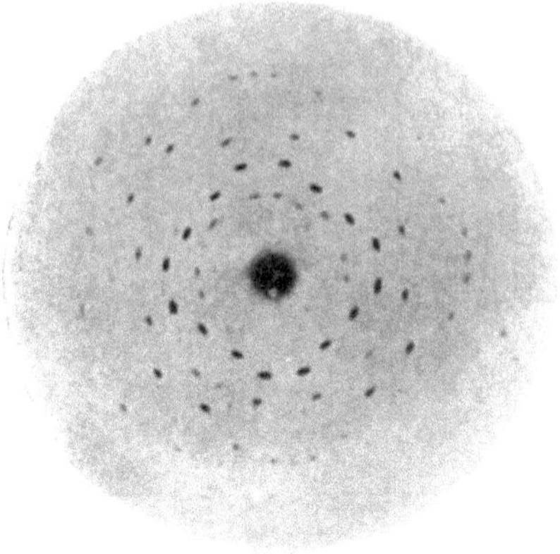

Fig. 81.

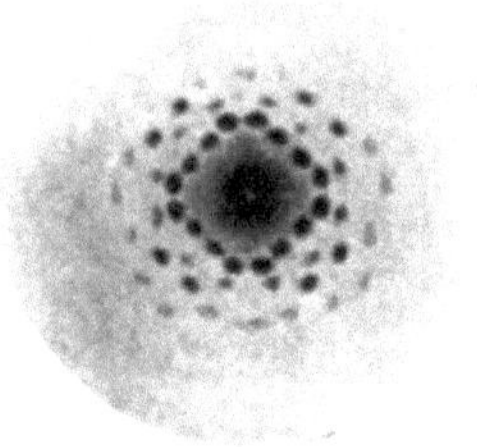

Fig. 82.

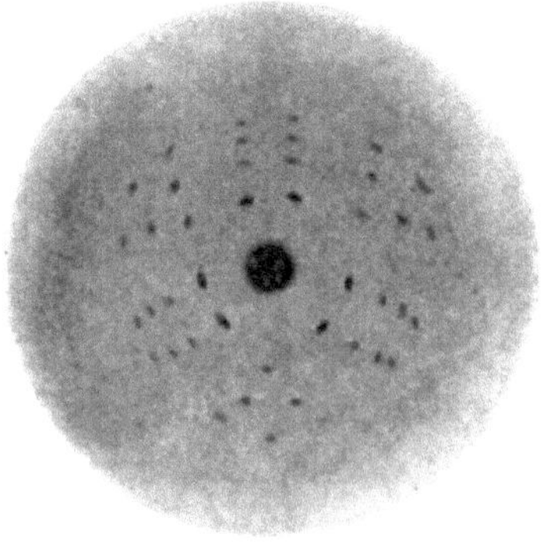

Fig. 84.

Laue Photographs.

[*To face p.* 73.

the wave-lengths of *Röntgen rays*, these rays should be *diffracted by interference* within the crystals. This diffraction will follow the laws derived from purely geometrical considerations in the previous sections, for the *atoms* corresponding to the *lattice points* must obviously act as *diffraction centres* with respect to electromagnetic waves.

An *experimental proof* of an *interference of Röntgen rays* produced by *crystals* would therefore, on the one hand, demonstrate the lattice structure of crystals, and, on the other hand, confirm the conjecture that Röntgen rays are electromagnetic waves with a wave-length of the order of 10^{-9} cm.

Starting out with this idea, *Laue*, in 1912, was the first to investigate the action of crystals on Röntgen rays, and actually to demonstrate experimentally the suspected phenomena. In these researches, which were carried out by Laue in collaboration with *Friedrich* and *Knipping*, a narrow parallel pencil of rays was separated from the radiation of a Röntgen tube by means of lead screens with adjustable slits. The rays fell on a plate of crystal about 1 cm. square and $\frac{1}{2}$ mm. thick, and a photographic plate was mounted some centimetres behind the crystal.

Now, from a parallel bundle of incident rays, a stationary space-lattice diffracts rays only in quite definite directions, which, apart from the direction of incidence, depend only on the ordinal numbers of the diffraction. [These directions are defined by § 104 eqns. (8), in which λ is to be given the value following from § 104 eqn. (9).] The points at which the directions of these diffracted rays *meet* the plane of the photographic plate must thus be *blackened* on the plate.

Experiment actually gave these results. In Figs. 81 and 82 (on Plate I) are reproduced the so-called *Laue photographs* which were obtained with a crystal of *zinc blende* (ZnS). In the exposure shown in Fig. 81, the distance of the plate from the crystal amounted to 3·5 cm., whereas it was only 1 cm. in the exposure reproduced in Fig. 82. On that account, the spots appear more closely packed in the second photograph.

In both exposures, the rays met a *face of the cube at right angles*. Their direction was therefore that of an edge of the cube, and this explains the *fourfold symmetry* to be observed in the photographs. An axis of symmetry in a crystal is termed n-fold, when through this axis n planes can be drawn, with respect to which the crystal is symmetrical. Now it is easy to see that each *edge of a cube* in a regular crystal represents a *fourfold axis of symmetry*. For, through each edge, four planes of symmetry can be drawn, the lines along which they intersect a plane normal to the edge being represented in Fig. 83; two of these intersecting lines are again edges, and two face diagonals. Thus the photographs exhibit a fourfold symmetry, provided that the incident pencil of rays be perpendicular to a face of the cube. Corresponding to Fig. 83, four lines of symmetry are to be recognized, and hence all spots (provided they do not lie just on a line of symmetry) are repeated *eight times* in all. This is actually to be seen in Figs. 81 and 82.

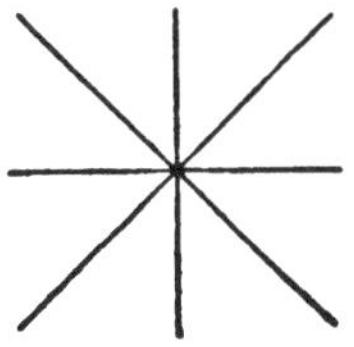

FIG. 83.

If the ray falls perpendicularly on an *octohedral* face cut from a regular crystal, geometrical considerations (into which we cannot enter here) show that its direction is that of a trigonal axis of the cube—a threefold axis of symmetry. An exposure made in this way is reproduced in Fig. 84 (on Plate I), in which each spot occurs *six times*.

In 1913, *W. H.* and *W. L. Bragg* (father and son) deduced from the Laue photographs most important conclusions on the *structure of crystals*. Although the space-lattice nature of crystals had been proved beyond doubt, yet, before crystals could be actually utilized for accurate measurements of wave-lengths, the exact lattice structure of at least *one* crystal had to be determined, and in this connection there were two questions to be solved.

The first question concerns a possible *centring of the space-lattice*. A cubic lattice may be simple, space-centred, or finally, face-centred. The symmetry relations are the same in all three cases. A cubic lattice (such as was considered in §§ 104 and 105) is *space-centred* when, in all the elementary cubes whose edges are given by the lattice constant, an additional lattice point is situated at the centre of the cube; each elementary cube then contains nine instead of eight lattice points. If, on the other hand, the interior of the cube is unoccupied, whilst lattice points are situated at the centres of the six faces (Fig. 85), a *face-centred* lattice results.

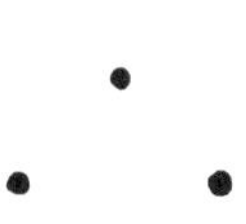

FIG. 85.

The second fundamental question of crystal structure arises from the fact that most crystals are composed of *different kinds of atoms*, *e.g.*, rock-salt is made up of sodium and chlorine atoms. It might well be expected that each kind of atom would form a separate space-lattice, but the question was still unsolved as to how the different space-lattices were combined in the crystal.

Very ingenious reasoning, which we cannot discuss here,[1] enabled W. H. and W. L. Bragg to answer these questions for numerous crystals, by investigating the Laue photographs. The arrangement and number of the interference spots differ, according as to whether and how the space-lattice is centred. Moreover, the relative positions of the combined space-lattices of the various kinds of atoms could be determined, on the grounds of the fact that the behaviour of an atom with respect to Röntgen rays depends on the number of electrons contained in the atom, and thus, broadly speaking, on the *atomic weight*. Hence a lattice plane of atoms reflects Röntgen rays in a different manner according as the two species of atom have much the same, or widely differing atomic weights. Thus, in spite of a presumably identical crystal structure, the different alkali halides, for example, should give rather different Laue photographs. [*E.g.*, in potassium chloride (sylvine) the two atomic weights only differ slightly (39 and 35·5), but in potassium bromide they are very different (39 and 80).] A comparison of the photographs of various alkali halides thus rendered possible a recognition of the structure of crystals of these compounds. In a later section we shall discuss at length the general results of crystal analysis; for the moment, we shall only refer to the structure of the alkali halides, because one of them, rock-salt (NaCl), is the principal crystal used in the measurement of the wave-lengths of Röntgen rays.

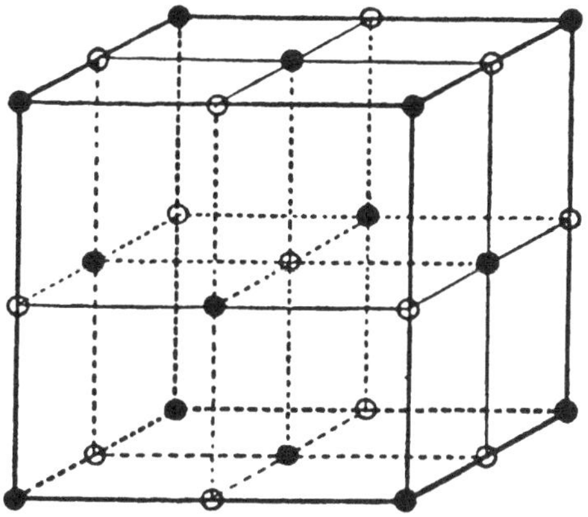

FIG. 86.—Structure of a rock-salt crystal.

In a *crystal of rock-salt*, the atoms form a *simple cubic lattice* in the sense of the considerations of the earlier sections (§§ 104 and 105), if we disregard the difference between the Na and Cl atoms. For the Na atoms by themselves form a

[1] *Cf.*, for instance, Sir W. H. and Prof. W. L. Bragg's "*X-Rays and Crystal Structure*" (London: Bell, 1924).

face-centred lattice, as do the Cl atoms likewise, and the two lattices are combined in such a way that the points of the one lattice lie at the mid-points of the edges of the other lattice (*cf.* Fig. 86, where the two kinds of atoms are distinguished by black and by white circles).

Now once the structure of rock-salt crystals is known, the *distance between the lattice planes* can immediately be calculated, taking the faces of the cubes as lattice planes. Denoting this distance by d, we see that each cube, the length of whose edges is d, contains a total of eight atoms, which are situated at its corners. Since, however, each corner belongs to the eight cubes which meet at it, the number of atoms contained in a cubic centimetre is given by

$$(1) \qquad Z = \frac{1}{d^3}.$$

On the other hand, if M be the so-called *molecular weight* of rock-salt, ϱ its *density*, and L the *Loschmidt number*,

$$(2) \qquad \varrho = \frac{M}{L} \cdot \frac{Z}{2}.$$

We have to divide by two, because it is the mass of *two* atoms (one Na and one Cl atom) taken together that is equal to the molecular weight divided by the Loschmidt number. Combining eqns. (1) and (2), we find the distance between the lattice planes to be given by [2]

$$(3) \qquad d = \sqrt[3]{\frac{M}{2L\varrho}}.$$

The molecular weight of NaCl is equal to the sum of the atomic weights of Na (23·00) and Cl (35·46), *i.e.*,

$$(4) \qquad M = 58{\cdot}46.$$

The value found for the density of rock-salt from very accurate measurements is

$$(5) \qquad \varrho = 2{\cdot}164.$$

Hence, by eqn. (3) and § 88 eqn. (6),

$$(6) \qquad d = 2{\cdot}814 \times 10^{-8} \text{ cm}.$$

This value has been derived with an accuracy of about 1/3%, depending chiefly on the uncertainty occasioned by the Loschmidt number. Since, however, Röntgen-spectroscopic measurements can be carried out with a much greater accu-

[2] In other lattice structures the numerical factor in eqn. (3) would, of course, be different.

racy, an *ideal value* of $2{\cdot}81400 \times 10^{-8}$ cm. is assigned to d for such purposes.

As soon as a fundamental constant in the theory of crystals is thus evaluated, very exact *measurements of the wave-lengths* of Röntgen rays at once become possible. These may be most simply carried out by means of the *method of the rotating crystal*, devised by *W. H.* and *W. L. Bragg* in 1913. This method is based, on the one hand, on the law of reflection at the lattice planes derived in the previous section and likewise due to the Braggs, and, on the other hand, on the fact that, as far as the *reflection of Röntgen rays by crystals* is concerned, practically only the set of lattice planes *parallel to the front face of the crystal* requires consideration.[3]

Accordingly, a very narrow pencil of Röntgen rays from a small lead diaphragm is directed upon a crystal, which is mounted on a goniometer table, rotating about a vertical axis, in such a way that the axis of rotation lies in the front face of the crystal.[4] The crystal is very slowly rotated by hand, or by clockwork. According to § 105 eqn. (13), rays of the wave-length λ are then reflected only for quite definite glancing angles, *viz.*, only for those angles which, for integral (and low) values of the ordinal number n, satisfy the equation

$$2d \sin\theta = n\lambda . \tag{7}$$

As the crystal rotates, the ray glides along its face; the advantage of this method over one where the crystal is kept in the same position is that the disturbing effects of the unavoidable defects in the crystal face are thereby eliminated.

Now it follows from the geometrical considerations of § 105 that rays of the same wave-length are *focussed* at points on the circumference of a circle, whose centre lies on the axis of rotation, and which passes through the slit. If, therefore, a *photographic film* be placed round this circle, and if the radiation contains only rays of definite wave-lengths, *line spectra* of several orders must be formed on the film. The spectrum of the first order can at once be recognized, for it is far more intense than that of the second order, while the latter is more intense than the spectrum of the third order, and so on.

In general, only the *spectrum of the first order* is utilized. The wave-lengths of the lines contained in it are readily

[3] Naturally, the face itself is always a lattice plane.

[4] In *Seeman's* apparatus a sharp edge is fixed directly against a part of the crystal face that is as perfect as possible. Edge and face together thus form a fine slit.

found by constructing a *scale* of wave-lengths in accordance with eqn. (7) (wherein we put $n = 1$). The crystal of rock-salt may, of course, be replaced by any other, quite arbitrary crystal, even of unknown structure, provided only that one line can be identified with certainty, because of its origin always being the same.[5] A comparison of the positions of this line in the scales of the two crystals immediately enables the scale of the crystal used to be converted into that of rock-salt.

By means of the photographic method, a spectroscopic accuracy of measurement of almost one-millionth part of the measured value can be obtained. With this accuracy the wave-lengths of *different* lines can be compared, and, as was mentioned above, this necessitates the use of an ideal value of the rock-salt constant, which cannot be so accurately determined. The *photographic* method is due to *de Broglie* (1913); W. H. and W. L. Bragg were then still working with a so-called *ionization spectrometer.* This instrument consists of a lead cylinder filled with gas, and provided with a narrow slit through which Röntgen rays can pass into the gas. The cylinder is rotated around the crystal along the circle on which rays of equal wave-length are focussed, in such a way that the window always remains on the circle. Rays entering the window ionize the gas, and its resulting conductivity can be accurately measured. More recently the ionization method has been very much improved, and it is principally used for measurements of intensity in Röntgen spectra, for the intensity of the Röntgen rays is approximately proportional to the exactly measurable ionization. It may also be remarked that wave-lengths can be determined by the perfected ionization method with an accuracy of one in ten thousand.

If now we consider a small region of the spectrum extending from a wave-length λ to a wave-length $(\lambda + \Delta\lambda)$, the *spectral resolution* of this region is defined by the difference quotient $\Delta\theta/\Delta\lambda$, and hence approximately by the differential coefficient. It follows from eqn. (7) that

$$\frac{d\theta}{d\lambda} = \frac{n}{2\,d\cos\theta}. \tag{8}$$

The resolving power thus increases with the ordinal number, but this advantage is counterbalanced by the rapid decrease in intensity with increasing ordinal number. Nevertheless, close doublets are often measured in the second order

[5] *Cf.* the remarks in § 108 on the subject of characteristic rays.

spectra, sometimes even in the third, fourth, or fifth orders. Moreover, as eqn. (8) shows, the resolution increases with the glancing angle θ, and it becomes especially strong when the angle θ approaches the value of $90°$, so that $\sin \theta$ is almost equal to unity. According to eqn. (7), however, this is the case when twice the distance between the lattice planes becomes nearly equal to an integral multiple of the wave-length of the region to be resolved. The resolution of a given small region, such as a doublet, is therefore best effected with a crystal which satisfies this relation for as small an ordinal number n as possible.

Since neither $\sin \theta$ can be greater, nor n smaller than unity, it follows from eqn. (7) that the greatest wave-length which can ever be diffracted by a crystal is equal to twice the distance between the lattice planes, which in a rock-salt crystal is approximately equal to $5{\cdot}6 \times 10^{-8}$ cm.[6] The unit of wave-length used in Röntgen spectroscopy is the so-called *X-unit* (X.U.).[7] This is equal to one-thousandth of the optical spectroscopic unit, the Ångström unit, which, in its turn, is equal to the one-hundred millionth part of a centimetre [§ 86 eqn. (1)]. Hence

$$(9) \qquad 1 \text{ X.U.} = 0{\cdot}001 \text{ Å.U.} = 10^{-11} \text{ cm.}$$

Thus with a rock-salt crystal, only wave-lengths up to about 5,000 X.U. can be measured. For the measurement of larger wave-lengths, other crystals with larger distances between the lattice planes are used, such as selenite or potassium ferrocyanide, in which the lattice plane distances are $7{\cdot}6 \times 10^{-8}$ and $8{\cdot}5 \times 10^{-8}$ cm. respectively. By means of these, wave-lengths as large as 13,000 X.U. can be measured. Apart from rock-salt, the principal crystal used in wave-length determinations is *calcite*, in which the distance between the lattice planes is $3{\cdot}029 \times 10^{-8}$ cm.[8] It may also be observed that soft Röntgen rays (*i.e.*, rays of large wave-length) are very strongly absorbed by gases even at quite low pressure, so that, in the determination of such wave-lengths, the whole path of the rays has to be *in vacuo*.

With very *hard* rays the method of the rotating crystal cannot be used. Hard rays penetrate too deeply into the crystal, and thereby produce a strong broadening effect on

[6] If the face of a regular crystal is not that of a cube, the lattice plane distance is less than d.

[7] The name is due to the fact that Röntgen rays are also called X-rays.

[8] Proposals have also been made to refer all wave-lengths to an idealized calcite constant ($d = 3{\cdot}02900 \times 10^{-8}$ cm.) instead of to the idealized rock-salt constant.

the lines. Hence, for very hard rays another method is used, which was developed by *Rutherford* and *Andrade*, in order to measure the wave-lengths of the extremely hard γ-rays. A widely diverging beam of rays falls on a thin plate of crystal, and the rays are reflected (at least, in a regular crystal) principally by the lattice planes that are normal to the crystal face struck by the rays. The crystal is fixed in one position, and behind it is placed a slit, which singles out from the emerging beam rays reflected at a certain glancing angle, and hence of a quite definite wave-length. By means of this method, into the details of which we cannot enter here, wave-lengths as small as 50 X.U. have been measured.

Finally, the last method of Röntgen ray analysis that we shall mention is the *crystal powder* method, developed on the one hand by *Debye* and *Scherrer*, and on the other hand by *Hull*. In this form, the time succession of diffraction centres in the rotating crystal method is replaced by a corresponding succession in space. The powder, in fact, consists of numerous minute crystals mingled together in all possible orientations. Hence, when a narrow pencil of rays strikes a small tube filled with the powder, reflection occurs at every possible angle, so that in this way also, the incident radiation is resolved into spectra.

§ 107. Duane and Hunt's Equation.

It follows from *Einstein's law of light-quanta* that, in the *generation of Röntgen rays by the arrest of cathode rays*, a single light-quantum must be produced from the kinetic energy of a single arrested electron. This kinetic energy is equal to the product of the elementary quantum of electricity (e) and the tube potential (V), the latter being given by the potential difference between cathode and anti-cathode.[1] Provided that the kinetic energy of the electron be *completely* transformed into a light-quantum, the following relation should hold :—

$$h\nu = eV, \tag{1}$$

where ν denotes the frequency of the Röntgen radiation generated. Hence, for the wave-length we have

$$\lambda = \frac{hc}{eV}. \tag{2}$$

If we measure in volts, the value of V must be multiplied by 300 ;[2] and if we express the wave-lengths in X-units,

[1] *Cf.* § 91.
[2] *Cf.* § 64 eqn. (10).

these must be multiplied by 10^{11}. Hence, giving the universal constants h, c, and e their known values, we find

(3) $$\lambda V = 1{\cdot}234 \times 10^7, \ (\lambda \text{ in X.U., } V \text{ in volts}).$$

This equation, however, only holds for the *limiting case* in which the kinetic energy of the arrested electron is *completely* transformed into a light-quantum and does not appear in any other form. In general, of course, this is not the case; for a portion of the kinetic energy is used up in other ways, and hence, in general, the *light-quantum* is *smaller* than the kinetic energy, whilst it can never be larger, according to the Law of the Conservation of Energy.

Thus, on the one hand, eqn. (3) determines the *largest possible frequency* or, what amounts to the same thing, the smallest possible wave-length, for a given tube potential; on the other hand, it determines the *lowest tube potential* necessary for the production of rays of a given wave-length in the continuous Röntgen spectrum. Eqn. (3) is generally known as *Duane* and *Hunt's* equation, after the investigators who first derived it (in 1915) from Einstein's law and confirmed it experimentally.

This equation forms an exact, quantitative expression for the observation made soon after the discovery of Röntgen rays, that the hardness of the rays increases with the tube potential. Later, more exact observations have shown that the *continuous Röntgen spectrum* exhibits a *sharp boundary towards the side of short wave-lengths.* According to eqn. (3), this boundary must be displaced in the direction of decreasing wave-length with increasing tube-potential, and eqn. (3) is therefore also known as Duane and Hunt's *displacement law.* Measurements carried out first by Duane and Hunt, and later by other investigators, have proved eqn. (3) to be satisfied with the greatest accuracy, as regards the limiting wave-length; the measurements covered a range from about 4,500 to 170,000 volts.[3] Indeed, the accuracy is so great that a determination of the elementary quantum of action can be based upon this displacement (assuming the value of e to be known).[4]

[3] In accordance with eqn. (3), the corresponding limiting wave-lengths lie between about 80 and 3,000 X.U.

[4] *Cf.* the comprehensive report by E. Wagner in the *Jahrb. d. Radioakt. u. Elektronik,* **16** (1920), p. 190.—It should, however, be borne in mind that the value of the Loschmidt number is also implicitly contained in the wave-length [in virtue of § 106 eqn. (3)]. Since the wave-length is inversely proportional to the cube root of the Loschmidt number, while the latter is equal to the quotient of Faraday's constant and the elementary quantum of electricity, it is not really the quotient e/h that occurs in eqn. (2), but the quotient e^3/h.

Eqn. (3) determines, as already explained, the minimum potential necessary for the generation of a given wave-length, or the so-called *excitation potential*. This may be experimentally determined with great accuracy. If the tube potential be gradually increased, the given wave-length is found to appear, though very weakly, at a quite definite potential, and, as soon as that minimum value is exceeded, the intensity of the radiation in the region of that wave-length is found to increase *linearly* with the potential. The intensity is determined by means of an ionization chamber (*cf.* § 106), the conductivity of the ionized gas being measured by the deflections of an electrometer. In Fig. 87, which reproduces the results of experiments made by Duane and Hunt in 1915, the dependence of the measured values of the ionization upon the potential, for various wave-lengths, is represented by black dots. As the figure shows, almost straight lines may be drawn through the dots, the intersections of these lines with the axis of abscissæ giving the excitation potentials for the wave-lengths in question. In this way, from each of the curves, Duane and Hunt were able to *calculate*, according to eqn. (2), the *elementary quantum of action*.[5] This method of the direct determination of h has subsequently been greatly improved. For example, *Duane* and *Blake*[6] obtained in later measurements a value of $(6{\cdot}557 \pm 0{\cdot}013) \times 10^{-27}$ erg sec., while the value derived by other methods, and accepted at present as the most correct, is $(6{\cdot}545 \pm 0{\cdot}012) \times 10^{-27}$ erg sec.

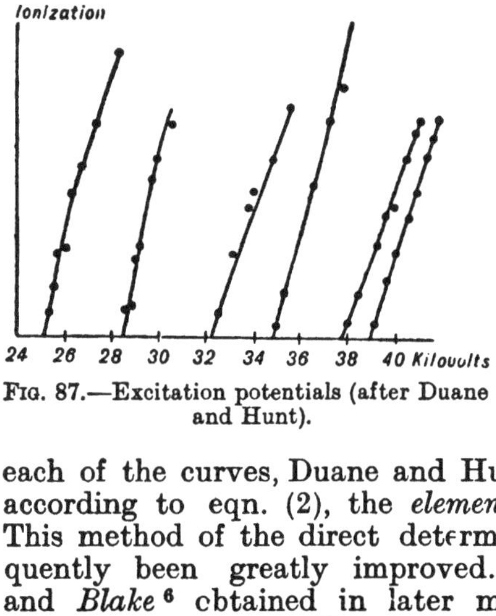

FIG. 87.—Excitation potentials (after Duane and Hunt).

Duane and Hunt's measurements thus represent a direct experimental confirmation of the quantum hypothesis. It must, however, be remembered that the relations based on the quantum hypothesis refer *only to the frequency* of the excited Röntgen radiation, but make no predictions on the

[5] From left to right, the six lines in Fig. 87 correspond to the wave-lengths 488, 424, 377, 345, 318, and 308 X.U. The values of h calculated from the six lines are: 6·44, 6·37, 6·34, 6·38, 6·41 and $6{\cdot}39 \times 10^{-27}$ erg sec.

[6] *Phys. Rev.*, **10** (1917), pp. 93 and 624.

subject of the energy relations. It is by no means implied that the kinetic energy of each and every arrested electron is transformed into a light-quantum. As is shown by measurements on the heat generated by Röntgen rays, only about *two thousandths of the energy* of the cathode rays in a Röntgen tube are transformed into energy of Röntgen radiation. In the few cases, however, in which a corpuscular quantum of energy is changed into a light-quantum, Einstein's relation must be obeyed, as, indeed, is proved by Duane and Hunt's measurements.

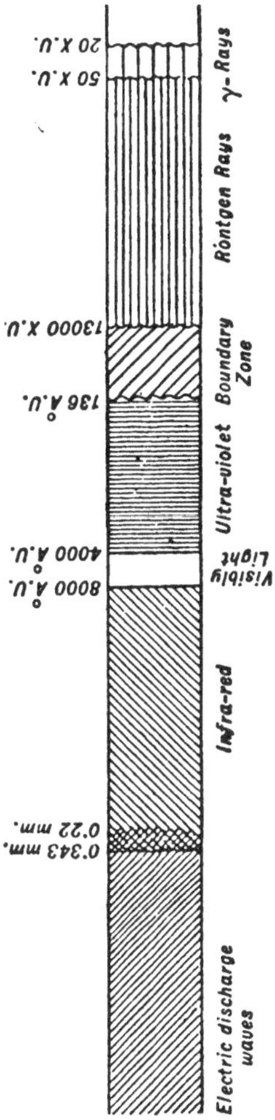

Fig. 88.—Complete spectrum.

Just like ultra-violet rays, Röntgen rays may, of course, also give rise to the so-called *photo-electric effect*, consisting in the *liberation of electrons* from the surface of irradiated bodies (*cf.* § 91). In consequence of their magnetic deflectability, these electrons can be isolated from the secondary Röntgen radiation which is emitted by a body subjected to the action of primary Röntgen rays. They were discovered by *Dorn* in 1900, in the form of so-called *secondary cathode rays*. Owing to the much greater frequency of the generating light, their velocity is much larger than that of the electrons liberated by ultra-violet light. In the case of the secondary cathode rays, Einstein's law of light-quanta is again found to be accurately obeyed; the kinetic energy of a liberated electron is equal to a light-quantum of the generating radiation, less the work expended in the liberation.

The quantum relations holding for Röntgen rays also enable an *indirect determination of Röntgen wave-lengths* to be made, which is of special importance in those regions of the spectrum where a direct determination by means of crystals is not possible. Thus, on the ground of Duane and Hunt's equations, very large wave-lengths up to 40,000 X.U. could be determined by measuring the tube potential.[7] Furthermore, the equation of the photo-electric effect renders possible an indirect determination of wave-lengths, as will later be discussed at length. In virtue of this effect it has been possible to measure, on the one hand, long waves up to about 40,000 X.U. in length,[8] and, on the other hand (as we shall see in detail later), wave-lengths as short as 20 X.U. in the region of the γ-rays.

Thus, whereas the Röntgen rays measured directly by means of *crystals* extend over a region of some *eight octaves* (from about 50 to about 13,000 X.U.), the above-mentioned indirect methods have resulted in the addition of approximately one octave on the side of shorter, and of approximately two octaves on the side of longer wave-lengths. Hence, in the present state of experimental research, the Röntgen rays, including the measured γ-rays, comprise some *eleven octaves*. About one octave from the far side of the long wave-length boundary of this region lies the shortest wave-length hitherto measured in the ultra-violet—136 Å.U. (= 136,000 X.U.). We can, however, scarcely speak of a true gap, because (as we shall discuss more fully later) the atomic spectra of the elements establish a complete connection between ultra-violet rays and Röntgen rays.

The whole spectrum of electromagnetic waves is (in continuation of § 86) represented in Fig. 88. The unit of scale is an octave, so that, as regards wave-lengths, the scale is logarithmic. From the largest wave-length measured in the infra-red (0·343 mm.) down to the smallest known wave-length (20 X.U.), the spectrum extends over some *twenty-six octaves*. Of these about eight lie in the infra-red, about five in the ultra-violet, and about eleven are covered by Röntgen rays (in the wider sense), whilst the whole of the rays visible to the human eye are comprised in the narrow limits of a single octave.[9]

[7] *Cf.* H. M. Dadourian, *Phys. Rev.*, **14** (1919), p. 234.

[8] *Cf.* § 108.

[9] In Fig. 88 an important discovery made in 1923 is taken into account. In that year *Nichols* and *Tear* succeeded in bridging the gap which had hitherto separated the region of electric discharge waves from that of infra-red rays. Using electric discharges, they were able to produce waves with wave-lengths as small as 0·22 mm., whereas the wave-lengths of the longest known infra-red rays are 0·343 mm. [*Proc. Nat. Acad.*, Washington, **9** (1923), pp. 211–214]. *Cf.* § 86.

Chapter XIV

THEORY OF THE ELEMENTS

§ 108. Moseley's Law, and the Natural Sequence of the Elements.

In 1905 *Barkla* made the important discovery of *characteristic Röntgen radiation.* As Barkla observed, when primary Röntgen rays impinge on a body, the *secondary* Röntgen radiation emitted by it consists chiefly of *homogeneous rays*, the *hardness* of which is found to be a *characteristic property of the atoms* of the body. Barkla measured the hardness of the rays by means of their *absorption* in aluminium plates, the absorption coefficient representing a constant characteristic of the element in question.[1] He found the radiation from chemical compounds to be composed of the rays of the elements contained in the compounds. He further observed that the hardness of the characteristic rays *increases with the atomic weight*, and, moreover, that the majority of the elements appeared to possess not merely one, but *two* absorption coefficients differing widely in value. Barkla therefore distinguished between a harder *K-radiation* and a softer *L-radiation.* He attributed quite rightly to the limitations of experimental methods the fact that only one species of rays could be demonstrated in the case of many elements. For, on the one hand, the *K*-rays become so hard with increasing atomic weight that, in the state of Röntgen technique at that time, it was impossible to excite them ; on the other hand, the *L*-radiation becomes so soft with decreasing atomic weight that it could not be demonstrated by the methods then known. Furthermore, as Barkla discovered, characteristic rays are generated only by primary rays which are *harder* than the characteristic rays themselves.[2]

[1] The reciprocal value of the absorption coefficient represents the depth (in centimetres) to which the rays penetrate in aluminium before their intensity is diminished in the ratio $e : 1$.

[2] This law is analogous to the well-known Stokes's rule, according to which the wave-lengths of fluorescent light are greater than those of the exciting light. On that account, the characteristic radiation was also called by Barkla fluorescent radiation.

The invention of the method of the *rotating crystal* rendered possible an exact analysis of the characteristic radiation, and led to its resolution into a *line spectrum*, the exact measurement of *wave-lengths* taking the place of the inaccurate determination of hardness. We owe the foundation of *Röntgen spectroscopy* to *Moseley* who, in 1913, by means of a specially constructed vacuum spectrograph, resolved the K- and the L-radiations of numerous elements into distinct sharp lines, whose wave-lengths he determined accurately. Moseley found that both the K- and the L-series are of the same type for the different elements. In each series one line was found to be particularly strong, and this line he called the K_a- and the L_a- line respectively. More recent Röntgen-spectroscopic researches (which we shall discuss more fully in a later section) have resulted both in the observation of new lines in both series, and also in the discovery of two further series. These, on account of their softness, have actually been demonstrated only in elements of high and the highest atomic weights, and are known as the M- and the N-series.[3]

The increase in hardness with the atomic weight had been established by Barkla, but Moseley succeeded in finding a much more exact expression for this relation in a law which, discovered by him, was destined to be of fundamental importance for the modern *theory of the elements*. As Moseley discovered, the *elements* can be arranged in a *consecutively numbered series*, in such a way that the *square root of the frequency of a definite Röntgen line increases linearly with the number of the element.*

The "*natural*" *sequence of the elements* thus obtained agrees on the whole, as Moseley recognized, with the series formed by arranging the elements in order of *increasing atomic weight*. Whereas, however, no regular relation exists between the atomic weight numbers, in the natural sequence the displacement of the Röntgen lines from element to element takes place with the greatest exactness, so that the existence of any gap in the series is clearly revealed by too large a jump.

In Fig. 89, some photographs taken by Moseley of the K- series for a sequence of elements are reproduced. In addition to the K_a- line, a second, weaker, (K_β-) line can be recognized. (It was not until spectroscopic methods had been perfected that further lines were first observed.) The

[3] The N-series is of longer wave-length than the M-series, and the wave-lengths of the M-series are greater than those of the L-series. *Cf.* § 117.

photographs are so arranged that points lying vertically with respect to each other correspond to the same wave-length. The wave-lengths increase from left to right, and the frequencies from right to left, so that, in Fig. 89, the element with the lower number lies above the element with the higher number. The photographs refer to the elements calcium, titanium, vanadium, chromium, manganese, iron, cobalt, nickel, copper, and zinc. Zinc was used alloyed with copper, in the form of brass.[4] The exposures were made with interchangeable anti-cathodes of the elements to be investigated, so that the characteristic radiation was directly generated by the cathode rays.[5] The photographs show with wonderful clearness that the given elements follow the natural sequence, but that there is an element missing between calcium and titanium. (This is scandium, which at the time was not available to Moseley, because of its rarity.) It may also be noticed—and this is most remarkable—that cobalt precedes nickel in the natural sequence, although it has the higher atomic weight of 58·97 as against 58·68. Brass shows both the zinc and the copper lines, and cobalt appears to be contaminated by nickel and iron.

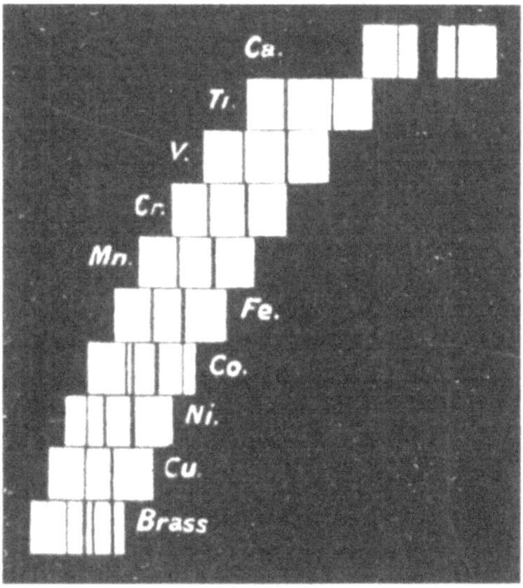

FIG. 89.—*K*-series (after Moseley).

Direct Röntgen-spectroscopic measurements, carried out with the help of crystals (as will be more fully described in a later section) have made it possible to establish the natural

[4] The alloy was used because zinc is not sufficiently resistant to heat.

[5] Moseley utilized a slit near the anti-cathode, so that the emerging pencil of rays was widely divergent. In this way he was able to photograph both *K*-lines in one exposure, without rotating the crystal, provided the latter was suitably placed. The subsequent introduction of the method of the rotating crystal resulted in a considerable improvement in the photographs.

sequence of the elements from sodium to the highest element *uranium*. Furthermore, indirect measurements of characteristic wave-lengths in the boundary region between Röntgen rays and ultra-violet rays have enabled us (as we shall discuss more fully later) to follow the course of the spectra from sodium down to the element with the second smallest atomic weight, helium.

The natural sequence, which undoubtedly commences with hydrogen in the first and helium in the second place, is thus *unequivocally* established. Including uranium, it comprises *ninety-two places*, only *five* of which belong to *unknown elements* (Nos. 43, 61, 75, 85, 87). The natural sequence of the elements obtained in this manner is represented by Table III, in which, for each element, the *atomic number* or *ordinal number*, which determines its place in the natural sequence, the conventional name, the chemical symbol, and the atomic weight [6] relative to oxygen = 16·000 are given. This table, it is true, includes elements for which it has not been possible to make Röntgen-spectroscopic photographs. These elements comprise the rare gases, mercury (which could not be applied as an anti-cathode material either in the metallic form or in compounds), and some particularly rare elements (thulium, radium, polonium, actinium, protoactinium). Apart from the chemical periodicity, the positions of these elements in the natural sequence follow from their atomic weights.

TABLE III

The Natural Sequence of the Elements

1	Hydrogen	H	1·008	17	Chlorine	Cl	35·46
2	Helium	He	4·00	18	Argon	A	39·88
3	Lithium	Li	6·94	19	Potassium	K	39·10
4	Beryllium	Be	9·02	20	Calcium	Ca	40·07
5	Boron	B	10·82	21	Scandium	Sc	45·10
6	Carbon	C	12·00	22	Titanium	Ti	48·1
7	Nitrogen	N	14·008	23	Vanadium	V	51·0
8	Oxygen	O	16·000	24	Chromium	Cr	52·01
9	Fluorine	F	19·00	25	Manganese	Mn	54·93
10	Neon	Ne	20·2	26	Iron	Fe	55·84
11	Sodium	Na	23·00	27	Cobalt	Co	58·97
12	Magnesium	Mg	24·32	28	Nickel	Ni	58·68
13	Aluminium	Al	26·97	29	Copper	Cu	63·57
14	Silicon	Si	28·06	30	Zinc	Zn	65·37
15	Phosphorus	P	31·04	31	Gallium	Ga	69·72
16	Sulphur	S	32·07	32	Germanium	Ge	72·60

[6] The atomic weights are those given in the fifth report of the German Atomic Weight Commission (November, 1924).

TABLE III—*continued.*

33	Arsenic . . .	As	74·96	64	Gadolinium .	Gd	157·3
34	Selenium . .	Se	79·2	65	Terbium . .	Tb	159·2
35	Bromine . . .	Br	79·92	66	Dysprosium .	Dy	162·5
36	Krypton . . .	Kr	82·9	67	Holmium . .	Ho	163·5
37	Rubidium . .	Rb	85·5	68	Erbium . . .	Er	167·7
38	Strontium . .	Sr	87·6	69	Thulium . .	Tm	169·4
39	Yttrium . . .	Y	89·0	70	Ytterbium . .	Yb	173·5
40	Zirconium . .	Zr	91·2	71	Lutecium . .	Lu	175·0
41	Columbium . .	Cb	93·5	72	Hafnium . .	Hf	178·3
42	Molybdenum .	Mo	96·0	73	Tantalum . .	Ta	181·5
43	—	—	—	74	Tungsten . .	W	184·0
44	Ruthenium . .	Ru	101·7	75	—	—	—
45	Rhodium . .	Rh	102·9	76	Osmium. . .	Os	190·9
46	Palladium . .	Pd	106·7	77	Iridium . . .	Ir	193·1
47	Silver . . .	Ag	107·88	78	Platinum . .	Pt	195·2
48	Cadmium . .	Cd	112·4	79	Gold . . .	Au	197·2
49	Indium . . .	In	114·8	80	Mercury. . .	Hg	200·6
50	Tin	Sn	118·7	81	Thallium . .	Tl	204·4
51	Antimony . .	Sb	121·8	82	Lead . . .	Pb	207·2
52	Tellurium . .	Te	127·5	83	Bismuth . .	Bi	209·0
53	Iodine . . .	I	126·92	84	Polonium . .	Po	(210)
54	Xenon . . .	Xe	130·2	85	—	—	—
55	Cæsium . . .	Cs	132·8	86	Radium emanation . . .	Ra Em	222
56	Barium . . .	Ba	137·4				
57	Lanthanum . .	La	138·9	87	—	—	—
58	Cerium . . .	Ce	140·2	88	Radium . .	Ra	226·0
59	Praseodymium .	Pr	140·9	89	Actinium . .	Ac	(226)
60	Neodymium . .	Nd	144·3	90	Thorium . .	Th	232·1
61	—	—	—	91	Protoactinium .	Pa	(230)
62	Samarium . .	Sm	150·4	92	Uranium . .	U	238·2
63	Europium . .	Eu	152·0				

In the natural sequence, the atomic weight increases throughout, except in four instances. The heavier element argon precedes the lighter element potassium (Nos. 18 and 19); cobalt comes before nickel (Nos. 27 and 28) as already mentioned; tellurium occurs before iodine (Nos. 52 and 53); and, finally, thorium precedes protoactinium (Nos. 90 and 91).

Now, for hydrogen and helium, the atomic number agrees with the nuclear charge number, as follows from our considerations of the line spectra of hydrogen and ionized helium. It therefore seems plausible to *identify* quite generally the *atomic numbers* of the elements with their *nuclear charge numbers*, which, referred to the elementary quantum as unit, give the positive charge of the atomic nucleus, and at the same time the *number of electrons revolving around the nucleus* in the *neutral* atom. The strongest support for this conception, proposed by *van den Broek* in 1913, is to be found in Bohr's general formula for the frequency of a spectral line in a hydrogen-analogue [§ 95 eqn. (2)]. For, according to this formula, the square root of the frequency of a spectral line due to a definite transition increases linearly with the *nuclear charge number*, while, in accordance with Moseley's law, it increases linearly with the *atomic number*. Hence, if the atomic number agrees with the nuclear charge number for *one* element (as with helium), the agreement must obviously hold for *all* the elements. The correctness of van den Broek's assumption is also proved by the numerical relations of chemical periodicity which we shall discuss in a later section.

Moreover, this hypothesis has been directly confirmed by experimental researches on the *passage of α-rays through metal foils*. As we have already mentioned in an earlier section (§ 90), α-rays are deflected by the atomic nucleus in their passage through matter. The so-called *scattering* of the α-rays thus produced therefore depends on the nuclear charge number of the metal traversed. From measurements carried out on the scattering produced by metal foils *Chadwick*, in 1920, was able in this manner to calculate the nuclear charge number of three metals with an accuracy of 1%. For platinum, silver, and copper, Chadwick found the values 77·4, 46·3, and 29·3. Within the limits of error of 1%, these values do, indeed, agree very well with the atomic numbers 78, 47, and 29, given in Table III.

§ 109. The Periodic System.

In the natural sequence of the elements, experience shows that *six elements* occupy a *very special position*, while simple numerical relations exist between their atomic numbers. These elements are, in fact, absolutely *passive* from the *chemical* point of view ; in contrast to all the other elements, they form *no chemical compounds* whatsoever. They are the so-called *rare gases*,[1] *viz.*, helium, neon, argon, krypton, xenon, and radium emanation.[2] Their atomic numbers are

2, 10, 18, 36, 54, 86.

The differences between these numbers are

8, 8, 18, 18, 32,

[1] The rare gases (apart from radium emanation) were discovered by *Ramsay* in tho atmosphere, towards the end of the nineteenth century.

[2] The "International" Committee on Chemical Elements has proposed the names *radon*, *thoron*, and *actinon* for the three radioactive emanations. *Radon* thus replaces *niton*, but the term *radium emanation* is also still widely used.

and we may suspect in advance a fundamental reason for the fact that the first atomic number and the resulting differences between the successive numbers are of the form $2z^2$ ($z = 1, 2, 3, 4$).

From the point of view of modern atomic theory, *chemical activity* must obviously be referred to *electrical forces* proceeding from the atoms. It is these electrostatic forces which lead to a combination of oppositely charged atoms, and thus to the *formation of molecules.* Since, in the neutral atom, the number of electrons revolving around the nucleus is equal to the nuclear charge number, an atom can only become chemically active when the number of electrons surrounding its nucleus becomes either more or less than the nuclear charge number. In the first case, the atom as a whole becomes electrically negative ; in the second case, it becomes positive ; moreover, the charge is determined by the number of electrons gained or lost. It must therefore be concluded from the chemical passivity of the rare gases that, in their atoms, the *arrangement of the electrons about the nucleus* is generally *stable,* in the sense that electrons are neither removed from the atomic system, nor are new ones introduced into it—a state of affairs that appears possible in all other atoms.

Thus the rare gases to a certain extent represent landmarks in the natural sequence of the elements, and divide the series into *seven periods.* Of these, the first period contains the first *two* elements (hydrogen and helium) ; the second and third periods each contain *eight* elements, Nos. 3 to 10 (lithium to neon), and Nos. 11 to 18 (sodium to argon). The fourth and fifth periods each comprise *eighteen* elements, Nos. 19 to 36 (potassium to krypton), and Nos. 37 to 54 (rubidium to xenon). The sixth period includes *thirty-two* elements, Nos. 55 to 86 (cæsium to radium emanation). The seventh period, beginning with the unknown element No. 87, apparently breaks off in the sixth place at uranium ; and it must remain an open question whether elements with still higher nuclear charge are not capable of existence, have already disintegrated, or have not yet been discovered because of their rarity.[3]

As long ago as 1869, *Lothar Meyer* and *Mendeléeff* simultaneously, but independently, made the important discovery that, if the elements be arranged in order of increasing atomic weight, characteristic physical and chemical properties recur *periodically,* and that, if the successive periods be written

[3] *Cf.* § 113.

TABLE IV

The Periodic System of the Elements

	I		II		III		IV		V		VI		VII		VIII	
	A	B	A	B	A	B	A	B	A	B	A	B	A	B	A	B
1	1 H															2 He
2	3 Li		4 Be			5 B		6 C		7 N		8 O		9 F		10 Ne
3	11 Na		12 Mg			13 Al		14 Si		15 P		16 S		17 Cl		18 A
4	19 K	29 Cu	20 Ca	30 Zn	21 Sc	31 Ga	22 Ti	32 Ge	23 V	33 As	24 Cr	34 Se	25 Mn	35 Br	26 Fe 27 Co 28 Ni	36 Kr
5	37 Rb	47 Ag	38 Sr	48 Cd	39 Y	49 In	40 Zr	50 Sn	41 Cb	51 Sb	42 Mo	52 Te	43	53 I	44 Ru 45 Rh 46 Pd	54 Xe
6	55 Cs	79 Au	56 Ba	80 Hg	57—71 rare earths	81 Tl	72 Hf	82 Pb	73 Ta	83 Bi	74 W	84 Po	75	85	76 Os 77 Ir 78 Pt	86 RaEm
7	87		88 Ra		89 Ac		90 Th		91 Pa		92 U					

down in successive horizontal rows, an arrangement is possible in which elements occurring in the same vertical column are distinguished by similar properties. The further development of the so-called *periodic system* resulted in a distinction between *eight vertical groups*, each with two sub-groups. An exact representation, however, was not possible until the knowledge of atomic numbers also led to an exact recognition of the number of places in the periods.

Experience shows that the table of the periodic system is to be constructed in the following manner (Table IV.). Eight vertical columns are formed, each with two sub-groups. The groups corresponding to the vertical columns are denoted by I, II, *etc.*, up to VIII; the sub-groups may be distinguished in the table by the letters A and B. The two elements in the first period are now placed in sub-groups I A and VIII B. The eight elements each of the second and the third period are to be allotted to the eight groups in such a way that the first two in each period fall in A columns, and the last six in B columns. In the fourth and fifth periods, each containing 18 elements, we proceed as follows: we place the first eight elements in the eight A sub-groups, and the last eight in the eight B sub-groups, whilst the ninth and tenth elements are allotted to sub-group VIII A, so that, in each of the fourth and fifth periods, this sub-group contains a *triad*. In the sixth period we follow the same arrangement as in the fourth and fifth, but with the difference that in sub-group III A a *group of fifteen elements* is to be placed, instead of a single element. Thus, in the sixth period, one of the 16 positions in the eight A and eight B sub-groups will contain 3 elements, and another 15 elements. Finally, in the seventh period, the six elements are to be placed in the A sub-group.

The three triads consist of the *iron*, the *palladium*, and the *platinum metals*. The group of fifteen elements is known as the *rare earth* group. Furthermore, the periodic system indicates to us the chemical nature of the five unknown elements. The 43rd and 75th elements must be homologues of manganese, the 61st element a rare earth, the 85th a halogen, and the 87th an alkali metal.

§ 110. Isotopy.

As we have seen in the previous considerations, the position of an element in the periodic system depends solely on its *nuclear charge*. Furthermore, we must assume the *nuclei* of the elements to be *built up* of protons and electrons. Hence, provided only that the *resultant nuclear charge*—the *algebraical sum* of all the positive and negative elementary charges forming the nucleus—be the same in two atoms, these must exhibit the same chemical behaviour, even though the number of protons and nuclear electrons differs in the two. Whereas, however, the nuclear charge, and thus the *chemical character*, appears to be determined by the *difference* in the numbers of protons and electrons in the nucleus, the *mass* of the atom, and therefore the so-called *atomic weight*, depends *only* on the number of *protons*, since the mass of an electron can be neglected in comparison with that of a proton. It is thus quite possible for two substances despite *different atomic weights* and *different nuclear structures*, to agree in their chemical and in many important physical properties to such an extent that they would really represent only *different species of one and the same element*.

The first phenomenon of this kind to attract the notice of physicists was the *inseparability* of certain *radioactive elements*. When, for instance, salts of *thorium* and *ionium* (the parent substance of radium) were in any way mixed together, it was found quite impossible to separate them again by any method whatsoever; moreover, the *spectra* of thorium and ionium were observed to be quite identical. Similarly, radium and mesothorium 1 were found to be inseparable, as also were lead and radium D (*cf.* § 113).

All these cases proved that substances differing in their radioactive behaviour could yet be identical in chemical and also in physical respects. Recognition of this fact led *Soddy*, in 1910, to the belief that also the *inactive* (non-radioactive) *elements* might represent *mixtures of several, chemically inseparable species*. Soddy further pointed out that this assumption would solve a great difficulty which still faced the atomic theory at that time. This difficulty arose from the *deviations* of the so-called *atomic weights from whole numbers*.

If atoms are actually built up of protons (*i.e.*, hydrogen nuclei and electrons), and if the mass of the electrons can be neglected in comparison with that of the protons, then all atomic weight numbers must obviously be *integral*. Now, although this is the case with many elements, and especially with the most abundant ones, yet in other elements such as chlorine, the deviations from whole-number values are so considerable that they could not at first be reconciled with that necessary consequence of atomic theory. This difficulty, however, vanishes at once if, according to Soddy's assumption, the elements really represent *mixed elements*. For, in that case, only the significance of an average weight would attach to the atomic weight, and its constancy would be explained on the grounds that any mixture ratio once formed (*e.g.*, in times when the earth was in a gaseous state), must remain unaltered for ever, in view of the inseparability of the mixed atoms.

This conception, suggested by Soddy in 1910, was experimentally confirmed in 1913 by *J. J. Thomson*, who showed by means of *positive ray analysis* that the rare gas *neon* must really contain *two atomic species* of *different atomic weights* (20 and 22). Accurate experiments made in 1914 also confirmed the opinion, expressed by Soddy, that the *atomic weight of lead* must differ in value, according as to whether we are dealing with " ordinary " lead or with lead which has been obtained from uranium minerals or from thorium minerals. (As we shall show in § 113, both uranium and thorium are gradually transformed into lead.) Whereas the atomic weight of ordinary lead is found to be 207·2, atomic weights as low as 206·05 have actually been determined for lead from uranium minerals, and as high as 207·9 for lead from thorium minerals.[1] The term *isotopes*, suggested by Soddy, soon came into general use to denote substances which have different atomic weights, but exhibit the same chemical properties.[2]

Researches on isotopes, properly speaking, began in 1919 with *Aston's* development of positive ray analysis into the so-called *mass-spectroscopy*, by means of an artifice already described (*cf.* § 90). Using his new method, Aston was first able to prove, with greater accuracy, that the rare gas *neon* does actually consist of *two isotopes* with the exactly *integral atomic weights* 20·0 and 22·0. Shortly afterwards, he was also able to prove that *chlorine*, in which the deviation from a whole-number value had always been felt to be particularly embarrassing, is a mixture of two isotopes with the exactly integral atomic weights 35·0 and 37·0. It follows from the mixture weight (35·46) that the ratio of the lighter to the heavier gas is about three to one, as can also be recognised from the intensity of the lines in the mass-spectra.

[1] *Cf.* Hönigschmid, *Zeitschr. f. Electrochemie*, 24 (1918), p. 163, and 25 (1919), p. 91. *Cf.* also § 113, note 7.

[2] *Isos topos* in Greek means " the same place." The name was suggested by the fact that isotopes occupy the same place in the periodic system of the elements.

By means of his mass-spectrograph, Aston has investigated numerous other elements for isotopes. This method, however, has necessarily been limited to such elements as can be introduced, in a gaseous state at ordinary temperature into a Geissler tube, whether as element or in the form of a chemical compound. Subsequently, the analysis of non-volatile substances has also been successfully carried out, by introducing them on the *anode* of the discharge tube and investigating the anode or positive rays produced. Using this method, *Aston*, *G. P. Thomson*, and *Dempster* have made further analyses.[3]

In Fig. 90 (Plate II), Aston's *mass-spectra* of neon, chlorine, argon, and krypton are reproduced, showing most clearly the integral nature of the pure atomic weights.[4] Aston's method is sufficiently delicate to be able to detect an isotope present only to the extent of 1 in 1,000 in a mixture. The degree of blackening of the individual lines enables quantitative conclusions to be drawn on the mixture ratios of the isotopes ; and it is possible in this way to determine the mixture weight by a mass-spectroscopic method [5] (though Dempster's method gives more accurate results).

In Table V are reproduced the results of researches (up to the end of 1924) on the subject of isotopes in the non-radioactive elements ; it contains 53 elements and 105 atomic species or, if the bracketed, less certain, results given

[3] In Dempster's method the anode rays, which were deflected only by a magnetic field, were caught on a metal plate connected to an electrometer. In this way the relative quantities of the isotopes in the mixed element could be fairly accurately estimated.

[4] Lines of known atoms or groups of atoms serve as reference lines in these spectrograms, especially the lines of carbon (12) ; the C-complexes, CH, CH_2, CH_3, CH_4 (13, 14, 15, 16) ; the C_2-complexes (24, 25, 26, 27, 28, 29, 30) ; O (16), OH (17), OH_2 (18), CO (28), and CO_2 (44). Spectrum I shows, besides these lines, which almost always occur, the two neon lines very clearly at 20 and 22. In spectra II, III, and IV can be seen the lines of Cl_{35} and Cl_{37}, as well as lines at 36 and 38 due to $Cl_{35}H$ and $Cl_{37}H$. We may also notice "second order lines" at 17·5 and 18·5. These are produced by chlorine ions with double charges, so that the specific charge is then twice as large, which would also be the case if the charge remained single and the mass were halved. Lines 63 and 65 are due to $COCl_{35}$ and $COCl_{37}$. Spectra V and VI show not only the first order lines of argon, but also, quite clearly, a second order line at 20 and a third order line at 13·3. Spectrum VIII is that of krypton, showing plainly five strong lines at 80, 82, 83, 84, 86 and a sixth, weaker, line at 78. These lines also occur in the second order in the mass-spectrum of krypton. An explanation of spectrum VII may well be omitted here. For details of Aston's mass-spectroscopy, *vide* his book, "*Isotopes*" (London : Arnold, 1924).

[5] In the case of boron, it was actually recognized on these grounds that its "atomic weight" must be lower than that previously accepted. Accurate, direct chemical determinations of the atomic weight later proved that the atomic weight of boron, as stated, was really one unit too high in the first decimal place.

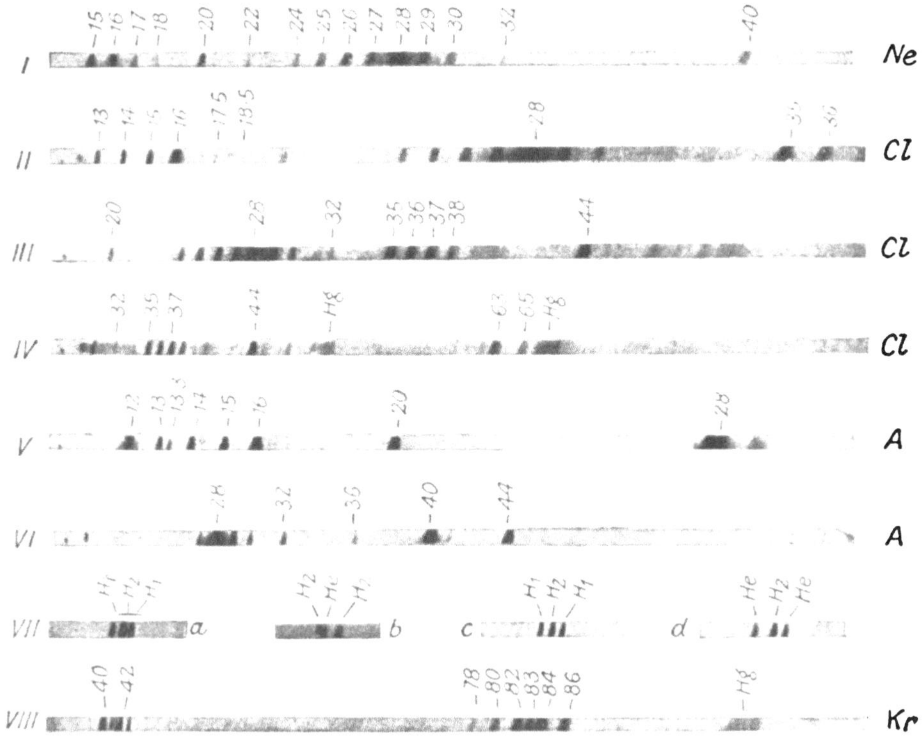

Fig. 90.—Mass-spectra (after Aston).

[*To face p.* 96.

TABLE V

Isotopes

Atomic Number.	Element.	Atomic Weight.	No. of Isotopes.	Masses of Isotopes, in order of relative amounts in mixture.
1	Hydrogen .	1·008	1	1·008
2	Helium . .	4·00	1	4
3	Lithium .	6·94	2	7, 6
4	Beryllium .	9·02	1	9
5	Boron . .	10·82	2	11, 10
6	Carbon . .	12·00	1	12
7	Nitrogen .	14·008	1	14
8	Oxygen . .	16·000	1	16
9	Fluorine .	19·00	1	19
10	Neon . .	20·2	2 (3)	20, 22, 21
11	Sodium . .	23·00	1	23
12	Magnesium .	24·32	3	24, 25, 26
13	Aluminium .	26·97	1	27
14	Silicon . .	28·06	3	28, 29, 30
15	Phosphorus .	31·04	1	31
16	Sulphur .	32·07	1	32
17	Chlorine .	35·46	2	35, 37
18	Argon . .	39·88	2	40, 36
19	Potassium .	39·10	2	39, 41
20	Calcium .	40·07	2	40, 44
21	Scandium .	45·10	1	45
22	Titanium .	48·1	1 (2)	48, (50)
23	Vanadium .	51·0	1	51
24	Chromium .	52·01	1	52
25	Manganese .	54·93	1	55
26	Iron . .	55·84	2	56, 54
27	Cobalt . .	58·97	1	59
28	Nickel . .	58·68	2	58, 60
29	Copper . .	63·57	2	63, 65
30	Zinc . .	65·37	4	64, 66, 68, 70
31	Gallium .	69·72	2	69, 71
32	Germanium .	72·60	3	74, 72, 70
33	Arsenic . .	74·96	1	75
34	Selenium .	79·2	6	80, 78, 76, 82, 77, 74
35	Bromine .	79·92	2	79, 81
36	Krypton .	82·9	6	84, 86, 82, 83, 80, 78
37	Rubidium .	85·5	2	85, 87

TABLE V—*continued.*

Atomic Number.	Element.	Atomic Weight.	No. of Isotopes.	Masses of Isotopes, in order of relative amounts in mixture.
38	Strontium .	87·6	2	88, 86
39	Yttrium .	89·0	1	89
40	Zirconium .	91·2	3 (4)	90, 94, 92, (96)
47	Silver . .	107·88	2	107, 109
48	Cadmium .	112·4	6	114, 112, 110, 113, 111, 116
49	Indium .	114·8	1	115
50	Tin .	118·7	7 (8)	120, 118, 116, 124, 119, 117, 122, (121)
51	Antimony .	121·8	2	121, 123.
52	Tellurium .	127·5	3	128, 130, 126
53	Iodine . .	126·92	1	127
54	Xenon . .	130·2	9	129, 132, 131, 134, 136, 128, 130, 126, 124
55	Cæsium . .	132·8	1	133
56	Barium . .	137·4	1 (2)	138, (136)
57	Lanthanium .	138·9	1	139
58	Cerium . .	140·2	2	140, 142
59	Praseodymium	140·9	1	141
60	Neodymium .	144·3	3 (4)	142, 144, 146, (145)
80	Mercury .	200·6	(6)	197–200, 202, 204
83	Bismuth .	209·0	1	209

in the table be taken into account, actually 115 atomic species. Table V is supplemented by Table VI, referring to the ten radioactive elements which, together with two gaps, occupy the last twelve places in the periodic system. Of these elements, 44 species are known. Thus, in the present state of research a total of 66 out of the 87 known elements, or more than three-quarters (including all the elements with ordinal numbers from 1 to 40), have been investigated for isotopes, and in the process about 165 atomic species have been discovered. Since there are still 21 elements to be investigated, the total number of atomic species may well exceed two hundred.

Of the elements investigated for isotopes, 24 have been

TABLE VI

Radioactive Isotopes (including the Inactive Species of Elements Nos. 81–92)

Atomic Number.	Element.	Number of Isotopes.
81	Thallium . .	4
82	Lead . . .	8
83	Bismuth . .	5
84	Polonium . .	7
86	Radium emanation .	3
88	Radium . .	4
89	Actinium . .	2
90	Thorium . .	6
91	Protoactinium .	3
92	Uranium . .	2

proved to be single, or so-called *pure elements*, at least in the sense that, if they do contain any admixture of isotopes, the quantity of these is too small to be detected.[6] As might be expected, the atomic weights of the pure elements are very approximately integral. The small deviations can be explained on the basis of the so-called *packing effect*, which we shall discuss in a later section (§ 116).[7]

The number of isotopes is especially large in the case of xenon (9) lead (8), and tin (7 or 8). The difference in weight between the lightest and the heaviest isotope amounts in selenium, krypton, and tin, to eight units, and in xenon even to twelve units. The difference is relatively largest for lithium and neon, where the heavier isotope is respectively 16 and 10% heavier than the lighter.[8]

The instances of so-called *isobares* are also noteworthy, *viz.*, those in which atomic species differing in chemical

[6] Hydrogen and oxygen have also been proved to be pure elements by diffusion experiments.

[7] The large deviation with cæsium is, however, striking.

[8] Since the relative difference in the atomic weights of two isotopes is generally not very large, a gradual displacement of the mixture ratio in the course of time would be scarcely noticeable. For example, as a simple calculation shows, an alteration of the mixture ratio by 1% in bromine, which contains two isotopes, 79 and 81, in approximately equal quantities, would only cause a change in the mixture weight of two units in the second decimal place (0·81 — 0·79 = 0·02).

character have the same atomic weight. Such isobares occur, *e.g.*, for an atomic weight of 40 in argon and calcium, for an atomic weight of 70 in zinc and germanium, *etc.*

Since isotopes are identical in all properties determined by the electrons surrounding the nucleus, it is plain that a *separation of isotopes* can only be effected by such methods as are based on differences in nuclear properties. On the other hand, the ordinary chemical methods of separation will be of no avail, especially those based on differences in solubility and vapour pressure, in which the principal part is taken by the outer electrons. One of the main possible ways in which isotopes may be separated utilizes the fact that the *velocity of evaporation* depends on the atomic mass. *Brönsted* and *Hevesy* have actually succeeded, by this method of successive evaporations, in preparing *two species of mercury* which differ in density by 1 in 2,000, so that their mixture weights differ by 0·1 units. The same investigators, using the same process, have been able to prepare *two species of chlorine* differing in their mixture weights by more than two units in the second decimal place.[9] It is, however, too much to hope at present for a preparation of pure isotopes in this manner, for the yield becomes smaller and smaller as the separation proceeds.[10]

The differences in nuclear mass and structure between isotopes also find expression, to a small degree, in *spectral* properties. The conception of the nucleus as a point charge is only approximately correct and the electrical field surrounding the nucleus may exhibit small differences, with varying composition of the nucleus, despite equality of the nuclear charge number. To this is most probably due the fact established by *Aronberg* and by *Merton*,[11] that the wavelength of the spectrum line of *lead* at 4,058 Å.U. proves to be greater by 0·01 Å.U. for uranium lead, and smaller by 0·002 Å.U. for thorium lead, than in the case of common lead. The phenomenon of isotopy finds far stronger expression in molecular spectra than in optical spectra, as we shall later discuss at length.

[9] For further details, *vide* Hevesy and Paneth's "*Lehrbuch der Radioaktivität*" (Leipzig: Barth, 1923), Ch. xix. (English translation: Oxford, Clarendon Press).

[10] At the present time there does not yet seem to be any possibility of adapting the analytical process of the positive ray method to a method of preparation in which the ions of the isotopes generated in the discharge tube might be separated, in virtue of their different deflections in both an electric and a magnetic field, and then collected.

[11] T. R. Merton, *Proc. Roy. Soc.*, *A* 100 (1921), p. 84.

§ 111. Disintegration of the Nucleus, and Transformation of the Elements.

Some few years after the first observation of radioactive radiation, closer investigation led to a new discovery of far greater moment, *viz.*, the discovery of the *transformation of elements*. In 1900 *Rutherford* established the surprising fact that the radioactive element thorium is constantly evolving a gas that is itself radioactive, and which he called *thorium emanation* [1]; and a similar process was shortly afterwards discovered by *Dorn* to be going on with *radium*. We know now that thorium emanation and radium emanation are isotopes of the rare gas whose nuclear charge number is 86; the elementary nature of the emanation was manifest from its *characteristic spectrum*.

In 1902 an explanation of the formation of the emanations was put forward by *Rutherford* and *Soddy* in a theory distinguished by great simplicity, but representing a complete break with the hitherto deeply rooted and fundamental conceptions of physics and chemistry. Rutherford and Soddy's theory, which is known as the *disintegration theory*, is in fact based on the assumption that the atoms of radioactive substances undergo a series of successive disintegrations; and that the expulsion of the projected particles in the form of radioactive corpuscular radiation changes the chemical character of the residual atom.[2]

From the point of view of the nuclear theory of the atom later developed, we must obviously refer the processes of disintegration to the *atomic nucleus* itself; and then, since the *nuclear charge number* determines the *chemical character*, the phenomenon of the *transformation of radioactive elements*

[1] Rutherford's discovery resulted from the fact, previously observed by M. and Mme. Curie, that bodies enclosed in the same space as radium themselves become radioactive. This phenomenon, which at the time was termed "induced radioactivity," was more fully investigated by Rutherford in the case of thorium, and it was then that he discovered that the "induced activity" is due to the emanation.

[2] The disintegration theory thus really opened up absolutely new territory in natural philosophy, for, in the processes of nature previously allocated to physics, the molecules remain unchanged, whilst chemical processes were defined as being those which occur inside the molecules, without, however, producing any changes in the atoms. But the phenomena of radioactivity were referred by Rutherford's disintegration theory to processes occurring *inside the atoms themselves*, and thereby altering them. Moreover, radioactivity was shown to be a property of the atoms themselves by the fact that it was found to be quite independent of the kind of chemical compound concerned, and also that it is entirely unaffected by temperature changes of a thousand degrees, whereas the velocity of most chemical reactions is doubled by a rise in temperature of only ten degrees.

at once becomes intelligible. The expulsion of an α- or a β-particle by the disintegrating atom distinguishes between two varieties of transformation, known briefly as the α- and the β-transformation. Since an α-particle possesses a positive charge of two elementary quanta and a mass of four hydrogen atoms (§ 90), an *alpha-transformation reduces the atomic number by two*, and the *atomic weight by four*. On the other hand, since β-particles are identical with electrons, a *beta-transformation raises the atomic number by one, without noticeably altering tl e atomic weight*. Hence, in the *periodic system of the elements*, an α-transformation *displaces* the element affected *through two places to the left*, whilst a β-transformation displaces it *through one place to the right.*

Actually, the emanation resulting, together with α-rays, directly from radium is found to have an atomic weight four units lower than that of radium,[3] *viz.*, 222 as compared with 226 ; and, whereas radium belongs to the second group of the periodic system, *i.e.*, the alkaline earth group, the emanation (a rare gas) belongs to the zero group, or, what amounts to the same thing, to the eighth group.

The *formation of helium during the disintegration of the emanation which is itself radioactive*, was first demonstrated in 1903 by *Ramsay* and *Soddy*, using a *spectroscopic method.* This was the first instance of the observation of the generation from another element of an element already known to chemists. The actual *identity of the atoms of helium formed* and the *α-particles* expelled during the disintegration of radioactive substances (at least, after neutralization) was proved in 1909 by a beautiful experiment carried out by *Rutherford* and *Royds.*

In this experiment, a substance emitting α-rays was contained in a closed glass tube, the wall of which was not thicker than about one-hundredth of a millimetre, so that the α-rays could pass through it. This tube was placed inside another, thick-walled, glass tube, which was likewise closed. Since no α-rays could be detected outside the larger tube, it was evident that all the α-particles passing through the thin glass wall were retained in the evacuated space between the two walls. In this evacuated space, however, Rutherford and Royds were actually able to demonstrate spectroscopically the presence of helium, after some time had elapsed.

On closer investigation, it was found that a preparation

[3] The density, and hence the atomic weight, of radium emanation were first determined by Gray and Ramsay in 1902, a measured volume of it being weighed by means of a special micro-balance.

containing one gram of radium (*e.g.*, in the form of 1·314 grams of $RaCl_2$) generates in a year about 167 mm.[3] of helium, corresponding to a mass of 0·0000298 gram (the density of helium being taken as 0·000179).[4] Since the mass of a helium atom is equal to the atomic weight (4·00) divided by the Loschmidt number, we obtain the number of helium atoms generated in a year by one gram of radium, on dividing the above-given number (0·0000298) by four, and then multiplying by the Loschmidt number ($6{\cdot}062 \times 10^{23}$).[5] In this way we find that one gram of radium produces $4{\cdot}52 \times 10^{18}$ helium atoms in a year.

Again, the accurate counts already mentioned (in § 90), *made by the electrical method*, have shown that a gram of radium emits $3{\cdot}72 \times 10^{10}$ α-particles in one second.[6] This number is four times as large if the preparation is in equilibrium with three disintegration products which are formed from radium, and also emit α-rays [*cf.* the remarks in § 112 in connection with eqn. (12)]. Since there are 86,400 seconds in a day, and 365 days in a year, the number of α-particles ejected by the radium preparation in a year is equal to $3{\cdot}72 \times 10^{10} \times 4 \times 365 \times 86{,}400$, or $4{\cdot}69 \times 10^{18}$. The good agreement between the calculated number of the α-particles emitted and the accumulated helium atoms formed an important empirical confirmation of the disintegration theory. Conversely, the observation of the *rate of formation of helium* from radium, together with a *count of the α-particles*, enables the *Loschmidt number* to be *determined directly*, although not very accurately.

In every radioactive substance, the ratio of the number of atoms disintegrating in a second to the total number of atoms of the substance present is found to represent a characteristic constant. This ratio is called the *disintegration constant*, on the essential assumption that the *number of atoms disintegrating* in an element of time is always *proportional to the number of atoms available*. Denoting the disintegration constant by λ and the number of atoms present by N, we therefore have for the number of atoms disintegrating in an element of time (Δt) the relation

$$(1) \qquad \Delta N = -\lambda N \Delta t\,.$$

[4] The density of monatomic helium is only twice that of diatomic hydrogen.

[5] Since, according to Avogadro's Law (to be discussed in a subsequent chapter), the number of molecules in 1 cm.[3] at normal pressure and 0° is the same for all gases, *viz.*, $2{\cdot}708 \times 10^{19}$, the number of monatomic helium molecules contained in 0·167 cm.[3] is found by simply multiplying together the two numbers just given.

[6] The exact determination of this number is due to Hess and Lawson (1918).

On replacing the differences by differentials and integrating, we obtain the equation

$$N = N_0 e^{-\lambda t}, \tag{2}$$

where N_0 denotes the number of atoms at the time $t = 0$. The time (T) in which N diminishes to one-half of N_0, half of the atoms originally present having by then disintegrated, is known as the *half-value period*, or period of half change. Its connection with the disintegration constant follows from the relation

$$\frac{1}{2} = e^{-\lambda t};$$

whence

$$T = \frac{\log_e 2}{\lambda} = \frac{0{\cdot}693}{\lambda}. \tag{3}$$

The reciprocal value of the disintegration constant is called the *period of average life ;* it is larger than the half-value period, being obtained on dividing the latter by 0·693.

The disintegration constant of *radium* may at once be calculated from the number of α-particles emitted in a second. For we obtain the number of atoms in one gram of radium simply by dividing the Loschmidt number by the atomic weight of radium (226). This gives $2{\cdot}68 \times 10^{21}$; and, since $3{\cdot}72 \times 10^{10}$ α-particles are emitted in one second, while the same number of radium atoms disintegrate, the value resulting for the disintegration constant of radium is

$$\lambda(\mathrm{Ra}) = \frac{3{\cdot}72 \times 10^{10}}{2{\cdot}68 \times 10^{21}} = 1{\cdot}39 \times 10^{-11}\ \mathrm{sec.}^{-1}. \tag{4}$$

It follows from eqn. (3) that the *half-value period of radium* is given by

$$T(\mathrm{Ra}) = 4{\cdot}99 \times 10^{10}\ \mathrm{sec.} = 1{,}580\ \mathrm{years}. \tag{5}$$

For many radioactive substances, the half-value period (to values as low as about 0·002 sec.) can also be derived from observations on the *diminution in intensity of the radioactive radiation effects.* Thus, denoting the intensity of the radioactive radiation measured by any apparatus[7] at an arbitrary time and at the time $t = 0$ by I and I_0 respectively, we have, as is also found experimentally, the relation

$$I = I_0 e^{-\lambda t}. \tag{6}$$

In the case of radium emanation, the value found in this

[7] The simplest method is to measure the conductivity acquired by air as a consequence of the ionizing effect of the radioactive radiation.

way for the exponential constant is precisely the same as that given for the disintegration constant by the scintillation method. At the same time, this evidently proves that the intensity of the radioactive radiation at any moment is proportional to the number of atoms of the radioactive substance in existence at that moment.

§ 112. Radioactive Equilibrium.

In the following considerations, we shall apply the fundamental equation of the disintegration theory [§ 111 eqn. (2)] to the special case in which the *product of the transformation* is itself radioactive, but with a much *smaller life period than the parent substance.* Such, for example, is the case with radium emanation and radium, the half-value periods being in the approximate ratio 1 : 150,000.

Let λ be the disintegration constant of the parent substance, and λ' that of the transformation product. We shall suppose N atoms of the former and N' atoms of the latter to be present at a given moment. At an arbitrary time $t = 0$, let the values of N and N' be represented by the constants N_0 and N_0'.

We now consider an arbitrarily chosen element of time Δt; during that interval, atoms of the transformation product will *disintegrate* and, *simultaneously,* new atoms of the product will *be generated* by the disintegration of the parent substance. The number of disintegrating atoms is $\lambda' N' \Delta t$; the number of newly generated atoms $\lambda N \Delta t$.

If now, as we assumed, the parent substance is very long-lived as compared with the transformation product, then, even after an interval considerably longer than the half-value period of the transformation product, the quotient N/N_0 will not differ noticeably from unity.[1] Hence, when we are considering the state of the transformation product, we may always replace the variable quantity N by the constant N_0. The change in the number of the atoms of the transformation product in an element of time (dt) is accordingly given by the relation

$$(1) \qquad \frac{dN'}{dt} = -\lambda' N' + \lambda N_0 .$$

In order to integrate this equation, it is best to begin by differentiating again with respect to the time; denoting the

[1] In a system composed of radium and emanation we might take one year as the time.

differential coefficient dN'/dt by φ, for the sake of brevity, we then obtain

$$(2) \qquad \frac{d\varphi}{dt} = -\lambda'\varphi .$$

It immediately follows from this that (replacing φ by the derivative dN'/dt)

$$(3) \qquad \frac{dN'}{dt} = C \,.\, e^{-\lambda' t} .$$

By means of eqn. (1), we can easily determine the value of the constant C, for, at the time $t = 0$, it must be equal to the differential coefficient dN'/dt. In accordance with eqn. (1) we have

$$(4) \qquad \left| \frac{dN'}{dt} \right|_{t=0} = -\lambda' N_0' + \lambda N_0 .$$

The right-hand side of this equation therefore represents the constant C, so that eqn. (3) may be written in the form

$$(5) \qquad \frac{dN'}{dt} = (-\lambda' N_0' + \lambda N_0) e^{-\lambda' t} .$$

Integrating this again, we have

$$(6) \qquad N' = N_0' e^{-\lambda' t} - \frac{\lambda}{\lambda'} N_0 e^{-\lambda' t} + C' .$$

The integration constant C' can be determined from the equality of the numbers N' and N_0' at the time $t = 0$. Introducing the time $t = 0$ into eqn. (6), we find

$$(7) \qquad N_0' = N_0' - \frac{\lambda}{\lambda'} N_0 + C' ,$$

whence

$$(8) \qquad C' = \frac{\lambda}{\lambda'} N_0 .$$

Eqn. (6) may therefore be written in the form

$$(9) \qquad N' = N_0' e^{-\lambda' t} + \frac{\lambda}{\lambda'} N_0 (1 - e^{-\lambda' t}) .$$

In this equation let us give t a value which is large compared with the half-value period of the transformation product, but small compared with the half-value period of the parent substance. For such a value of t, the expression $e^{-\lambda' t}$ may simply be put equal to zero. Eqn. (9) then becomes

$$(10) \qquad N' \lambda' = N_0 \lambda .$$

If, however, this relation be satisfied, then by eqn. (1)

$$(11) \qquad \frac{dN'}{dt} = 0.$$

The number of atoms of the transformation product must therefore be constant because, in a given time, just as many new atoms of the transformation product are generated as disintegrate. The system consisting of the parent substance and its transformation product is then said to be in *radioactive equilibrium.* Even if equilibrium did not exist at the time $t = 0$, it must nevertheless be attained after an interval which is large as compared with the half-value period of the transformation product, provided that, during this interval, there be no external interference with the system (such as the removal of the transformation product as formed).

We shall now suppose the transformation product in its turn to generate a new active substance, the number of atoms and the disintegration constant of which may be denoted by N'' and λ'', this substance again generating another active product, and so on. Then, when radioactive equilibrium exists, the following relation must hold between the parent substance and all the transformation products :—

$$(12) \qquad N_0 \lambda = N' \lambda' = N'' \lambda'' = N''' \lambda''', \textit{etc.}$$

Since the number of atoms of a radioactive substance disintegrating in unit time is given by the product of the disintegration constant and the number of atoms present, *equal numbers of atoms of all substances* which are in *radioactive equilibrium* with each other *disintegrate* in the same time. This explains why a radium preparation ejects *four times* as many α-particles as radium itself, for it contains, besides radium, three transformation products emitting α-rays (RaEm, RaA, RaC), in equilibrium with the radium.

Eqn. (12), moreover, permits us to derive the *disintegration constant of the long-lived parent substance* from the known disintegration constant of any transformation product. For example, it has been found that, in all uranium minerals, there is a content of $3{\cdot}4 \times 10^{-7}$ grams of *radium* for every gram of *uranium.* Since the atomic weights of uranium and radium are 238 and 226 respectively,[2] it follows that, in radioactive equilibrium, corresponding to n-atoms of uranium we have $n \times 3{\cdot}4 \times 10^{-7} \times 238/226$, or $n \times 3{\cdot}6 \times 10^{-7}$ atoms of radium. We then obtain the half-value period of uranium

[2] It is not necessary to take into consideration uranium II, which is always contained as an isotope in the mixed element uranium, for it only forms about 0·025 % of the whole element.

on dividing that of radium (1,580 years) by $3{\cdot}6 \times 10^{-7}$; the *half-value period of uranium* is thus found to be 4,400 million years. This value agrees well with that obtained directly by the scintillation method. Counts have shown that one gram of uranium emits $2{\cdot}3 \times 10^{4}$ α-particles in a second, only half of which, however, arise from uranium I, the other half being generated by uranium II, which, as we shall see in the next section, is always present in radioactive equilibrium with uranium I, though in much smaller amount. Since the atomic weight of uranium is 238, one gram contains about $2{\cdot}55 \times 10^{21}$ atoms. The disintegration constant is therefore found on dividing the half of $2{\cdot}3 \times 10^{4}$ by this number, the result being

$$\lambda\,(\mathrm{U}) = 4{\cdot}5 \times 10^{-18}\ \mathrm{sec.}^{-1}.$$

It follows from this, according to § 111 eqn. (3), that the half-value period of uranium amounts to $4{\cdot}9 \times 10^{9}$ years, in good agreement with the value derived indirectly.

The ratio between the half-value periods of *radium* and *radium emanation* is also found to obey eqn. (12) satisfactorily. One gram of radium is in equilibrium with $0{\cdot}663$ mm.^{3}, or $6{\cdot}56 \times 10^{-6}$ grams of emanation, this quantity, which serves as a unit of measure, being known as a "*curie.*"[3] The half-value period of emanation must hence be equal to that of radium multiplied by $6{\cdot}56 \times 10^{-6}$, and by the ratio between the atomic weights of radium and the emanation, *i.e.*, by 226/222.[4] In accordance with § 111 eqn. (5), we thus find for the half-value period of radium emanation $3{\cdot}33 \times 10^{5}$ sec., while direct observation on the decrease in the intensity of the radiation gives a value of $3{\cdot}29 \times 10^{5}$ sec., or 3·81 days.

Finally, we will apply eqn. (9) to the special case in which, at a given moment which we choose as zero on the time scale, the *transformation product is separated from the longer-lived parent substance* with which it is supposed, till then, to have been in equilibrium. From the moment of separation, we have therefore to distinguish between two preparations. In the first, the transformation product which was completely removed at the time $t = 0$ continues to form, whilst in the

[3] The unit of measure used for radium emanation concentrations is the *Mache* unit; this corresponds to an emanation content of $3{\cdot}64 \times 10^{-10}$ curie/litre. (The quantity of emanation then contained in one litre will produce, on transference to an electroscope, a saturation current of 10^{-3} electrostatic units a second.) The strongest therapeutic springs have an emanation content of 2,000 or 3,000 Mache units.

[4] For n radium atoms are in equilibrium with $n \times 6{\cdot}56 \times 10^{-6} \times 226/222$ emanation atoms.

second preparation, which contains no parent substance, the number of atoms of the transformation product in it decreases continuously, in consequence of their progressive disintegration. Let N_1' and N_2' be the numbers of atoms of the disintegration product in the two preparations at any given time.

It follows from eqn. (9) that, since N_1' is to be made equal to zero at the time $t = 0$, the number N_1' is given by

$$(13) \qquad N_1' = \frac{\lambda}{\lambda'} N_0 (1 - e^{-\lambda' t}).$$

On the other hand, at the time $t = 0$ the number N_2' is equal

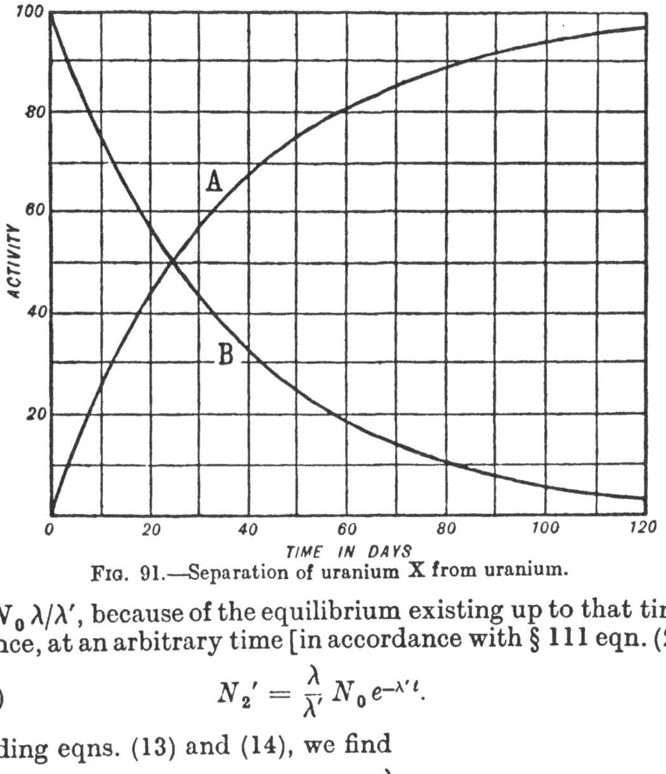

FIG. 91.—Separation of uranium X from uranium.

to $N_0 \lambda/\lambda'$, because of the equilibrium existing up to that time. Hence, at an arbitrary time [in accordance with § 111 eqn. (2)],

$$(14) \qquad N_2' = \frac{\lambda}{\lambda'} N_0 e^{-\lambda' t}.$$

Adding eqns. (13) and (14), we find

$$(15) \qquad N_1' + N_2' = \frac{\lambda}{\lambda'} N_0 ;$$

i.e., the *sum of the atoms of the transformation product in the two preparations* is *not affected by the separation*, but remains *constant*.

An instructive example of the fulfilment of this law is afforded by the *separation of uranium X from uranium.* Uranium X is of far shorter life than uranium, its half-value period only amounting to 23·8 days.[5] Since uranium emits α-rays, but uranium X β-rays, the intensity of the β-radiation emitted by the separated preparations forms a measure of the number of uranium X atoms present.[6] In Fig. 91 curve A shows the *increasing* intensity of the β-radiation from the "recovering" uranium preparation, and curve B the *decreasing* intensity in the separated uranium X preparation. At every instant the sum of the ordinates of the two curves has the same value, namely, the initial value (supposed equal to one hundred).

§ 113. The Transformation Series.

At the present time, 40 radioactive elements,[1] or, as they are briefly termed, "*radioelements*," with the most various life periods are known in physics; the half-value periods range from 10^{-11} sec. to 10^{12} years. Apart from potassium and rubidium, from which a weak β-radiation has been detected, radioelements are distributed over the last twelve places in the periodic system with the ordinal numbers 81 to 92, omitting, however, the 85th and 87th places.

Exact investigations have shown that the radioelements (apart from potassium and rubidium) can be arranged in *three transformation series*, in such a way that each element in such a series arises from the preceding element through either an α- or a β-transformation. The three series are known as the uranium-radium series, the actinium series, and the thorium series, though the second is obviously only a branch of the first.

The *uranium-radium series* is represented in Table VII, in which elements with the same atomic weight are situated in the same row, and elements with the same atomic number in the same column; the groups of the periodic system corresponding to the atomic numbers are also shown. An α-transformation, in which the atomic weight decreases by four, is indicated by an oblique arrow, and a β-transformation, in which the atomic weight remains unaltered, is shown by a horizontal arrow. The half-value period is given under each element (here, and in the two following tables, years,

[5] The small admixtures of uranium X_2 and uranium II can be disregarded, and we may speak simply of uranium and uranium X. *Cf.* § 113.

[6] The α-radiation can be cut out by absorption.

[1] Not including the stable end products of the transformations.

The Uranium-Radium Series

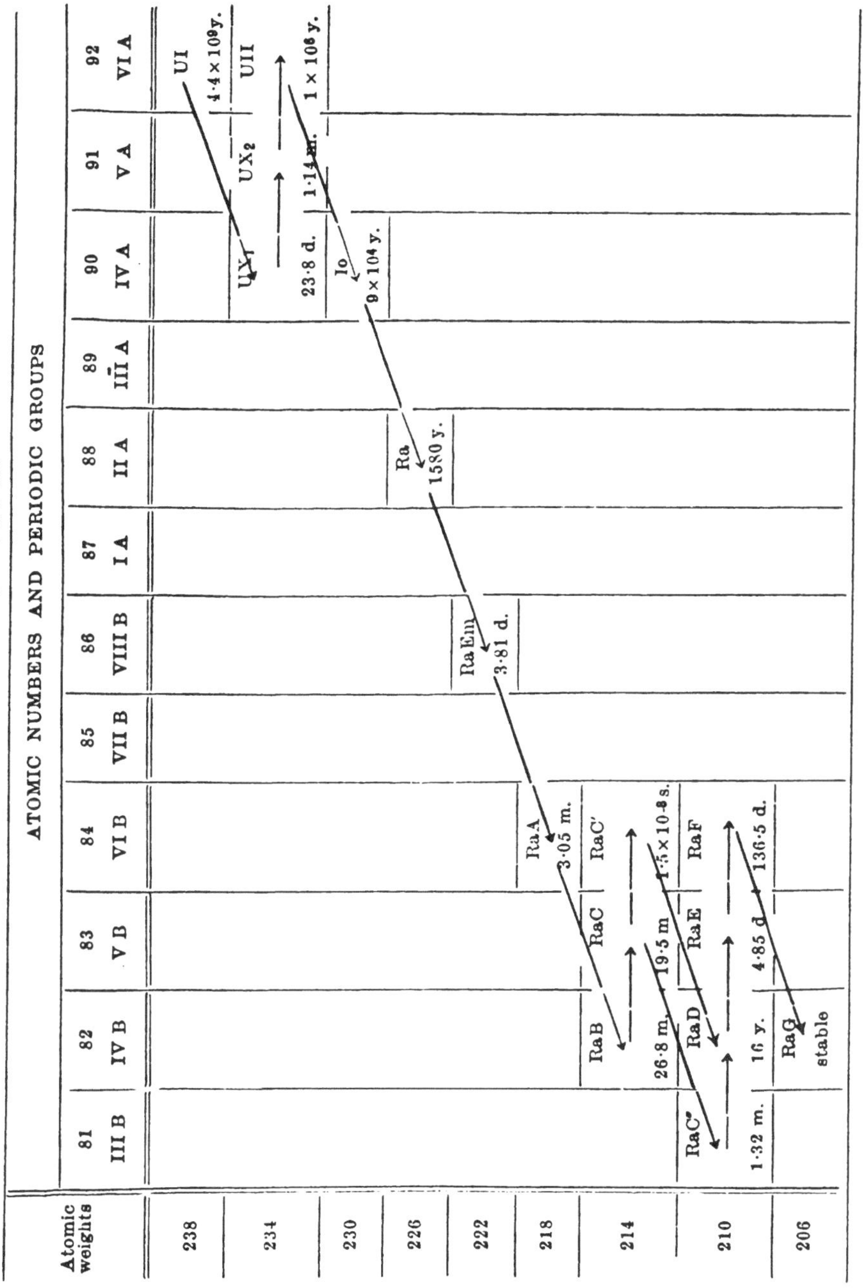

days, hours, minutes, and seconds are denoted respectively by the letters *y*, *d*, *h*, *m*, and *s*).

As can be seen from Table VII, *uranium* is a *mixed element* with *two isotopes*, the heavier of which undergoes one α- and two β-transformations to form the lighter. One gram of uranium I, however, is in equilibrium with only 0·25 mgm. of uranium II, and, indeed, with only $1{\cdot}5 \times 10^{-11}$ grams of uranium X_1, and $4{\cdot}9 \times 10^{-16}$ grams of uranium X_2. Hence from the point of view of practical chemistry, the mixed element can be regarded as being pure uranium I. Uranium II cannot be separated from uranium I, with which it is isotopic; on the other hand, uranium X_1 and uranium X_2, which are isotopes of thorium and protoactinium, can be separated by suitable precipitation methods.

An α-transformation leads from uranium II to the long-lived parent substance of radium, *ionium*. Inasmuch as this is an isotope of thorium, it cannot be prepared in a pure state, but only in the form of an enriched thorium preparation, to the extent of 30%.[2] Radium, which has already been obtained in some hundreds of grams from minerals, and which has also been prepared in a chemically pure state, gives rise to *radium emanation* on an α-transformation. Since this is a rare gas, not chemically combined but only occluded in radium salts, it can easily be liberated on heating.

Radium emanation produces the so-called *short-lived active deposit of radium*, consisting of five substances, RaA, RaB, RaC, RaC′ and RaC″, whose genetic relations can be seen from Table VII. These substances are all solid, and are mostly charged positively immediately on their production, so that they are readily deposited on negatively charged plates. The five elements are so short-lived that they can only be observed in mixtures, though the proportions of the mixtures may be varied at will. Analyses of the curves representing the decrease in activity of such mixtures have made it possible to determine the half-value periods of the individual constituents of the short-lived active deposit. The smallest of the half-value periods have been indirectly determined on the basis of the relation between the life period and the range of the emitted α-rays, which we shall discuss in the following section.

At radium C, the transformation series *branches*, a β-transformation leading to the extraordinarily short-lived RaC′, whereas an α-transformation leads to RaC″. The α-trans-

[2] Whereas the atomic weight of pure thorium is 232·1, the combining weight of the thorium preparation richest in ionium is 231·5.

formation, however, is much rarer than the β-transformation; of 10,000 RaC atoms, only 4 undergo the transformation into RaC″.

When a sufficiently large quantity of emanation (such as half a cubic millimetre) disintegrates inside a closed glass vessel, a fine deposit which has the properties of a *species of lead* (known as *radium D*) is formed on the walls. The transformation law shows that radium D must result from an α-transformation of RaC′ (which emits α-rays), and also by a β-transformation from RaC″. Radium D has the comparatively large half-value period of 16 years; it changes after emitting a weak β-radiation into an isotope of bismuth, known as radium E, from which another β-transformation gives rise to RaF, known as *polonium*. Polonium emits α-rays, and when it has decayed no new rays can be detected. Hence the element produced from polonium after an α-transformation and known as RaG, which (according to the transformation law) must be a *species of lead*, representing the so-called *uranium lead*, must be assumed to be the *end product of the uranium-radium series*.

Quite analogous to the radium series is the *actinium series*, represented in Table VIII. There is no doubt about the sequence from protoactinium to the presumably final product actinium D, isotopic with lead. *Protoactinium* is a very long-lived element, with a half-value period of 12,000 years; the disintegration product, arising from an α-transformation, is *actinium*, whose half-value period is much smaller, amounting only to 20 years. Hitherto it has not been possible to prepare either protoactinium or actinium in a pure state, for they have not yet been completely separated from their homologues, tantalum and lanthanum. One β- and two α-transformations lead from actinium to *actinium emanation*, whose life period, however, only amounts to a few seconds. Actinium emanation gives rise to the *active deposit* of actinium. The sequence of transformations in this deposit much resembles that of the radium active deposit. Whereas, however, almost the whole of the RaC atoms undergo a β-transformation, just the converse is the case with the atoms of AcC. Of 10,000 AcC atoms, 9,984 emit α-rays and form AcC″, whilst only 16 undergo the β-transformation into AcC′. AcC′ and AcC″ lead directly to the final stable product, AcD, whereas the corresponding product radium D is weakly radioactive.

Evidence that the *actinium series* probably represents a *branch of the uranium series* is to be found in the observed

TABLE VIII

The Actinium Series

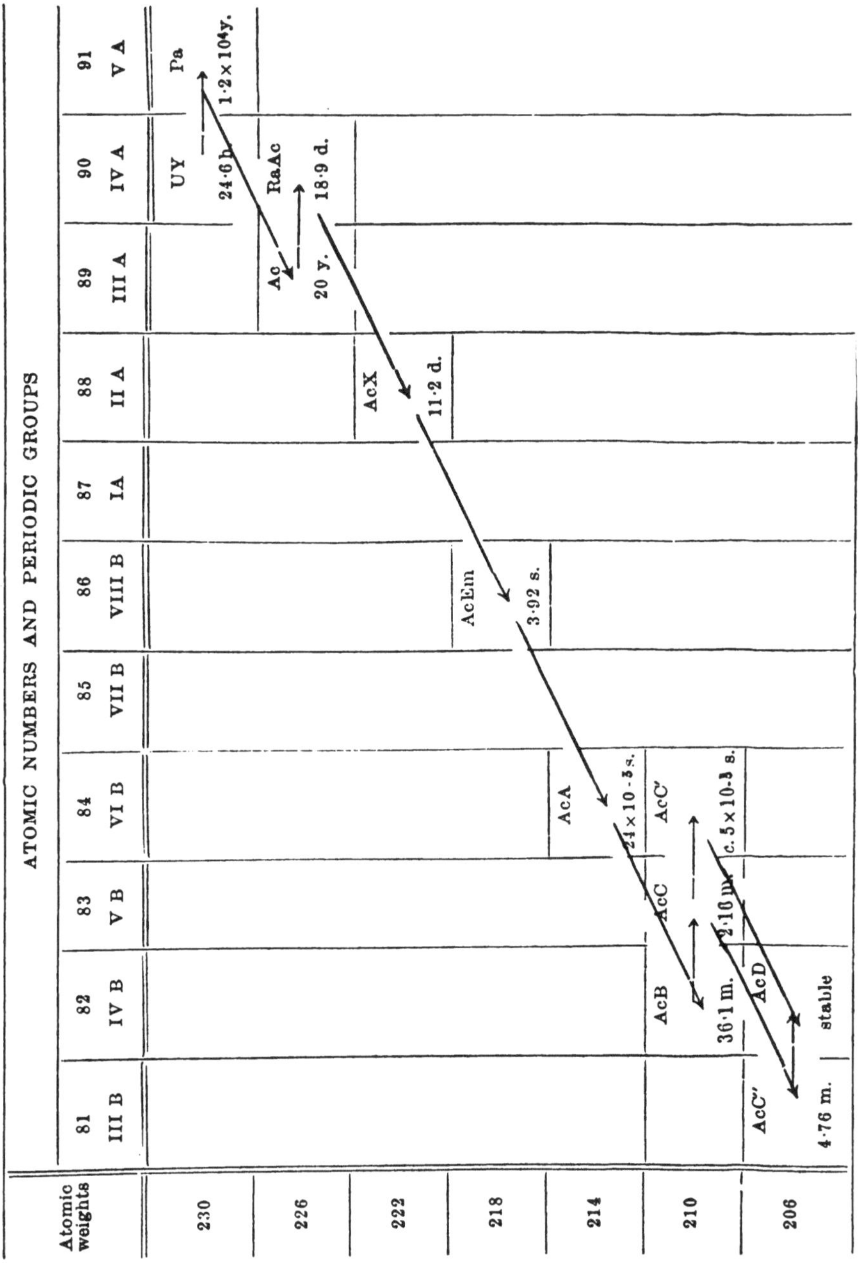

fact that the analogous members of the actinium series occur in all uranium minerals to the same extent, being responsible for about three *per cent.* of the total activity. This means that the uranium series branches in such a way that 97 out of every 100 atoms follow the transformation into radium, and 3 the transformation into actinium. At what point the branching occurs is, however, not yet certain. On the grounds of the transformation law alone, it is probable that the parent substance of protoactinium is *uranium Y*, an isotope of thorium, emitting β-rays, which is present in uranium and contributes three *per cent.* towards the activity, though the relative masses are only in the ratio of 2×10^{-14} to 1. It is further possible that uranium Y, like ionium, is produced from uranium II by an α-transformation, so that a branching involving two α-transformations would occur at this radio-element. The atomic weights of the actinium series are also only hypothetical, for no single member of the series has yet been isolated in the pure form in sufficient quantity to allow of an experimental determination of its atomic weight.[3]

Furthermore, the position of *uranium Z*, which is obviously due to a branching in the uranium series, is not yet certain. It is an isotope of protoactinium, has a half-value period of 6·7 hours, and contributes about 0·35 per cent. towards the activity of uranium, in which it is always present, though the mass ratio is only 5×10^{-16} to 1.

The complete *independence of the thorium series from the uranium series* is indicated by the fact that both uranium minerals almost free from thorium, and thorium minerals almost free from uranium, are known. The thorium series is shown in Table IX.

From the initial substance of the thorium series, *thorium*, an α-transformation leads to the relatively long-lived *mesothorium* 1. Since its half-value period amounts to 6·7 years, and in all thorium minerals there are about 3×10^{-10} grams of mesothorium 1 to 1 gram of thorium, it follows from the law of radioactive equilibrium that the half-value period of thorium must be about 2×10^{10} years. Pure mesothorium 1 cannot be prepared from minerals, because it is an isotope of radium, and all thorium minerals always contain a little uranium, and thus radium.[4] The pure substance can only be

[3] Minute quantities of pure AcX have been prepared by the method of recoil (to be described in the following section—§ 114). An experimental atomic weight determination is first to be hoped for with protoactinium, as soon as a larger quantity of this has been obtained in a pure state.

[4] Since 1 gram of thorium is in equilibrium with 3×10^{-10} grams of mesothorium, and 1 gram of uranium with $3{\cdot}4 \times 10^{-7}$ grams of radium, the

TABLE IX
The Thorium Series

Atomic weights	81 III B	82 IV B	83 V B	84 VI B	85 VII B	86 VIII B	87 I A	88 II A	89 III A	90 IV A
	ATOMIC NUMBERS AND PERIODIC GROUPS									
232										Th c. 2×10^{10} y.
228								MsTh 1 6·7 y.	MsTh 2 6·2 h.	RdTh 1·90 y.
224								ThX 3·64 d.		
220						ThEm 54·5 s.				
216				ThA 0·14 s.						
212		ThB 10·6 h.	ThC 60·8 m.	ThC′ c. 10^{-11} s.						
208	ThC″ 3·20 m.	ThD stable								

obtained in minute quantities by the disintegration of thorium salts which have been previously purified from uranium and its transformation products.

From mesothorium 1, two β-transformations[5] lead through the very short-lived mesothorium 2 to *radiothorium*, which is an isotope of thorium with an atomic weight smaller by four units, and with a half-value period of nearly two years. From radiothorium, two α-transformations lead to the extremely short-lived *thorium emanation*, which in its turn gives rise to the *active deposit of thorium*. Considerable analogy exists between this and the deposits of radium and actinium, the final product here being thorium D, which is isotopic with lead. Approximately 65% of thorium C undergoes a β-transformation and 35% an α-transformation. About one ten-thousandth of the atoms of ThC disintegrate with the emission of α-rays of abnormally long range (11·3 cm.), so that at least a three-fold branching of the thorium series must be assumed to occur at ThC.

The *final product* of all three transformation series is represented by different isotopes of *lead*. Consequently, both in uranium and in thorium minerals, appreciable quantities of lead must accumulate in the course of the ages ; and hence a determination of the lead content enables us to fix the *age of the minerals*, and therewith the age of the *geological formations* in which the minerals are found. Since $n \times 1{\cdot}5 \times 10^{-10}$ atoms out of n uranium atoms disintegrate in the course of one year, while the ratio of the atomic weight of lead to that of uranium is 206 : 238, 1 gram of uranium must produce annually $1{\cdot}3 \times 10^{-10}$ grams of uranium lead. We therefore obtain the approximate age (in years) of a mineral on multiplying the amount of uranium lead, corresponding to 1 gram of uranium, by the reciprocal of $1{\cdot}3 \times 10^{-10}$ or, roughly, by 8,000 million.[6] A ten *per cent.* content of uranium lead thus indicates an age of about 800 million years. Determinations of age must, however, be based only on the content of uranium lead (RaG), from which a common contamination of ordinary lead must be distinguished. Inasmuch, however, as the atomic weights of uranium lead

quantities of mesothorium and radium contained in a thorium mineral will be approximately the same when uranium is present in the mineral to the extent of even only one in a thousand of the thorium content.

[5] It is true that no radiation can be detected from mesothorium 1 ; but since the atomic number of mesothorium 2 is 89, a β-particle must be ejected during the transformation into mesothorium 2.

[6] In an accurate calculation it must be remembered that the uranium has also decreased in amount in the course of time ; allowing for this, the ages found for the minerals are somewhat smaller.

and ordinary lead are different (206·0 and 207·2), a determination of the atomic weight of the mixture of RaG and ordinary lead at once enables us to calculate the proportions of the two substances.[7]

In this way it has been found that the *oldest mineral known* is a zircon from Mozambique, a 21% content of uranium lead showing the age to be 1,500 *million years*. The youngest uranium mineral appears to be a uraninite occurring in the carboniferous system, with a uranium lead content of about 4% and an age of 320 million years.[8]

Whether the uranium and thorium series, both of which presumably end with lead, begin with elements of still higher atomic weight than uranium and thorium is an open question. But it is unlikely that such higher elements can be found, for, if they were shorter-lived than uranium or thorium, then, in accordance with what has just been said about the age of minerals, they would have disintegrated long ago, seeing that a formation of new elements after the solidification of the earth is scarcely conceivable. On the other hand, if their life periods were sufficiently long, they must occur in recognizable quantities in uranium and thorium minerals, as a consequence of radioactive equilibrium, in which case they could hardly have escaped detection up till now. It does not, however, appear impossible for elements with higher atomic weights than that of uranium to exist, if they belong to transformation series other than those described, which may have hitherto eluded discovery simply because of the rarity of their members.

Members of such transformation series, as yet unknown, are possibly represented by those isotopes of *potassium* and *rubidium* to which is due the β-radiation discovered in these two metals by *Campbell* in 1907. It is true that, both with potassium and with rubidium, it is still an open question whether one of the two known isotopes is radioactive, or whether the activity is caused by a third unknown isotope. If the former be the case, it may be estimated from the hardness of the β-rays emitted that the half-value period of rubidium (which radiates the more strongly) is about a

[7] For example, whereas the Morogoro ore found in East Africa gives an atomic weight of 206·05, and can thus only contain uranium lead, it follows from the atomic weight of 206·4, found for the lead in uraninite from North Carolina, that 70 % of the lead has been formed from uranium, and 30 % is due to an impurity.

[8] Thorium minerals are frequently of secondary origin, so that the ages found for them cannot always be ascribed to the geological formations in which they occur. The ages calculated for thorium minerals range from 10 million to 600 million years.

hundred thousand million years, while the half-value period of potassium is about a billion years.[9]

It was not until after the first observations of radioactivity in 1896 that by far the greater number of the radioelements were discovered; before that date only uranium and thorium were known, apart from potassium and rubidium. Of the radioelements discovered after that time, five represent newly discovered elements in the sense that they are the principal representatives of places in the periodic system which had previously been empty. These five are *radium* and *polonium*, discovered in 1898 by *M.* and *Mme. Curie*; *actinium*, discovered by *Debierne* and *Giesel* in 1899; *emanation*, discovered in 1900 by *Rutherford* and *Dorn*; and finally, *protoactinium*, discovered by *Hahn* and *Meitner* and by *Soddy* in 1918.[10] The genetic relations between the radioelements became evident on the establishment (in 1913) by *Soddy* and by *Fajans* of the law of α- and β-transformations —even before the discovery of the natural sequence of the elements.

§ 114. Radioactive Radiation.

The most striking phenomenon due to the disintegration of the nucleus, which, indeed, first drew the attention of physicists to that disintegration, is the *radioactive radiation* emitted by the disintegrating atoms, part of which is *corpuscular* in nature and part of the nature of *electromagnetic waves*. The radioactive rays arise either as so-called *primary rays from the atomic nucleus*, or as so-called *secondary rays* consequent on the interaction of the primary rays and the *electron groups*, surrounding the atomic nucleus.

Foremost among the primary rays are the α- and β-particles ejected by the disintegrating atomic nucleus. The *initial velocities of the α-particles* represent *constants* characteristic of the radioelements emitting them. Expressed in cm./sec., the initial velocities lie between $1{\cdot}40 \times 10^9$ for uranium I, and $2{\cdot}06 \times 10^9$ for thorium C′; the abnormally rapid α-rays, which, as already mentioned, are emitted by about a ten-thousandth part of the disintegrating thorium C atoms, have a velocity of $2{\cdot}3 \times 10^9$.

As was mentioned in an earlier section (§ 90), the *range* of the α-rays in any gas is *proportional to the cube of the initial*

[9] For the relation between the hardness of the β-rays and the life period, *vide* § 114.

[10] Before protoactinium, its short-lived isotope uranium X_2 (formerly called "brevium") had already been discovered.

velocity ; the same relation also exists between the distance yet to be covered at any moment and the constantly decreasing velocity at that moment. The ranges in air at atmospheric pressure and 0° likewise represent constants characteristic of the radioelements which emit α-rays ; they vary from 2·53 cm. for uranium I to 8·17 cm. for thorium C′. The exceptionally rapid α-rays emitted by a ten-thousandth part of the disintegrating thorium C atoms have a range of 11·3 cm. Furthermore, the range varies directly as the absolute temperature of the gas, and inversely as its pressure, so that the range of α-rays *in vacuo* is found to be extremely large. Referred to RaC′, the range in air at 0° and atmospheric pressure is connected with the initial velocity by the relation

(1) $$v^3 = 1{\cdot}076 \times 10^{27} R_0 ,$$

or,

(2) $$v = 1{\cdot}025 \times 10^9 R_0{}^{1/3} .$$

The range at 15°, which is frequently given, is obtained from R_0 on multiplying by 1·055.

A simple relation connects the ranges of α-rayers with their *life-periods.* As was found by *Geiger* and *Nuttall* in 1911, the *logarithms of the disintegration constants* are *linear functions of the logarithms of the ranges,* within each transformation series. We have the relation

(3) $$\log \lambda = A + B \log R_0 ,$$

where B is a universal constant, but A has a definite value for each of the three transformation series (*cf.* Fig. 92). With logarithms to the base 10, B is equal to 60 and, for the uranium-radium series, A is equal to $(-\ 41{\cdot}5)$. By means of eqn. (3) it has been possible to determine the half-value periods of the exceedingly short-lived radioelements. That eqn. (3) is not generally valid, however, can be seen from the position of AcX in Fig. 92, the range corresponding to the thorium rather than to the actinium series.[1]

Whereas the ranges are taken as characteristic constants for the radioelements emitting α-rays, the *absorption coefficients in aluminium* are generally taken for those elements which emit β-rays. Since the intensity of the β-rays diminishes in accordance with an exponential law, as experience shows, it decreases from I_0 to I, on passing through a layer of thickness d, according to the relation

(4) $$I = I_0 e^{-\mu d} ,$$

[1] As Fig. 92 shows, Geiger and Nuttall's equation does not hold in general for the thorium and the actinium series nearly so accurately as for the uranium radium series.

where μ denotes the absorption coefficient. The value of d for which I is equal to half of I_0 is known as the *half-value thickness*, or half-value layer; it is connected with the absorption coefficient by the relation [2]

$$(5) \qquad D = \frac{\log_e 2}{\mu}.$$

Expressed in reciprocal centimetres, the absorption co-

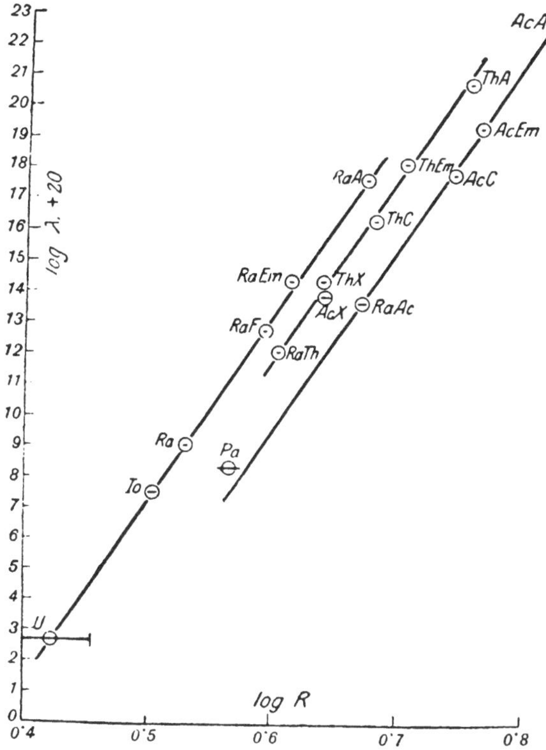

FIG. 92.—Relation between range and disintegration constant. (*Zeitschr. f. Phys.*, vol. 8. Springer, Berlin.)

efficients in aluminium lie between 13 (for RaB and RaC) and 5,500 (for RaD); the half-value thicknesses between $5{\cdot}13 \times 10^{-2}$ and $1{\cdot}26 \times 10^{-1}$ cm. Naturally, the smaller the absorption coefficient and the larger the half-value thickness, the harder and thus the more penetrating are the rays.

[2] The same relation exists between half-value thickness and absorption coefficient as between half-value period and disintegration constant.

Many radioelements possess *several different values for the absorption coefficient*,[3] which shows that the β-radiation emitted by them is composed of rays of different velocities. This is still more clearly demonstrated by the *magnetic spectra*, which are based on the different deflections of rays of different velocities in a magnetic field. Thus, if the β-radiation as deflected by a magnet be allowed to fall on a photographic plate (which is thereby blackened), with a suitable arrangement of the apparatus sharp black lines will be formed on the plate (Fig. 93). Each of these lines corresponds to a definite initial velocity, which can be easily calculated,[4] but some of them are due to the secondary β-rays (which we shall discuss later), only those of higher velocity being of primary origin.[5] The initial velocities of the β-rays lie between $9{\cdot}9 \times 10^9$ and $2{\cdot}994 \times 10^{10}$ cm./sec., amounting to between 33 and 99·8% of the velocity of light. The fastest β-rays have been found in the magnetic spectrum of radium C. In general, large values of the initial velocity are associated with small half-value periods.[6]

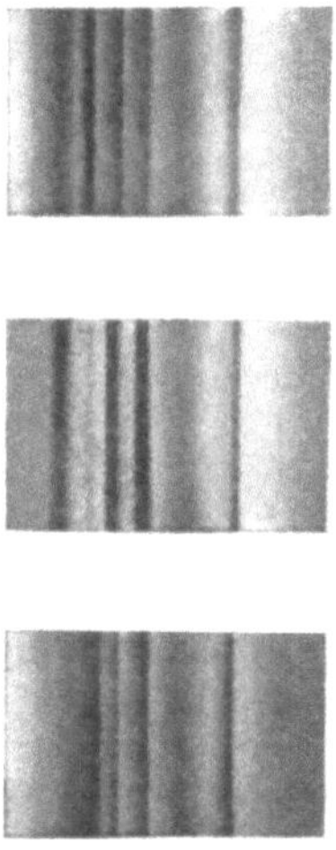

FIG. 93. — Magnetic spectra of β-rays.

By measuring the charge imparted by β-rays to an electroscope, the *number of β-particles emitted in a second* by a β-rayer can be calculated. It is found that this number is always *equal to the number of atoms undergoing a β-transformation* in a second. For example, it has been shown that the quantities of the β-raying elements RaB and RaC in equilibrium with a gram of radium emit together 7×10^{10} β-particles in a second. Half of this number, however, is nearly identical with the number of β-particles expelled by a gram of radium (apart from transformation products), which, as already mentioned, amounts to $3{\cdot}72 \times 10^{10}$. (The agreement is not perfect, because the

[3] *E.g.*, radium B has three μ-values: 890, 77, and 13·1.

[4] *Cf.* § 89 eqn. (1).

[5] The same applies to widely differing values of the absorption coefficient.

[6] Efforts to establish a quantitative relation between the disintegration constant and the absorption coefficient (on the lines of Geiger and Nuttall's equation) have so far been unsuccessful, principally because an exact separation of primary from secondary lines in the magnetic β-ray spectrum has not yet been effected.

primary β-particles are accompanied by secondary particles.) The emission of negative β-particles produces on the preparation ejecting them strong *positive charges*, up to 150,000 volts with a suitable experimental arrangement.

Primary γ-rays are generated when the expelled α- or β-particles lose kinetic energy owing to a "*braking*" *effect* while still *within the nucleus.* Like β-rays, γ-rays are also generally characterized by their *absorption coefficients in aluminium*; the hardest γ-rays (from thorium C″) have an absorption coefficient of 0·0916 $cm.^{-1}$, while the softest rays (from ionium) have a coefficient of 1,088 $cm.^{-1}$. The half-value thickness in air for the hardest γ-rays is some 150 metres, whereas for the fastest β-rays it is only about 1 metre.

In the case of soft and moderately hard γ-rays, the *wave-lengths* can be determined by the crystal method (*cf.* § 108). Thus, more than 20 separate lines have been detected in the combined γ-ray spectrum of the mixture of RaB and RaC, the wave-lengths lying between 71 and 1,365 X.U. It has only been possible to determine smaller wave-lengths indirectly, values as low as 20 X.U. having been calculated; this will be discussed more fully in a later section (§ 118). Estimations lead to a probable wave-length of only 5 X.U. for the hardest rays. It follows from Duane and Hunt's equation that the potential difference necessary to generate rays of such hardness in Röntgen tubes would be about two million volts, or nearly seven times as large as the highest potential yet produced. Only the hardest of the γ-rays, however, are likely to be primary rays.

The *secondary radioactive rays*, emanating from the portions of the atom lying outside the nucleus, are either β- or γ-rays. *Secondary β-rays* are the result of a *photo-electric* effect produced in the groups of electrons surrounding the nucleus by the primary γ-rays emitted from the latter. They are, of course, to be observed not only in such radioelements as undergo a β-transformation, but also in those which experience an α-transformation; in point of fact, a weak β-radiation of secondary origin can be detected in the α-raying elements, radium, radioactinium, and radiothorium.[7] *Secondary γ-rays* due to primary β- or primary α-rays, are simply the *characteristic rays* of the elements concerned, which are also to be found in their Röntgen spectra.

[7] Secondary β-rays can probably be produced by primary corpuscular rays as well as by primary γ-rays. The so-called "δ-rays," consisting of slow electrons and emitted by all α-rayers, are very likely of this origin.

A part of the corpuscular radiation emitted by radioactive substances is formed by the so-called *recoil rays*. In accordance with the Principle of the Conservation of Momentum, every atom which expels an α- or a β-particle must experience a recoil with a velocity the ratio of which to that of the expelled α- or β-particle is the same as that of the particle's mass to the mass of the remainder of the atom. For example, if an atom of radium A with an atomic weight of 218 expels an α-particle of mass 4 with a velocity of $1{\cdot}69 \times 10^9$ cm./sec., the atom of radium B resulting from the transformation must recoil with a velocity equal to that of the α-particle just given multiplied by 4/214. This velocity is about 3×10^7 cm./sec., and it is large enough to render the recoil rays recognizable.[8]

Among the effects of radioactive rays, the most striking are the *ionization of the air* and the continuous *development of heat*. The *number of pairs of ions* produced by an α-particle in the air is found to be proportional to the initial kinetic energy of the α-particle, and hence, by eqn. (2), to the square of the cube root of the range. Denoting the number of pairs of ions by k, we have (relatively to the data for radium C′)

$$k = 6{\cdot}25 \times 10^4 R_0^{2/3}. \tag{6}$$

According to this, the long-range α-particles of radium C′ each produce some 220,000 pairs of ions ; and even the slow α-particles of uranium I give rise to about 120,000. The ionization effect of β-rays is considerably less ; on complete absorption of β-radiation in air, a fast β-particle produces only about 10,000 pairs of ions.[9] γ-rays also have an ionizing action. The total effect of the γ-rays emitted by any quantity of RaC is approximately as great as that of the β-rays emitted by that quantity.[10]

The ionization effect of radioactive radiation allows of the *detection of extraordinarily small and unweighable amounts* of the radioelements. The detection, and also the measure-

[8] Like α-rays, these recoil rays have a range, which for rays of RaB in air amounts to 0·14 mm. By catching the recoil rays, it has been possible to prepare some radioelements, such as AcX, in a state of purity, but only in minute quantities. The detection of recoil rays on the emission of β-particles is far more difficult, seeing that, on account of its 7,000 times smaller mass, the momentum of a β-particle, despite its 20 times larger velocity, is much smaller than that of an α-particle. Nevertheless, the recoil rays of RaC atoms produced by the β-disintegration of RaB can be observed.

[9] One β-particle in air at normal pressure generates, on an average, about 70 pairs of ions along one centimetre of its path.

[10] The ionization effect of γ-rays, referred to one centimetre of path, is in itself much less than that of β-rays. This is, however, outweighed by the much longer paths of the γ-rays.

ment, of such minute quantities is most simply performed by observing the time required for the leaves of a charged electroscope to collapse, when the air is made conducting by the rays from the radioactive substance.[11] By means of this method it is possible to detect even the billionth part of a gram of a radioactive substance.[12]

The *heat* which is continuously evolved by an active preparation enclosed in a vessel originates in the kinetic energy of the α- and β-rays and the recoil rays in their collision with the walls of the vessel, and also in the absorbed electromagnetic energy of the γ-rays. It may be calculated from the characteristic constants of the rays, and from the disintegration constants of the radium series, that the heat evolved in an hour by a gram of radium, together with its short-lived transformation products, must amount to 137 calories. Of this quantity of heat 90·1% is due to the α-rays, 3·4% to the β-rays, 4·7% to the γ-rays, and 1·8% to the recoil rays. In good agreement with the calculated result, the hourly development of heat is found by experiment to be 136 calories.[13]

The total quantity of heat which would be evolved by a gram of radium on complete disintegration amounts to $3{\cdot}7 \times 10^9$ calories. It is, perhaps, of interest to note that this quantity is about a million times as large as that developed on the transformation of a gram of "electrolytic gas" into water, although this chemical reaction is marked by an unusually large evolution of heat.[14]

§ 115. The Disruption of the Elements.

In radioactive phenomena the disintegration of the nucleus represents a spontaneously occurring process which

[11] Strong radioactive preparations are measured by the intensity of the "saturation currents" produced by them.

[12] *E.g.*, only about three-millionths of a milligram of radium A can be obtained from a gram of radium.—An important application of the extraordinary sensitivity of radioactive measurements has been made by *Hevesy* in his method of *radioactive indicators*. Chemical processes in which lead or bismuth take part in only very minute quantities can be detected and measured by mixing with the lead or bismuth isotopic radioelements, which cannot be separated from them, but whose presence even in the smallest amounts can clearly be recognized by radioactive measurements.

[13] The continuous evolution of heat has also been detected in the cases of polonium, thorium, and uranium. A gram of uranium in equilibrium with its disintegration products develops about a ten-thousandth of a calorie in an hour.

[14] Even if the sun were composed exclusively of uranium and its transformation products, the heat resulting from the disintegration would only account for one-half of the actual quantity of heat radiated.

cannot be influenced in any way whatsoever. It was, therefore, a discovery of the greatest importance when, in 1919, *Rutherford* for the first time demonstrated an *artificial disruption of an element*, and that an element with a low atomic number.

Rutherford's discovery was based on observations on the origin of the H-rays which we have already mentioned in an earlier section (§ 90). Shortly after the detection of these rays, *Marsden*, in conjunction with *Lantsberry*, was surprised to find that a piece of nickel foil, coated with radium C, produced *scintillations* through *ordinary air* at a distance much greater than the range in air of the α-rays emitted from the nickel foil. For, whereas the range of these α-rays (actually coming from the disintegration product RaC′) amounts only to 7 cm., scintillations could be observed even at a distance of 40 cm. This was all the more unexpected, because even H-rays generated in hydrogen have a range in air of only 29 cm.

On following up this, at first mysterious, phenomenon, Rutherford made the important discovery that the number of scintillations depended on the substance filling the space between the nickel foil and the screen. When this space was evacuated or filled with carbon dioxide or oxygen instead of with air, there were no scintillations ; on the other hand, the number increased on replacing the air by pure nitrogen. At the same time, measurements of the electric and magnetic deflections proved that the long-range rays actually were H-rays, and thus consisted of hydrogen nuclei.

These facts led Rutherford to the important conclusion that the *α-particles effect a disruption of the nitrogen nuclei, hydrogen nuclei being expelled* from the disrupted nuclei.

It is found from calculations on scintillation counts that, on an average, only one single hydrogen nucleus is liberated for some hundred thousands of α-particles emitted. Since the amount of RaC′ in radioactive equilibrium with a gram of radium ejects $3{\cdot}72 \times 10^{10}$ α-particles in a second, a preparation of one gram of radium (of which only the RaC′ is of account for the disruption of the nitrogen)[1] will thus liberate about 10^{12} hydrogen nuclei annually. Hence, the hydrogen atoms arising on neutralization would, in a year, make up a volume of about a thousandth of a cubic millimetre.

Later experiments of Rutherford have shown that hydrogen nuclei are liberated by α-particles also from the atoms of *boron*, *fluorine*, *sodium*, *aluminium*, and *phosphorus ;* and,

[1] The ranges for the other α-ray transformation products are too small.

according to the most recent researches, the same probably applies to the atoms of *beryllium*, *magnesium*, and *silicon*.

Disruption of the nuclei of elements only succeeds when the α-rays have a range of at least 5 cm. (which is the case with RaC'). That the effect of the α-particles shot into the atomic nuclei is simply one of *liberation*, however, and that the process of disruption is essentially *explosive*, is shown by the fact that the *kinetic energy of an H-particle* is found to be *larger* than that of the α-particle itself. In the case of *aluminium*, from which H-rays with a range of up to 90 cm. can be produced, the kinetic energy of the H-particles exceeds that of the α-particles by some 50%. Moreover, it is found that the H-rays travel not only in the direction of the α-rays, but in part also in the opposite direction, and this likewise supports the assumption of a nuclear explosion produced by the α-particles.

§ 116. Structure of Atomic Nuclei.

In conjunction with the radioactive β-radiation, the H-rays released by nuclear disruption afford a convincing empirical proof of the accuracy of the modern conception, which regards all atoms as being composed of protons and electrons. On the other hand, the phenomenon of radioactive α-radiation appears clearly to prove that the nuclei are *not directly* built up of protons and electrons, but rather that these components of the nuclei are grouped together especially frequently in the form of *aggregates* with a mass of four hydrogen atoms and a positive charge of two elementary quanta.

Light is thrown on the essential *rôle* played by *alpha-particles* in the construction of the nuclei, not only by the phenomena of radioactivity, but also by the fact that it is just the *most common atomic species* which have an *atomic weight divisible by four*. To elements with an atomic weight and atomic number of the form

$$(1) \qquad A = 4n\,, \quad z = 2n$$

—and such elements are known for values of n from 3 to 10— we must obviously ascribe a *pure alpha-structure* (*cf.* the first horizontal row in Table X). The ability of positive α-particles alone to combine into nuclei of this kind is explained by the fact that the α-particles themselves each consist of four protons and two electrons, and are thus able to attract one another electrostatically, when close enough and suitably arranged, despite their total positive charge.

Atomic weights divisible by four are represented, up to the end of the natural sequence, by comparatively common atomic species whose *atomic numbers* are *always even.* From the atomic weight 44 onwards, however (and to a certain extent, in consequence of isobares, even at 40), the *atomic weight* is *more than twice the atomic number.* If $4n$ be the atomic weight and $2n'$ the atomic number, the difference $(n - n')$ increases as we traverse the natural sequence, until it finally becomes equal to 13 (thorium has an atomic weight of $232 = 4 \times 58$ and an atomic number of $90 = 2 \times 45$).

This fact may be most readily explained on the hypothesis that the nuclei of these atoms also contain, besides positive α-particles of charge $(+ 2e)$, *neutralized alpha-particles* which, without affecting the nuclear charge, nevertheless contribute four units to the atomic weight. This hypothesis, due to *Miss Lise Meitner*, is principally supported by the experimental fact that, in the radioactive *disintegration series*, an α-rayer is frequently followed by two β-rayers, or a β-rayer by an α- and a β-rayer (*cf.* the previous Tables VII–IX in § 113).

A survey of the hypothetical construction of atomic nuclei from positive and neutralized α-particles is given in Table X. This refers to elements with even atomic numbers up to 42. Since the elements up to the atomic number 40 have already been investigated thoroughly for the existence of isotopes (*cf.* the earlier Table V in § 110), only the atomic species Mo_{96} is uncertain, although its existence is extremely probable, inasmuch as the " atomic weight " of molybdenum is an integer (96·0).[1]

Besides the elements with an atomic weight divisible by four, those whose atomic weight is of the form $(4n + 3)$, where the *atomic number* is *always odd*, are of especially frequent occurrence. In particular, elements with an atomic weight and ordinal number of the form

$$(2) \qquad A = 4n + 3, \quad z = 2n + 1,$$

n having values of from 1 to 9 with the exception of 3, are represented by comparatively common atomic species.[2] To many elements with an atomic weight divisible by four

[1] Representatives of higher series than those shown in Table X would be, for instance, the isotopes of 50 (Sn), with atomic weights of 116, 120, and 124, s being put equal to 4, 5, 6. For isotopes of 54 (Xe), with atomic weights of 128, 132, and 136, $s = 5, 6,$ or 7. For the isotope of 80 (Hg), with an atomic weight of 204, s is to be put equal to 11.

[2] These are : Li (3, 7), B (5, 11), F (9, 19), Na (11, 23), Al (13, 27), P (15, 31), Cl (17, 35), K (19, 39).

there corresponds an element with an atomic number greater by one, and an atomic weight higher by three. This indicates that the nuclei of such atoms perhaps contain, besides positive and neutralized α-particles, a single aggregate of mass 3 and charge $(+e)$.

TABLE X

Construction of Atomic Nuclei from α-Particles

(n α-particles, s of which are neutralized; below the chemical symbol are given the atomic number and atomic weight.)

$s=0$ $A=4n$ $z=2n$	$n=3$ C 6, 12	$n=4$ O 8, 16	$n=5$ Ne 10, 20	$n=6$ Mg 12, 24	$n=7$ Si 14, 28	$n=8$ S 16, 32	$n=9$ A 18, 36	$n=10$ Ca 20, 40
$s=1$ $A=4n$ $z=2n-2$	$n=10$ A 18, 40	$n=11$ Ca 20, 44	$n=12$ Ti 22, 48	$n=13$ Cr 24, 52	$n=14$ Fe 26, 56	$n=15$ Ni 28, 60	$n=16$ Zn 30, 64	
$s=2$ $A=4n$ $z=2n-4$	$n=17$ Zn 30, 68	$n=18$ Ge 32, 72	$n=19$ Se 34, 76	$n=20$ Kr 36, 80				
$s=3$ $A=4n$ $z=2n-6$	$n=20$ Se 34, 80	$n=21$ Kr 36, 84	$n=22$ Sr 38, 88	$n=23$ Zr 40, 92	$n=24$ Mo ? 42, 96			

Atomic species with an atomic weight of $(4n+1)$ or $(4n+2)$ are not so common as those with an atomic weight of $4n$ or $(4n+3)$. In this case also, the rule that the atomic number is even for an even atomic weight, and odd for an odd atomic weight, generally holds good ; there are, however, exceptions which occur principally near the beginning of the natural sequence of the elements.[3]

With respect to the number of isotopes, there exists a marked difference between elements of even and those of odd atomic number. Elements with *odd atomic numbers* have *few isotopes.* As the earlier Table V in § 110 shows us, they are either *pure elements*, or they consist of only *two isotopes*, the *atomic weights* of which are *odd* and *differ by two units.* (Apart from short-lived radioelements, the third and fifth elements, lithium and boron, in which the difference in atomic weights is only unity, are exceptions.) On the other

[3] Such exceptions are, on the one hand, Li_6, B_{10}, N_{14}, and on the other hand, Be_9, Mg_{25}, Si_{29}.

hand, elements with even atomic numbers generally have a larger number of isotopes, particularly for the higher atomic numbers.

Furthermore, it is found that elements of *even atomic number* are of *more frequent occurrence* than those of odd atomic number—obviously in consequence of a greater nuclear stability. This may be seen very clearly in Fig. 94, which shows the relation between the average percentage by weight of the elements in stone meteorites and the atomic numbers (drawn as abscissæ). It can, moreover, scarcely be chance that the five gaps in the natural sequence of the elements all belong to odd atomic numbers.

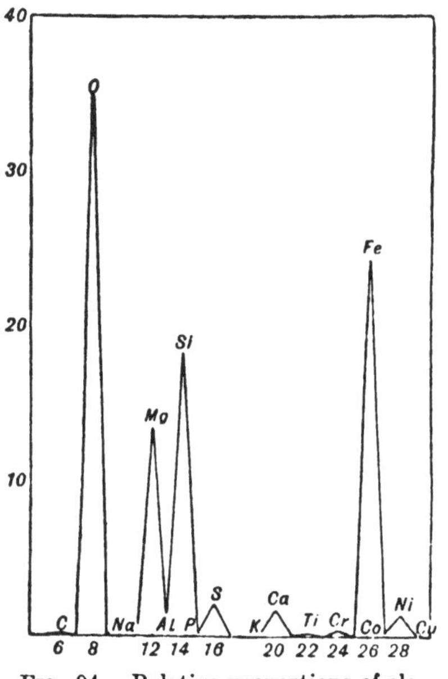

FIG. 94.—Relative proportions of elements in stone meteorites.

As regards the *linear dimensions of the atomic nuclei*, we can draw some conclusions from the earlier-mentioned experiments concerning the *scattering of alpha-rays* in their passage through metal foils (*vide* § 108). For, according to Coulomb's Law, the scattering depends on the size of the nucleus as well as on its charge, so that the experimental results enable us to derive an *upper limit for the radius of the nucleus* (on the assumption that it is spherical in form). In this way a value of 4×10^{-12} cm. has been calculated as the upper limit for the radius of the *gold nucleus*, or less than a thousandth of the radius of the atom. Again, it follows from the formula giving the *electromagnetic mass* of a spherical volume charge (§ 72) that, provided the mass of an electron is solely electromagnetic in origin, its radius must be about 2×10^{-13} cm., and that of a proton about 1×10^{-16} cm. If the mass of an electron were partly of another origin, the above values would represent lower limits, for, with a smaller radius, the electromagnetic mass of an electron would have to be larger than its actual mass. Thus the radius of the gold nucleus can be, at most, only about twenty times that of the electrons of which, together with protons, it is composed.

As a result of the small linear extent of a nucleus, its *internal energy* is, comparatively speaking, enormous. It is so large that the *equivalent mass* which may be ascribed to it, in accordance with the Principle of the Inertia of Energy (a consequence of the theory of relativity to be discussed in a later chapter), may have a *noticeable influence on the atomic mass ;* this equivalent mass is obtained by dividing the energy by c^2. In order to grasp this, we have only to remember that even the kinetic energy of a *single* α-particle expelled from an atomic nucleus amounts [according to § 90 eqn. (2)] to $1 \cdot 3 \times 10^{-5}$ ergs; the equivalent mass is $1 \cdot 4 \times 10^{-26}$ grams, or almost 1% of the mass of a hydrogen atom.

If, now, the internal energy of a nucleus be W, a positive or negative sign being taken according as energy is used up or liberated in "packing together" the components of the nucleus, the mass will appear larger in amount by W/c^2, as compared with the sum of the masses of the components. When the nucleus contains n hydrogen nuclei (the mass of the electrons may be disregarded), the *atomic weight* must thus be given by

$$(3) \qquad A = n \times 1 \cdot 008 + \frac{WL}{c^2},$$

where L, as usual, denotes the Loschmidt number.

Conversely, from eqn. (3) we can calculate the internal energy. Let us take, for example, *helium*, for which A is found to be 4·00 and n is to be put equal to 4 ; then

$$(4) \qquad W = -\frac{0 \cdot 03 \times c^2}{L} = -4 \cdot 5 \times 10^{-5} \text{ erg}.$$

The internal energy of an α-particle is therefore about *three times* as large as its kinetic energy for the largest velocity observed in α-rays.

It follows from the negative sign in eqn. (4) that on *building up a helium nucleus* from its components, *i.e.*, on forming helium from hydrogen, *energy is liberated.* Since there are $1 \cdot 5 \times 10^{23}$ atoms in a gram of helium, $6 \cdot 8 \times 10^{18}$ ergs or, approximately, $1 \cdot 6 \times 10^{11}$ calories would be liberated on forming one gram. The evolution of heat connected with the synthesis of one gram of helium from one gram of hydrogen would therefore be some five million times as large as that associated with the combustion of one gram of hydrogen to form water. In their great internal energy lies the explanation of the fact that the α-particles themselves remain intact even when they disrupt other atomic nuclei.

The extraordinary stability of helium nuclei is probably the cause of the especial part they play in the building-up of more complex nuclei. Obviously the *synthesis of these higher nuclei* from helium nuclei is connected with a readjustment of energy, but, from the circumstance that the pure elements and isotopes have almost integral atomic weights (relative to o = 16), we must conclude that this energy change is much less than that accompanying the synthesis of helium. Indeed, the deviations of the atomic weights from integral values are so small that we cannot even say whether, for example, in the synthesis of an oxygen nucleus from four helium nuclei, energy is used up or liberated.

The energy required for a nuclear synthesis is partially liberated again in a radioactive disintegration, or in an artificial nuclear disruption. Since the transformation of an atom of uranium into an atom of uranium lead is associated with the expulsion of eight α-particles with a kinetic energy whose equivalent mass, in each instance, is nearly 1% that of a hydrogen atom, the difference in atomic weight between uranium and uranium lead must at any rate be greater than 32, and perhaps 32·06 or more. Actually it is more than 32, being between 32·1 and 32·2.

Thus, by the discovery of isotopes, and by drawing attention to the packing effect, recent atomic researches have succeeded in explaining the deviations of the atomic weights from whole-number values. Modern physics is hence justified in regarding the atomic weights as a clear proof that all elements are *composed of hydrogen*, just as *Prout* did more than a hundred years ago (in 1815), in the sense that, as we know now, the two constituents of the lightest of all atoms, the hydrogen nucleus and its satellite, go to build up the atoms of *all* the elements.[4]

[4] The most recent researches on β-ray spectra seem to show that the structure of the nuclei is governed by *quantum relations*, just as is the arrangement of the electrons round the nucleus; *cf.* Ellis and Skinner, *Proc. Roy. Soc.*, *A*, **105** (1924), pp. 185–198. It is also worthy of note that, if positive and negative electricity be interchanged in an atomic model constructed on a quantum basis, a model is obtained whose linear dimensions are a thousand times as small, so that it may be regarded as a model of a nucleus. The cause of this reduction lies in the difference in mass of protons and electrons. The conception of the nucleus as an "inverted" Rutherford atomic model is due to *Lenz*.

CHAPTER XV

GENERAL THEORY OF SPECTRA AND OF ATOMIC STRUCTURE

§ 117. Röntgen Series.

IN laying the foundations of Röntgen spectroscopy, *Moseley* discovered an important relation which permits an *interpretation of the lines in Röntgen spectra* from the point of view of the *quantum theory* and Bohr's theory of spectra. This relation expresses that the *frequency* of the strongest line in the K-series, the K_a-line, can be represented by the formula

$$(1) \qquad \nu(K_a) = \frac{3}{4} R (z - a)^2 ,$$

where R denotes the Rydberg constant, z the nuclear charge number of the element concerned, and a is a constant with a value differing but little from 1·6 for the majority of the elements.

In terms of Bohr's theory of spectra, eqn. (1) finds an obvious explanation if we consider the K_a-line as being due to a *transition of an electron from a two-quantum to a one-quantum orbit* [inasmuch as 3/4 is equal to $(1/1^2 - 1/2^2)$]. Provided, however, that we regard the atom, in rough approximation, as a hydrogen-analogue, the number a is to be interpreted as a so-called *screening constant.* The nucleus behaves as though a part of its charge, *viz.*, ae, were screened off ; for not only is the electron that undergoes the transition attracted by the atomic nucleus, but it is also *repelled* by the other electrons which, as well as it, surround the nucleus. Since, however, the screening constant of the K_a-line has almost the same value for the most various elements, this obviously means that the transition giving rise to the K_a-line must occur in a region of the atom which is *similarly constituted* for all atoms of arbitrary nuclear charge numbers (provided these are not too small). Apart from elements with very low nuclear charge numbers, the arrangement of the electrons moving in one- and two-quantum orbits must therefore be approximately the same in all elements.

For the strongest line in the L-series, the so-called L_a-line, Moseley likewise found the relation

$$(2) \qquad \nu(L_a) = \frac{5}{36} R(z - b)^2 ,$$

where, again, b is a screening constant. Its value for most elements is about 3·5, but the variations in the values of the screening constant for the L_a-line are considerably greater than in the case of the K_a-line.

Eqn. (2) may evidently be interpreted as meaning that

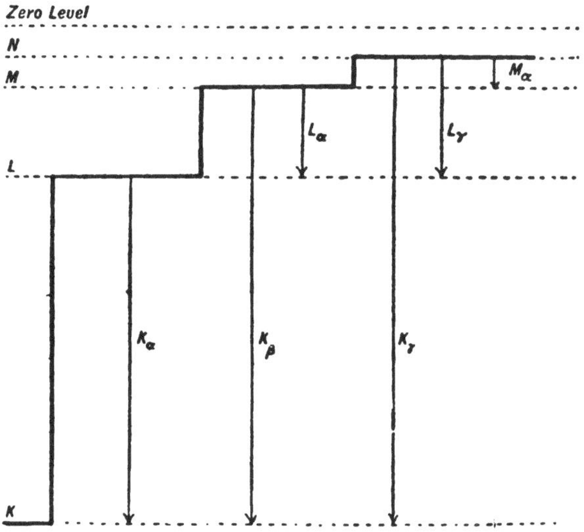

FIG. 95.—Diagrammatic scheme of the most important Röntgen lines.

the L_a-line arises from a transition of an electron *from a three-quantum to a two-quantum orbit.* The second strongest line in the K-series, generally known as the K_β-line, is due to a transition from a three-quantum to a one-quantum state, as follows from its wave-length. In like manner, the second strongest line in the L-series, usually denoted by L_γ,[1] arises from a transition from a four-quantum to a two-quantum state, and the strongest line in the M-series, the so-called M_a-line, from a transition from a four-quantum to a three-quantum state. The transition from the four-quantum to the one-quantum state gives rise to a weak line in the K-series, the so-called K_γ-line, whose wave-length is shorter than that of the K_β-line.

[1] The symbol L_β is used for another purpose.

A schematic representation of the origin of the above-named lines is given in Fig. 95, which is constructed according to the same principle as the earlier Fig. 63. The *one-quantum* level is called the *K-level*, the *two-quantum* the *L-level*, the *three-quantum* the *M-level*, the *four-quantum* the *N-level*, *etc*. The lines in the K-series are due to a transition to a K-level, those of the L-series to a transition to an L-level, and so on.

The diagrammatic representation leads at once to the following *combinations* giving the *frequencies* :—

$$(3) \qquad \begin{cases} K_\beta = K_\alpha + L_\alpha\,, \\ K_\gamma = K_\alpha + L_\gamma = K_\beta + M_\alpha\,, \\ L_\gamma = L_\alpha + M_\alpha\,. \end{cases}$$

These relations are found to be satisfied in great approximation; the small deviations observed are to be explained by the fine structure of Röntgen spectra, which we shall discuss in the next section.

As we pointed out in § 96, every *energy-level* may be associated with a so-called *term* obtained on dividing the energy, taken with a positive sign, by the product *hc*. This term represents a *reciprocal wave-length*, and the wave-length of the line emitted in the transition from one level to another is then given by the relation

$$(4) \qquad \frac{1}{\lambda} = T'' - T'\,,$$

where T'' and T''' denote the terms corresponding to the initial and final levels. On replacing the terms by their reciprocal values, the *wave-lengths characterizing the levels*, eqn. (4) becomes

$$(5) \qquad \frac{1}{\lambda} = \frac{1}{\lambda''} - \frac{1}{\lambda'}\,.$$

Each energy-level may also be defined by an equivalent *potential*, expressed in *volts*, which is calculated from λ' or λ'' by means of Duane and Hunt's equation [§ 107 eqn. (3)].

Soon after the discovery of Röntgen spectra, experiment led to the important rule that the *excitation potential* necessary for the generation of a definite line is always *equal to the characteristic potential of the final level*. Hence, it is not the value of λ but always that of λ'' which determines the excitation potential in accordance with Duane and Hunt's equation. The converse also holds good. If the potential be gradually increased until the value corresponding to λ'' in Duane and Hunt's equation be reached, then at this potential *all the lines* whose *common final level* is defined by the wave-length λ'' will be *simultaneously excited.*

From this fact *Kossel*, in 1914, first drew the important conclusion that the *emission of a Röntgen line* can only occur when an electron in the emitting atom has been *previously* removed from its normal orbit to *beyond the limits of the atom.* Thus, a *place in the atom* must first, as it were, *fall vacant.* If this vacant place corresponds to a quantity of energy $(-h\nu'')$, and if it be filled by an electron which was previously situated at an energy-level $(-h\nu')$, a line of frequency $(\nu'' - \nu')$ will appear. For the generation of this line, however, the energy $h\nu''$ must have been *previously expended.*

A convincing proof of the accuracy of Kossel's conception is afforded by the appearance of *absorption band heads in continuous Röntgen spectra.* If the *Röntgen line spectrum* of an element be *excited by primary continuous Röntgen radiation*, instead of by cathode rays, the *wave-lengths characteristic of the energy-levels in the line spectrum excited* manifest themselves as heads of absorption bands in the primary Röntgen spectrum.

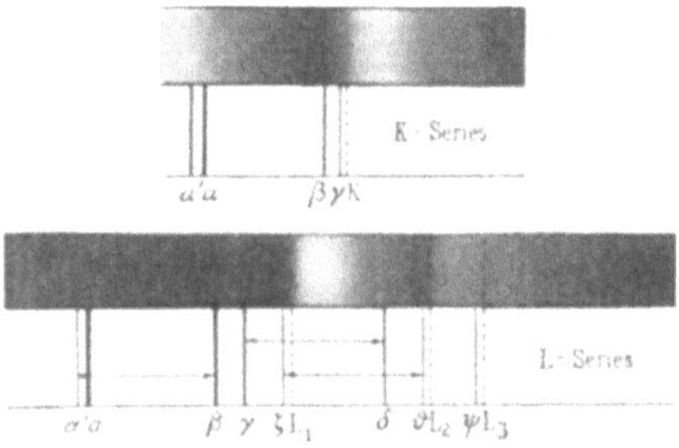

Fig. 96.—Absorption band heads in the continuous Röntgen spectrum.

The explanation of this phenomenon follows immediately from Kossel's conception. For, in accordance with Einstein's law of light-quanta, we must suppose *light-quanta* to be *absorbed from the primary continuous radiation* and to be *transformed into light-quanta of the secondary line radiation.* Hence, according to Kossel's theory, a line in the K-series can be excited only by primary rays of frequency *at least* as great as the characteristic frequency of the K-level. On the other hand, in the excitation of K-lines, no primary rays are absorbed, the wave-lengths of which are larger than the wave-length characterizing the K-level. Thus, we may expect to find the head of an absorption band at this wave-length, and likewise at the wave-lengths characteristic of the L- and M-levels.

These expectations are borne out by actual experiment.

In the upper photograph reproduced in Fig. 96, we see the absorption band head of the K-series in silver; beneath it are inserted the four lines of the corresponding K-series (their wave-lengths being larger). The primary Röntgen radiation was generated in a 40 kilovolt tube, so that, by Duane and Hunt's equation, the continuous Röntgen spectrum extended down to about 300 X.U. The sharp head to the absorption band, seen in the photograph, corresponds to a wave-length of 485 X.U. Since the wave-lengths of the absorption band heads are characteristic of the energy-levels themselves, thus corresponding to the difference between a level and the zero-line, the *heads of the absorption bands* also represent *series limits*, as is likewise seen from Fig. 96.

The problem of experimental Röntgen spectroscopy consists chiefly in the *exact determinations of wave-lengths,* both of absorption band heads and of individual emission lines. Much the most accurate method of measurement is the *direct* one based on *Bragg's equation of crystal reflection,* but this method can only be applied within a limited region of wave-lengths. The majority of wave-lengths are measured by means of a *rock-salt* or *calcite crystal,* which, as already mentioned in an earlier section (§ 106), can be used for wave-lengths up to about 5,000 X.U. Wave-lengths more than twice as large can be reached with the help of *selenite* or *mica,* which have a larger lattice constant. Exact measurements by means of crystals have been perfected, principally by *Siegbahn*, to such a degree that, at the present time, Röntgen wave-lengths can be determined with an accuracy of about a millionth part of the measured value. (That is to say, wave-lengths can be compared with one another, or referred to an idealized lattice constant with that accuracy.) [2]

In comparison with these direct precision methods, only a secondary value can be attached to the *indirect methods.* These are, however, of the greatest importance in the *boundary region* between the largest wave-lengths measurable with crystals and the shortest wave-lengths in the ultra-violet. The most important of these indirect methods is based on *Duane and Hunt's equation*, the wave-length of the *series limits* being thereby derived from the measured excitation potentials (defined by the values at which absorption suddenly commences).

In exceptional cases, the characteristic wave-lengths can also be measured with an *optical diffraction grating* when the *nuclear charge numbers* are *small*, because, with decreasing

[2] *Vide* § 106.

nuclear charge number, all series gradually shift over to the ultra-violet region.

As regards the *K-series*, the *series limit* is known from measurements of the *absorption band heads* for the elements between 12 (Mg) and 92 (U); for Mg it amounts to 9511·2 X.U., and for U to 107·5 X.U. The wave-lengths of the individual *lines* in the *K*-series are known for the elements from 11 (Na) to 60 (Nd) and, in addition, for 74 (W), 77 (Ir), and 78 (Pt). The extraordinary accuracy with which the

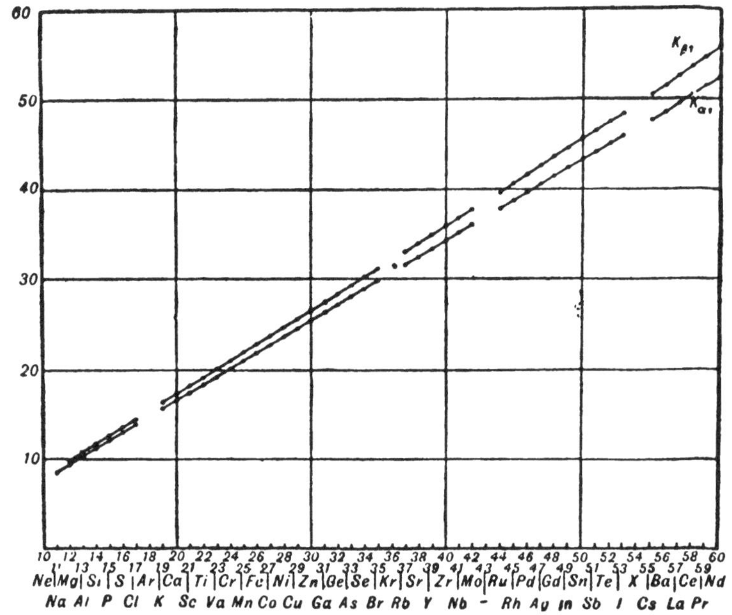

FIG. 97.—Linear relation between atomic number and $\sqrt{\nu}$ for the K_α-line.

linear relation between the square-root of the frequency of the K_α-line and the atomic number is satisfied is evident from Fig. 97.

The K_α-line for Na lies as high as 12,000 X.U.; for elements of still lower ordinal number, it lies in the boundary region between Röntgen rays and ultra-violet rays. By means of Duane and Hunt's equation, however, the *K*-limit has been determined in this boundary region also for 8 (O), 6 (C), 5 (B), 4 (Be), and 3 (Li); *e.g.*, for C it lies at 47, and for Li at 290 Å.U.

The limit (or, as we should rather say, having regard to

the fine structure to be discussed subsequently, the limits) of the *L-series* is known for the elements between 55 (Cs) and 92 (U). The emission lines of the *L*-series are known from 29 (Cu) to 92 (U). For Cu the wave-lengths of the *L*-lines are as large as 13,000 X.U.; for U they are about 1,000 X.U.

In 1921, *Millikan*, by means of his *high-vacuum spectrograph*, was able to observe and measure the *L*-series in the *extreme ultra-violet* for 13 (Al), 12 (Mg), and 11 (Na). One wave-length for Al amounts to 136·6 Å.U.; this also represents the shortest wave-length yet measured in the ultra-violet. Indirect measurements have enabled the *L-limit* to be determined also for 3 (Li), 4 (Be), 5 (B), 6 (C), 8 (O), and 9 (F).[3]

At the present time, the absorption band head in the *M*-series (discovered by *Siegbahn*) has been directly measured only in the cases of three elements, *viz.*, 83 (Bi), 90 (Th), and 92 (U). Wave-length determinations principally carried out by *Stenström* have been made for the elements between 66 (Dy) and 92 (U). It has also been possible to measure the *M*-limit indirectly for several elements, *e.g.*, for 42 (Mo), in which case it lies at about 34 Å.U.

The *N-series* was discovered by *Dolejsek* in 1922;[4] up till now measurements have only been made for 83 (Bi), 90 (Th), and 92 (U). The *N*-series is so soft that its lines, even for these highest elements, lie right on the boundary of the wave-lengths which can be determined by means of crystals.

§ 118. The Fine Structure of Röntgen Spectra.

On closer investigation of Röntgen spectra, it was soon found that the number of lines observed was too large for the simple assumption of a single *L*-level, a single *M*-level, *etc.*, to suffice for their explanation. It appeared necessary, in fact, to distinguish between *various modifications of the levels* (except for the *K*-level). This necessity was also rendered evident by researches on *absorption band heads*, which showed that the *L*-limit is *threefold* and the *M*-limit *fivefold* in those elements where they are observed. Only the *K*-limit appears to be single in every case (*vide* the earlier Fig. 96).

As soon as the numerous lines in the different series had been determined, it became one of the most important

[3] *Cf.* the summary by MacLennan in *Nature*, **113** (1924), p. 217.
[4] Dolejsek, *Zeitschr. f. Phys.*, **10** (1922), p. 129.

problems in Röntgen spectroscopy to refer the manifold of these lines, through their combinational relations, to a less complicated multiplicity of energy-levels, and thus to create a *scheme for the origin of Röntgen lines.* We owe the final solution of this difficult problem, obtained in 1922, principally to four investigators: *Smekal, Wentzel, Coster,* and *Bohr.*

According to Bohr, this scheme is most simply stated on the following lines. *Each level* is characterized by *three quantum numbers,* represented symbolically by n (k_1, k_2). The number n, the so-called *principal quantum number,* is naturally assumed to be 1 for the K-level, 2 for the L-level, 3 for the M-level, 4 for the N-level, 5 for the O-level, and so on. For a given principal number, the so-called *first subordinate quantum number* (k_1) can assume any integral value from unity to an upper limit, depending on the nuclear charge number of the atom, but in no case exceeding the principal number. Finally, provided k_1 is greater than unity, for a given pair of values of n and k_1 *two levels* are assumed to exist—one for which k_2 is put equal to k_1, known as the *normal level,* and a second in which k_2 has a value less by one than k_1, known as the *anomalous level.* When $k_1 = 1$, only the normal level, of course, can exist.

The largest number of possible levels associated with a principal number n is thus $(2n - 1)$, though this number is not necessarily realized in the atoms of all elements. In accordance with § 100, of two levels with the same principal quantum number we shall define the one with the larger first subordinate number as being the higher, *i.e.*, as lying nearer the zero-line; and, for equal values of n and k_1, the normal level shall always be higher than the anomalous one. For example, the five possible M-levels are thus represented in ascending order by the symbols

$$3\,(1, 1)\,;\quad 3\,(2, 1)\,;\quad 3\,(2, 2)\,;\quad 3\,(3, 2)\,;\quad 3\,(3, 3).$$

As stated by *Bohr,* the *selection principle of Röntgen lines* may be expressed in the following form: The only permissible transitions between the various energy-levels are those in which the *first subordinate number changes by unity,* whilst the *second subordinate number either remains unaltered or likewise changes by unity,* and these transitions lead to the generation of Röntgen lines. Hence the two conditions

$$\text{(1)}\qquad \begin{cases} \Delta k_1 = \pm 1 \\ \Delta k_2 = \begin{cases} 0 \\ \pm 1 \end{cases} \end{cases}$$

must be satisfied. (It is taken for granted that, in such a transition, the principal quantum number n changes.)

In Fig. 98 is reproduced the *scheme of Röntgen lines* as determined by the above-stated selection principle for *elements of the highest atomic numbers* (from emanation onwards). With these highest elements there are 1 K-level, 3 L-levels, 5 M-levels, 7 N-levels, 5 O-levels, and 3 P-levels.[1]

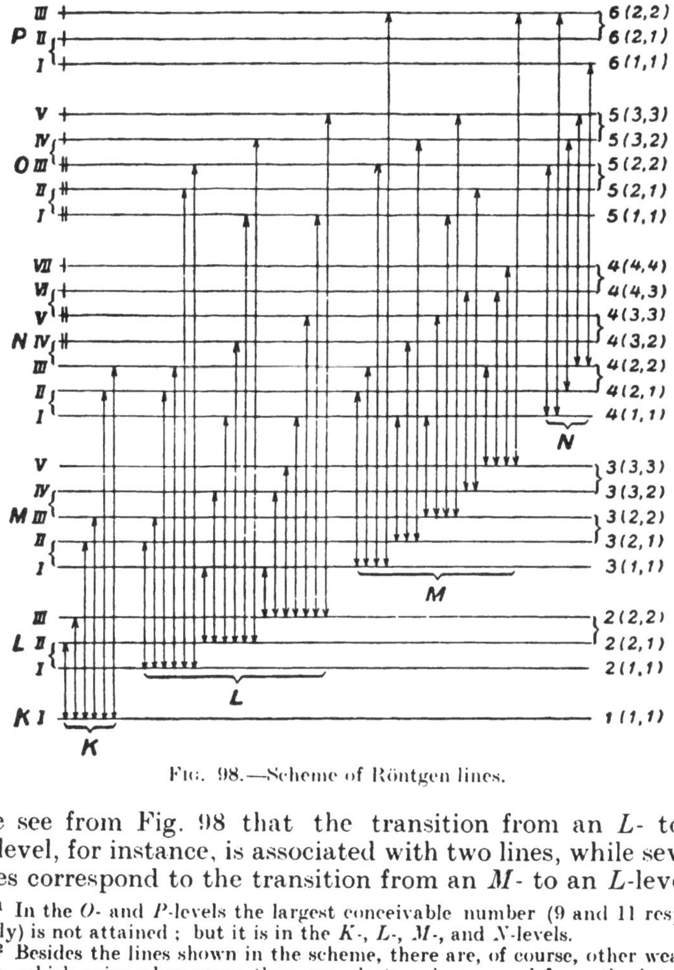

FIG. 98.—Scheme of Röntgen lines.

We see from Fig. 98 that the transition from an L- to a K-level, for instance, is associated with two lines, while seven lines correspond to the transition from an M- to an L-level.[2]

[1] In the O- and P-levels the largest conceivable number (9 and 11 respectively) is not attained; but it is in the K-, L-, M-, and N-levels.

[2] Besides the lines shown in the scheme, there are, of course, other weaker lines, which arise when more than one electron is removed from the interior of the atom. For instance, with each of the elements Mg, Al, Si, six different lines, due to a transition from a two-quantum to a one-quantum state, have been observed.

It is also evident from Fig. 98 that the individual groups of lines are resolved into *sub-groups* distinguished by the same final level, and thus by the same absorption band heads. In the figure single vertical dashes denote those levels which gradually lapse as we descend the natural sequence of the elements from emanation to xenon. In the case of the elements near krypton, other levels are also lacking, these being denoted in the figure by double vertical dashes. An example of a photograph of an L-series is to be found in Fig. 99, which gives the L-spectrum of rhodium, containing 11 lines.[3]

The individual *energy-levels* or, what amounts to the same thing, the individual *terms* are known at the present time with fair accuracy for most elements. They were determined either directly by measurements of absorption band heads or, with the help of the above-given scheme of lines, by measurements of the emission lines. The relations between the terms and the atomic numbers are shown in Fig. 100, where the ordinates represent the *square roots* of the terms referred to the Rydberg term as unit.[4] Only the portions of the curves drawn in full, however, correspond to *actually measured* values ; the numerical symbols and lines added underneath the figure will be considered later (§ 122).

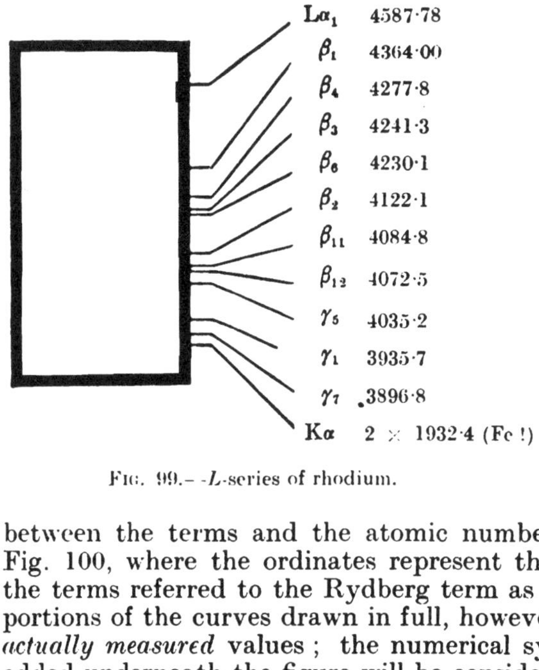

FIG. 99.—L-series of rhodium.

[3] The lines have been marked with the corresponding wave-lengths in X.U. As well as the rhodium lines, the K_α-line of iron is also to be seen in Fig. 99, iron being present as a small impurity in the rhodium. In Röntgen spectroscopy it is possible to detect impurities amounting to only from a hundredth to a tenth of 1 *per cent*.

[4] The terms, which are reciprocal to the wave-lengths, are therefore obtained on multiplying the squares of the ordinates by 109,737. The P-terms, which are still rather uncertain and have, of course, only been calculated for the highest of the elements, have been omitted in Fig. 100.

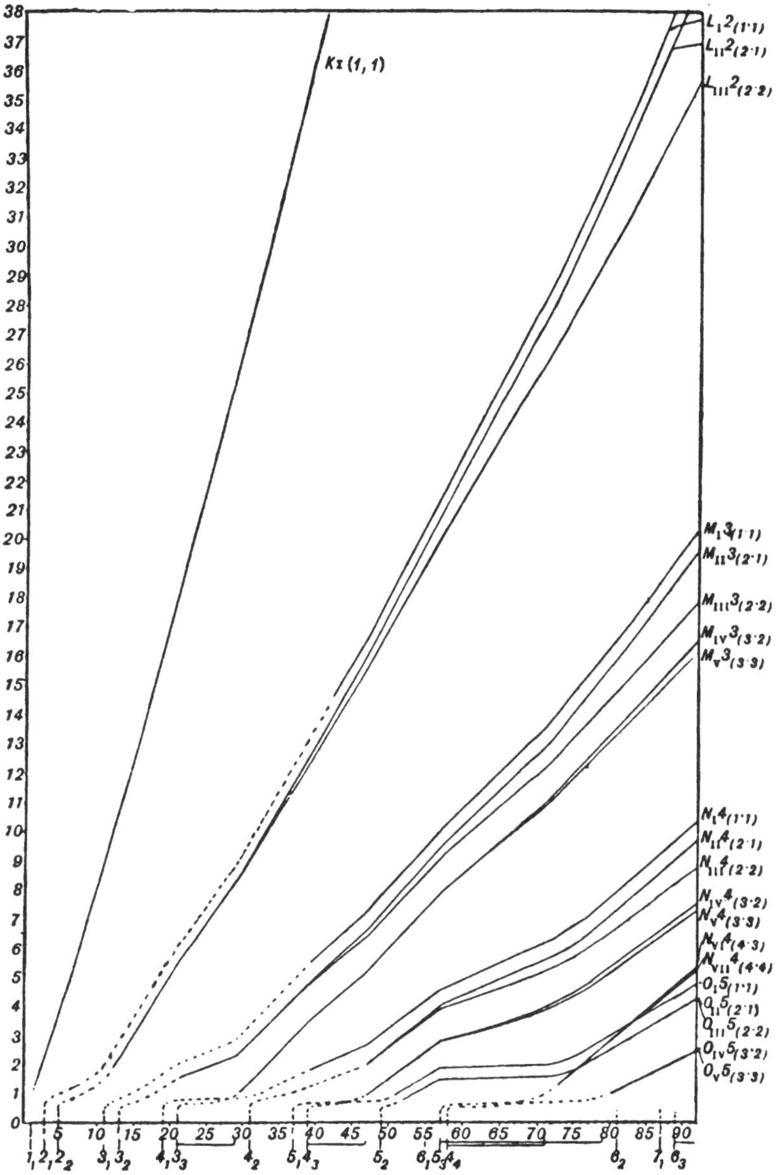

FIG. 100.—Röntgen terms of the elements.

The course of the various curves is more or less linear, in accordance with Moseley's Law; the larger the principal quantum number, the smaller is the angle between the curve and the axis of abscissæ.

If we neglect the second subordinate number, the manifold of terms appertaining to a definite principal number is explained by the *elliptical shape of the electronic orbits*, inasmuch as the definition of these requires a subordinate as well as a principal number. On the other hand, the difference between the normal and the anomalous terms is, on Bohr's view, to be referred to a possibility of more than one arrangement of the remaining electrons in atoms which have lost one electron.

Within every group characterized by one and the same value of the principal number, each pair of adjacent levels forms a so-called *level doublet.* There are, however, two kinds of doublets to be distinguished. The first is formed by two neighbouring levels with the same principal number and the same first subordinate number; in this case we speak of a *regular doublet.* The second kind of doublet is constituted by two adjacent levels for which the values of n and k_2 are the same, while the values of the first subordinate number differ by unity; we then speak of an *irregular doublet.* Within each group regular and irregular doublets alternate with one another. In the earlier Fig. 98, the regular doublets are marked by brackets on the right-hand side, and the irregular doublets by brackets on the left-hand side. The difference between the energy values in a doublet, divided by the elementary quantum of action, is known as its *frequency difference.*

Now, as *Sommerfeld* discovered in 1916, the frequency difference of a *regular* doublet is given by the simple relation

$$\Delta\nu = \Delta\nu_H (z - b)^4, \tag{2}$$

where $\Delta\nu_H$ denotes the so-called frequency difference of the hydrogen doublet [§ 100 eqn. (11)], z the atomic number, and b the relevant screening constant.[5]

Wave-length difference and frequency difference are connected by the simple relation

$$\Delta\lambda = -\frac{c}{\nu^2}\Delta\nu, \tag{3}$$

which is at once obtained by differentiating the equation

$$c = \lambda\nu.$$

[5] *E.g.*, or the doublet of the energy-levels $L\,(1, 1)$ and $L\,(2, 1)$, b is equal to 3·5.

Since, by Moseley's Law, the square root of the frequency is proportional to the atomic number less the screening constant, it follows from eqns. (2) and (3) that the *wave-length difference of a regular doublet* is almost *independent* of the *atomic number*.

A simple rule, discovered by *G. Hertz* in 1920, likewise holds good for *irregular doublets*: the *frequency difference* of irregular doublets proves to be almost *independent of the atomic number*. Hence, in Fig. 100, the curves of two terms constituting an *irregular doublet* have a nearly *parallel* course, whereas the curves of two terms forming a *regular doublet diverge* more and more as the atomic number increases.

In conclusion, we may just mention how the *wave-lengths* of the *γ-rays of a radioelement* can be derived from its *magnetic β-ray spectrum*, provided the *energy-levels* of an element *isotopic* with it are known from Röntgen-spectroscopic measurements. According to the fundamental relation obtained in an earlier section (§ 91), the kinetic energy of a secondary β-particle is given by the equation

$$E_{kin} = h\nu + W, \tag{4}$$

where ν is the frequency of the primary γ-rays and W the work necessary to remove the β-particle from its usual orbit "to infinity."

We shall now assume that the magnetic β-ray spectrum of a radioelement contains two lines corresponding to velocities v' and v''. The difference in these values is obviously explained by the fact that the β-particle is *liberated from a different level* in either case. If, however, the liberation is effected in each instance by γ-rays of the *same frequency*, we find, on applying eqn. (4) to each case and subtracting,

$$E_{kin}' - E_{kin}'' = W' - W''; \tag{5}$$

in calculating the kinetic energy, the dependency of the mass on the velocity must be taken into account [according to § 89 eqn. (6)].

Now if both secondary β-rays really were generated by primary γ-rays of the same frequency, we see that the energy difference $(W' - W'')$ divided by h gives the frequency of a known *Röntgen line* of the isotopic element. This being the case, however, we can immediately find out from what levels the secondary β-particle was liberated in the two instances, provided we know the energy-levels of the isotopic element. We have only to replace W in eqn. (4) by the energy of the level in question (*i.e.*, by the spectral term multiplied by $h\,c$), when, since the kinetic energy of the secondary β-particle is known, the frequency of the primary γ-ray can at once be calculated from eqn. (4).

By means of this method, devised by *Ellis* and by *Miss Lise Meitner*,[6] it has been found, for example, that two lines in the β-ray spectrum of *radiothorium*, corresponding to velocities of 51 and 47% of the velocity of light, are associated with values of the work of liberation, which are identical with the energy values defining the M_I- and L_I-levels in the isotope, thorium; it follows from this that the wave-length of the primary γ-rays is 14·6 X.U. To mention a second example, three lines are found in the magnetic spectrum of thorium B which are associated with velocities of 63·0, 71·4, and 72·1% of the velocity of light. These lines correspond to the values of the work of liberation from the K-, the L_{II}-, and the L_I- levels of the isotope, lead; and it follows that the wave-length of the primary γ-radiation has a value of 52 X.U.

§ 119. Excitation and Ionization of Atoms.

From the spectrum of the simplest of all atoms, hydrogen, it is clear that an electron revolving around the atomic nucleus can leave its usual orbit and revolve *temporarily in an orbit of higher quantum number*. Quite generally, we say that an atom is *excited* when one of its electrons describes an orbit *other* than its normal one, the electron then having a *higher energy* than usual. The *excitation* of an atom thus always demands an *introduction of energy*, while, on the *return of the atom to its normal state*, *energy is liberated*.

As subsequent spectroscopic considerations will show, the multiplicity of higher levels, to which the electron can temporarily be raised, is very great. Each of these levels, however, which are not present in the normal atom and can only be realized when the atom is excited, is associated with a quite definite energy value, just as is the normal state. Hence, from amongst the manifold of excited states there must be one state that is distinguished by its having the *smallest excess of energy over the normal energy value*. This state is termed, in a narrower sense of the word, simply the excited state, and its surplus energy is called the *excitation energy* of the atom. It therefore represents the smallest quantity of energy which a normal atom of the element

[6] C. D. Ellis, *Proc. Roy. Soc.*, *A* 99 (1921), p. 261; L. Meitner, *Zeitschr. f. Phys.*, 9 (1922), pp. 131 and 145.

concerned can ever take up in connection with a change of its internal energy.

The excitation of atoms is, of course, most readily effected with the *most loosely bound* electrons. These are obviously the *peripheral electrons*, which an atom loses when it enters into chemical combination as a positive ion, and which are therefore called the *valency electrons*. Since the elements in the first group of the periodic system are positively univalent, those in the second vertical column positively divalent, and so on, we must assume that the atoms of the elements in the first group each possess a single valency electron, those in the second group two valency electrons each, and so on.

In general, changes in the orbit of the peripheral valency electron are probably responsible for those variations of the internal atomic energy which, in accordance with Bohr's *frequency condition*, manifest themselves in the *optical spectra*. For, in the optical region, a light-quantum is roughly a thousand times smaller than in the region of Röntgen rays, so that optical lines can only be due to those electrons for which, because they are loosely bound, a change of orbit can be effected by a relatively small expenditure of energy. Moreover, it is easy to understand why the excitation of an atom should, in general, find expression only in the peripheral electrons, because it is only these electrons that, since they describe the highest-quantum orbits in the normal atom, will, on a transition to an orbit of still higher quantum number, find it unoccupied. An inner electron, on the other hand, must, on a transition to a higher-quantum orbit, generally find it already completely occupied (as follows from concepts of the grouping of electrons, to be discussed in greater detail later).

We must assume that an excited state of the atom is not only abnormal, but also extremely *unstable*. After a very short sojourn, which we shall later find to be of the order of 10^{-8} sec., the *return to the normal state* takes place. Furthermore, provided the surplus energy is not changed into kinetic energy, this return is accompanied by the *emission of a light-quantum* whose frequency, according to Bohr's condition, is equal to the quotient of the energy released and the elementary quantum of action. If the atom was not in the lowest of the excited states, but in one more highly excited, the return to the normal state may, of course, follow *by steps*, so that the emission spectrum then produced will contain lines for which both the initial and the final levels are possible only in an excited state.

Conversely, the excitation of an atom, or its transformation from an already excited to a more highly excited state, can also be effected by the *absorption of a light-quantum*, provided the product of the frequency of the light-quantum and the elementary quantum of action be equal to the difference in energy between the actually existing and a higher-quantum state.

A far more frequent way of exciting atoms than by absorption, however, is by supplying the requisite energy from the *heat store* of the body concerned. According to the fundamental concept of the kinetic theory of heat, incessant *collisions* occur in a gas between the rapidly moving molecules. By virtue of these collisions, *energy of translation* can of course be *changed into intra-atomic energy*, but when the atom struck was previously in the normal state, this can only happen if the relative kinetic energy is at least as large as the excitation energy. (In the monatomic vapours of the metals in the first two groups, which have been chiefly investigated, as well as in the rare gases, the concepts of atom and molecule are synonymous.) Now, when atoms thus excited by collisions return to the normal state with the emission of light-quanta, the well-known phenomenon of *thermo-luminescence* is observed. Moreover, the atoms of a gas can be excited, not only by sufficient heating, but also by collisions with rapidly moving ions *in electrical fields*, as is shown by the *luminescence of rarefied gases in Geissler tubes*.

The *excitation energy* in the narrower sense of the word, *i.e.*, the smallest quantity of energy which a previously normal atom can take up, may at once be derived from optical spectra, and also by other means, as we shall show in detail in the following sections. For the alkali metals its value lies between 2×10^{-12} and 3×10^{-12} ergs, being smallest for cæsium. For the elements of the second group in the periodic system, the excitation energy is two to three times as large as for cæsium ; for lead (in the fourth group) it is a little smaller than for cæsium, and for thallium (in the third group) it is about 20% smaller. The spectral lines corresponding to excitation in the narrower sense of the word lie in the optical region for most of the alkali metals and for Mg, Ca, Sr, and Ba, whereas for Zn, Cd, and Hg they fall in the ultra-violet.

The *mean energy* possessed by a monatomic *gas molecule* at an absolute temperature T, in virtue of its *translation*, is, as follows from the kinetic theory of gases (and as we shall show in detail in a subsequent section, *viz.*, § 148), equal to $3/2\ k\,T$, where k, the so-called Boltzmann constant, is equal to $1{\cdot}372 \times 10^{-16}$ ergs/deg.[1] At room temperature, for which T is about 300, the mean value of the translational energy of a monatomic molecule is hence about 6×10^{-14} ergs. Furthermore, according to a fundamental relation of molecular statistics, the *fraction* (α) of molecules whose translational energy is greater than x times the mean value is, quite generally, equal to e^{-x}.

Thus, for cæsium at room temperature, the fraction (α) of the atoms whose translational energy exceeds the excitational energy is about 10^{-13} (for x is approximately equal to 30). At a temperature of 1,200° C. (= 1,500° abs.) x becomes roughly equal to 6, and hence α to about 1/400. Whereas, therefore, at room temperature the fraction of atoms possessing a kinetic energy capable of exciting another atom by collision is entirely insignificant,[2] at 1,200° C. this fraction is large enough to produce a noticeable luminescence in the vapour.

Theory teaches us to expect that, when any vapour commences to glow with rising temperature, the first spectral line to be observed will be that corresponding to the transition from the lowest excited state to the normal state. The longer the wave-length of this line, the smaller will be the excitation energy, and the lower the temperature at which emission of the line will begin. As this temperature is exceeded, we must further expect that the transfer of translational energy will be sufficient to produce higher and higher excitations, so that new spectral lines will be continually appearing. In point of fact, all these theoretical conclusions are confirmed by experiment.

The energy necessary to remove a valency electron completely from its atom can also be derived from the optical spectrum (and by other methods which we shall discuss later). For, when the lines in the optical spectrum can be referred to terms, we have only to multiply by $h\,c$ the term corresponding to the lowest level, *i.e.*, the term with the largest numerical value (known as the *fundamental term*), in order to obtain this energy, the so-called *ionization energy*, in ergs.

For the alkali metals, and also for elements in the second group of the periodic system, the ionization energy proves

[1] In the case of molecules of diatomic gases, the factor 3/2 is replaced by the factor 5/2. Boltzmann's constant is the quotient of the gas constant and the Loschmidt number (*cf.* § 148).

[2] Actually, it is not the mean translational energy, but the mean value of the relative energy on collision, that comes in question. It follows from the kinetic theory of gases, however, that the two quantities only differ by an inconsiderable amount.

to be two or three times as large as the excitation energy. Hence, if we now refer the fraction α to the process of ionization instead of to that of excitation, as previously, we should expect to find the same value of α for an absolute temperature two or three times as high. Actually, however, ionization is to be observed for much smaller values of α than is excitation. For, on introducing the salts of a metal into a flame, the smallest ionization can at once be observed by a noticeable *electrical conductivity of the flame*, for the detection of which we possess in the galvanometer an extraordinarily sensitive instrument. As theory leads us to expect, the flame conductivity at a given temperature is greater the smaller the optical fundamental term of the metal concerned.

Ionized atoms can, of course, be excited in an entirely similar way to neutral atoms, by the removal of one of the remaining valency electrons from its normal orbit to a higher-quantum orbit. Thus, the elements may give, besides the spectra of their neutral atoms, *spectra of their ionized atoms*, the simplest example of which is afforded by the spectrum of ionized *helium*, described in an earlier section (§ 97). The spectra of ionized atoms obviously presuppose higher temperatures or stronger electrical fields than the spectra of neutral atoms. Whereas the spectra of neutral atoms are generated, apart from in Geissler tubes, principally in the electric arc, the spectra of ionized atoms are best produced by means of electric spark discharges. The two varieties of spectra are therefore commonly distinguished, from their modes of production, as *arc spectra* and *spark spectra*. According as we are dealing with single, double, or triple ionization, we speak broadly of a simple spark spectrum or of a spark spectrum of the second or third order respectively.

It is evident that the spark spectrum of an element of atomic number $(z + 1)$ must exhibit a strong resemblance to the arc spectrum of the element of atomic number z, except that, when the spectrum is approximately that of a "hydrogen analogue," the *terms* will appear *multiplied by four*, as was indeed the case with the simple example of ionized helium. Similarly, in spark spectra of the second and third orders, the terms appear multiplied by nine and by sixteen respectively (*cf.* the following section, § 120). The so-called *spectroscopic displacement law*, discovered by *Sommerfeld* and *Kossel* in 1919, holds quite generally. According to this law, the spark spectra of the elements in any group of the periodic system resemble in character the arc spectra of the elements in the previous group, especially as regards their fine structure (which we shall discuss later).

The easiest spark spectra to produce are those of the second group elements, a temperature of 2,000° C. being sufficient for this purpose, for their simple ions still contain one loosely bound electron. Arc and spark spectra of these elements are thus often to be observed together. On the other hand, it is extremely difficult to generate the spark spectra of the alkali metals, since their atoms in the neutral state only contain a single valency electron.

These conclusions from Bohr's theory of the atom provide an explanation of several apparently paradoxical results in *solar spectroscopy*, as was first pointed out in 1920 by the Indian physicist, *Saha*.[3] It had previously been quite inexplicable why sodium should appear very strongly in the solar spectrum, whereas potassium appears only weakly, and rubidium and cæsium are entirely absent. According to Saha's striking explanation, the alkali metals are present in the sun, just as in the earth. *To some extent*, however, they *remain unnoticeable*, because, in consequence of the high temperature, they are almost invariably *ionized*, and their spark spectra, the lines of which lie in the ultra-violet, escape spectroscopic observation. That the arc lines of Na occur strongly in the solar spectrum, those of K only weakly, and those of Rb and Cs not at all, follows from the fact that Rb and Cs have the smallest ionization energies, K having a somewhat higher, and Na a considerably larger value.[4] In contrast to the alkali metals, the alkaline earths are represented in the visible portion of the solar spectrum by the lines of their ionized atoms as well; for instance, the strong Fraunhofer lines *H* and *K* both belong to the spark spectrum of calcium.[5]

Moreover, Saha found in Bohr's theory a simple explanation for the remarkable fact that the heavier element calcium is to be observed in the spectrum of the sun's atmosphere at much greater heights than the much lighter element hydrogen. For the calcium lines lying in the visible region of wavelengths require a smaller excitation energy than the hydrogen lines lying in the same region; the hydrogen lines which are

[3] *Phil. Mag.*, **40** (1920), pp. 472 and 809, and **41** (1921), p. 267.

[4] *Cf.* the later Table XVII, in § 124.—Besides the differences in ionization energy, the abundance or rarity of the elements, it is true, also play a part.

[5] The earlier view that calcium is the most abundant element in the sun's atmosphere has, of course, been proved erroneous by Saha's researches.

readily excited belong to the ultra-violet.[6] In the higher regions of the sun's atmosphere, where the temperature is less than in the lower regions, the energy transferred by molecular collisions therefore suffices to excite the visible calcium lines, but not the visible hydrogen lines.

§ 120. Optical Spectra.

The optical spectra of the elements can be produced both by the emission and by the absorption of light-quanta. As was pointed out in the previous section, *absorption spectra* arise from the transition of an electron into a higher-quantum level, either from the normal orbit or from an orbit due to previous excitation. *Emission spectra* are produced by the return of the atom from an unstable to the normal state, whether this return takes place in one or in several stages.

Researches on optical spectra commenced with the observation by *Fraunhofer*, in 1814, of the absorption lines in the solar spectrum. In 1860 *Kirchhoff* and *Bunsen* founded *spectrum analysis* with their discovery of emission spectra. They proved that the bright lines in these spectra are characteristic of the elements, and the constant wave-lengths entirely independent of the temperature and chemical combination. Furthermore, they were able to demonstrate the coincidence of the Fraunhofer lines with the emission lines of certain elements, and to give a theoretical explanation of the former. The possibility of the existence of simple numerical relations between the wave-lengths of the lines in a spectrum was first pointed out by *Balmer* in 1885, in the particularly simple instance of the hydrogen spectrum (as has been discussed at length in an earlier section). About 1890, *Kayser* and *Runge* discovered simple series relations also in the spectra of the *alkali metals*; and, at about the same time, *Rydberg* succeeded in expressing these relations quantitatively by formulæ which exhibited a strong resemblance to Balmer's formula, and in which a fundamental part was played by the so-called Rydberg constant, already known from the hydrogen spectrum. Series relations similar to those for the alkali metals, for which they are most clearly marked, were subsequently established for many other elements as well.

The common mark of distinction of all optical *series* is

[6] The hydrogen line corresponding to the excitation energy, in the narrower sense of the word, is the first line in the Lyman series (1,216 Å.U.).

(as has already been shown by the simple example of the optical hydrogen series) the *decreasing intensity* with *decreasing wave-length*, together with the *convergence of the lines*, also with decreasing definition, towards a *limit* corresponding to the final level in the case of emission.

In those elements for which an arrangement of the optical spectral lines into series was possible, the most important series were soon recognized to be three in number, and these were distinguished as the *principal series*, the *first* or *diffuse subordinate series*, and the *second* or *sharp subordinate series*. The two subordinate series, which are clearly distinguished by the different sharpness of their lines, are found to have a *common limit* which, for the elements in the first two groups of the periodic system, always has a greater wave-length than the limit of the principal series. Subsequently, a fourth series of comparatively large wave-lengths was added to the three above-named series, and this is known as the *fundamental series* or, after one of its discoverers, as the *Bergmann series*.[1]

Later considerations have shown that the frequencies of the most important lines in the series named can be represented as differences between the members of *four series of terms*; these are distinguished by the initial letters of the words "sharp," "principal," "diffuse," and "fundamental" as *s*-, *p*-, *d*-, and *f*-terms.[2] The above four series arise on emission from the transitions represented in the following scheme:—

(1)	Principal series	$p \rightarrow s$,
	1st (diffuse) subordinate series .	$d \rightarrow p$,
	2nd (sharp) subordinate series .	$s \rightarrow p$,
	Bergmann series	$f \rightarrow d$.

Investigations on the separability of the terms in the Zeeman and the Stark effect, as well as considerations bearing on the selection principle, have led to a knowledge of the values possessed by the *subordinate quantum number k* for the various terms, this quantum number determining the shape, and thus the eccentricity, of the orbital ellipse described by the valency electron. The values are:

[1] The real discoverer of the "Bergmann series" was Saunders. The term "fundamental series" is, of course, very inappropriate.

[2] The *f*-terms are also often called *b*-terms (from the initial letter of the word Bergmann).

$$(2) \qquad \begin{cases} k = 1 \text{ for an } s\text{-term}, \\ k = 2 \text{ for a } p\text{-term}, \\ k = 3 \text{ for a } d\text{-term}, \\ k = 4 \text{ for an } f\text{-term}. \end{cases}$$

The terms for which k is to be put $= 5, 6, 7$ are known as the f'-, f''-, f'''- terms, and so on.

Observation shows that, the *lower* the *subordinate number of the initial state* on emission, the *sharper* is the spectral line produced. The lines in the second subordinate series are therefore the sharpest, those in the principal series being less sharp ; in the first (or diffuse) subordinate series the lines are, as the name indicates, ill defined, and in the Bergmann series they are least sharp of all.

Scheme (1) is a confirmation of the validity of the *selection principle*, discussed in an earlier section (§ 99), according to which only those spectroscopic combinations of terms are permissible where the *subordinate number changes by unity.* It is true that, in strong electrical fields, this principle does not hold, and lines then occur which appear to contradict it. For example, lines are observed in the generation of which the subordinate number remains unchanged ; frequently lines are seen which arise from a combination of two p-terms, or of two s-terms. In strong electrical fields combinations between d- and s-terms, and between f- and p-terms, may also occur, although the subordinate number then changes by two.[3]

It was found by *Rydberg* that the individual terms in the various series may be represented, in fair approximation, by the formula[4]

$$(3) \qquad \nu = \frac{R}{(m + a)^2},$$

where R denotes Rydberg's constant, and a is a constant which differs in value for the different succession of terms of an element. Furthermore, m denotes a variable whole number, its value commencing with 1 for the s-terms, 2 for the p-terms, 3 for the d-terms, and 4 for the f-terms.[5] The

[3] Series which are not in accord with the selection principle have been investigated chiefly by *Lenard* and *Stark*.

[4] The terms themselves are obtained on dividing eqn. (3) by c.

[5] Rydberg's formula was later improved by *Ritz*, the denominator being altered to

$$[m + a + \alpha f(1/m)],$$

where the function $f(1/m)$ is, in particular, either $1/m^2$ or $1/m$; α is a constant characteristic of all the terms of an element.

variable or *sequence number m*, however, must not be confused with the principal quantum number of the terms. It is only the most recent investigations of Bohr on the periodic system that have led to a knowledge of the principal quantum numbers of the terms, which we shall discuss in the following section.

The series relations appear most plainly with the elements of the first three groups, especially in that of the alkali metals. The *fundamental term* (*i.e.*, the numerically highest term, which corresponds to the lowest energy-level and measures the strength of the binding of the electron) is an *s*-term for the elements of the first and second groups. On the other hand, it has been shown that in the elements of the third group for which it has as yet been possible to arrange the lines in series, *viz.*, aluminium, gallium, indium, and thallium, the fundamental term is a *p*-term. This is confirmed by the fact that, in the absorption spectra given by the cold vapours of these metals (the atoms thus not being previously excited), only the two subordinate series occur, and the initial terms of the absorption lines in these are *p*-terms.

It has also been possible to derive series relations for some elements in the fourth, sixth, and seventh groups of the periodic system. Of rare gas spectra we are acquainted with the helium spectrum, which will be discussed at length in a later section (§ 125), and also with the *neon spectrum*, which is extraordinarily rich in lines. The latter spectrum has been analyzed by *Paschen*, who has succeeded in representing approximately 900 lines in it as combinations between four *s*-terms, ten *p*-terms, and twelve *d*-terms.[6]

Spark spectra of the first order have been investigated, apart from in helium, principally in the case of the *alkaline earths* (magnesium, calcium, strontium, and barium). The spark spectra of these metals do, in fact, exhibit a great similarity to the arc spectra of the corresponding alkali metals, the terms being quadrupled. The spark spectra of the alkali metals are just as rich in lines and as complicated as the arc spectra of the higher rare gases. With *aluminium Paschen* has discovered a *spark spectrum of the second order*,[7] and with *silicon Fowler* has obtained spark spectra of the second and of the *third orders*, as well as of the first order.[8] As is to be expected theoretically, the spectra of doubly ionized Al and of trebly ionized Si greatly resemble the arc

Paschen, *Ann. der Phys.*, 60 (1919), p. 405 ; and 63 (1920 p. 201.
[7] *Ann. der Phys.*, 71 (1923), p. 142.
[8] *Proc. Roy. Soc.*, *A* 103 (1923), p. 413.

spectrum of sodium, the terms being multiplied by nine and by sixteen respectively.

Both in arc and in spark spectra, the majority of the terms appear as *doublets* or *triplets*. In the arc spectra of the first and third groups, and in the spark spectra of the second group, all terms except the invariably single *s*-terms are *doublets*. In the arc spectra of the second group and the spark spectra of the third group, there are *two systems of terms:* one system of *simple terms* and, in addition, a system in which all the terms except the single *s*-terms are *triplets ;* the terms of the two systems occur in combination with one another. The rule appears to hold, quite generally, that with an *odd number of electrons* the terms are *doublets* (in the higher groups of the periodic system also quadruplets, sextets, or octets) ; whereas, with an *even number of electrons*, the terms are *singlets* or *triplets* (or else quintets, septets, or nonets). It is, however, just in the neutral atoms of an odd group, and in the simple ions of an even group, that the number of electrons surrounding the atomic nucleus is odd, and *vice versâ*.

In doublets and triplets the individual terms are denoted, as usual, by the addition of the suffixes 1, 2 or 1, 2, 3 respectively, the *lower suffix* indicating the *smaller term* and thus the *higher energy-level*. The system of simple terms occurring with an even number of electrons is distinguished by capital italic letters (S, P, D, F). The previously mentioned sequence number of the terms (which commences with the *s*-terms[9] at 1, with the *p*-terms at 2, with the *d*-terms at 3, and with the *f*-terms at 4) is customarily written before the capital or small letters. Thus we speak of a $3P$-, or $2s$-, or $5p_2$-, or $4d_3$-term, *etc.*

When two terms belonging to a doublet or a triplet are combined with any third term, the *stronger line* always corresponds to the one of the two terms with the *smaller suffix*. *E.g.*, the line $2p_1 \rightarrow 1s$ is always stronger than the line $2p_2 \rightarrow 1s$. The *frequency difference* of multiple terms invariably decreases with increasing sequence number, and converges to zero. For the same sequence number, the frequency difference of the *p*-doublet is always greater than that of the *d*-doublet, and this, in its turn, is greater than that of the *f*-doublet ; the same holds good for triplets.

Within one and the same group of the periodic system,

[9] The sequence numbers of the *s*-terms are frequently given the values 1·5, 2·5, 3·5 . . . instead of 1, 2, 3 . . . because the constant in Rydberg's equation (3) is approximately equal to 1/2 for the *s*-terms.

higher periods are always associated with larger frequency differences. With low atomic numbers, however, the terms do not yet appear multiple. Thus, with lithium and sodium the d-term is still simple, with magnesium it is not yet triple, and it is not until barium is reached that the f-term appears as a triplet.

The formation of spectroscopic combinations between the individual terms is governed by a *selection rule* formulated by *Sommerfeld* and *Landé*. So-called *inner quantum numbers* are allotted to the terms in accordance with the following scheme :—

$$(4) \quad \begin{cases} s & p_1\, p_2\, p_3 & d_1\, d_2\, d_3 & f_1\, f_2\, f_3 & S\, P\, D\, F \\ 1 & 2\ \ 1\ \ 0 & 3\ \ 2\ \ 1 & 4\ \ 3\ \ 2 & 0\ \ 1\ \ 2\ \ 3. \end{cases}$$

The selection rule then states that only those transitions occur in which the *inner quantum number* either *changes by unity* or *remains unaltered ;* the transition $0 \rightarrow 0$, however, is excluded.

By way of illustrating this rule, let us consider the principal series and the two subordinate series in the *arc spectra of the alkali metals*. According to Sommerfeld's selection principle, the lines in the *principal series* must be *doublets*, for the s-term is simple and, because its inner quantum number is 1, transitions from both the p_1- and the p_2-levels are possible, the inner quantum numbers of these levels being 2 and 1 respectively. Since the p_1-term corresponds to the higher level, the line $p_1 \rightarrow s$ has the larger frequency and also, in view of what has been said above, the greater intensity. Moreover, with increasing sequence number, the frequency difference of the p-doublet will, as we have seen, become steadily smaller. The lines in the principal series are therefore composed of constantly narrowing doublets, in which the line with the *smaller wave-length* has the *greater intensity*. The series limit, which is represented by the simple s-term, is the same for both components of the doublet.

In the second or *sharp subordinate series*, which, in emission spectra, is due to transitions from a simple s-level to a double p-level, the lines are obviously again doublets. Since, however, the p-term now represents the final level (instead of the initial level as before), the smaller frequency corresponds to the line belonging to the p_1-term. Hence, in contrast to the principal series, in the doublets of the sharp subordinate series the *line of greater wave-length* is the *stronger*. Since the constant term for the sharp subordinate series is double and the variable term simple, the *frequency difference of the doublets* is *constant* for the whole series, so that the series actually has two series limits.

In the lines of the *diffuse subordinate series*, Sommerfeld's selection principle shows that *three* transitions are always possible, *viz.*, $d_1 \rightarrow p_1$, $d_2 \rightarrow p_1$, and $d_2 \rightarrow p_2$. According to the rule given, the lines of greatest intensity must correspond to the transition $d_1 \rightarrow p_1$, for the suffixes are then the smallest. This transition is associated with a larger frequency than the transition $d_2 \rightarrow p_1$, but with a smaller frequency than the transition $d_2 \rightarrow p_2$, because, as we have seen, the frequency difference of the d-doublet term is always less than that of the p-doublet term (*cf.* Fig. 101, where the different intensities of the lines are also symbolically represented). The lines in the diffuse subordinate series are therefore *triplets*, in which the central line exhibits the greatest intensity.[10]

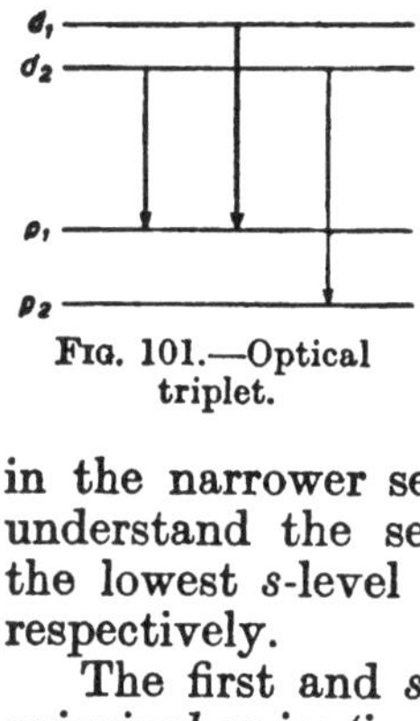

FIG. 101.—Optical triplet.

In every spectrum there is, of course, not merely one single principal series, a single diffuse and a single sharp subordinate series, but there is a multiplicity of such series; for example, the final level of a principal emission series may be any arbitrary s-term, and thus not only $1s$, but also $2s$ or $3s$, and so on. By principal and subordinate series in the narrower senses of the words, however, we always understand the series whose final levels on emission are the lowest s-level ($1s$) and the lowest p-doublet level ($2p$) respectively.

The first and *strongest line* of *longest wave-length* in the *principal series* (in the narrower sense of the word) lies in the *visible* region of the spectrum for all the alkali metals except cæsium. The wave-lengths are given in Table XI, from which it may also be seen that the frequency difference of the doublet increases with increasing atomic number. (With lithium the p-term is not yet double, and hence the line is simple.) For sodium the first line of the principal series lies in the yellow; it is the well-known Fraunhofer D-line, the component (D_2) of shorter wave-length being stronger (about twice as strong) than the other component (D_1), in accordance with the previous statements. For Li, K, and Rb the first

[10] As is immediately obvious, however, the series limit is only double; the frequency difference is that of the p-doublet.

TABLE XI

The First Lines in the Principal Series of the Alkali Metals
(Wave-lengths in Å.U.)

Li	Na	K	Rb	Cs
6708·2	D_1 5895·930	7698·97	7947·6	8943·6
—	D_2 5889·963	7664·94	7800·2	8521·1

line lies in the red, whereas for Cs it is situated in the infra-red close to the border of the visible region.[11]

Since the first line of the principal series is determined by the difference of the terms $1s$ and $2p$, the frequency of this line is equal to the difference between the limiting frequency of the principal series and the common limiting frequency of the two subordinate series. This rule is known as *Rydberg and Schuster's rule*, these two investigators having established it long before the birth of Bohr's theory.

In the elements of the first group there are no further levels between the $2p_2$-level and the $1s$-level, which represents the normal state, and therefore the frequency of the line $2p_2 \rightarrow 1s$, multiplied by the elementary quantum of action, gives the *excitation energy* of the atom. In accordance with the selection rule, it is a property of the $2p_1$-level as well as of the $2p_2$-level that, on the emission of light, *only a transition from that level back to the normal state is possible.* An atom which *absorbs* light of one of these two wave-lengths, and thereby becomes excited, can thus *emit* light only of the *same* wave-length on its return to the normal state. The double line $2p \rightarrow 1s$ is therefore also termed the *resonance line* of the alkali metals.

The *second line* in the principal series (*i.e.*, the doublet $3p \rightarrow 1s$) is situated in the ultra-violet in the cases of Li and Na ; for K it is the well-known violet doublet ; for Rb it is likewise a doublet in the violet, but of greater wave-length ; and, finally, for Cs it is the well-known blue doublet, from which cæsium takes its name.[12] This second line also shows clearly the increase in width of the line doublet with the atomic number.

[11] By way of comparison, it may be noted that the first line in the "principal series" of hydrogen lies far in the ultra-violet, at 1,216 Å.U.

[12] *Cæsius* is the Latin word for blue.

The *principal series in the arc spectrum of sodium*, in which it has been possible to measure no less than 57 lines, is given as an example of an optical series in Table XII.

TABLE XII

The Principal Series of Sodium.

($mp \rightarrow 1s$; wave-lengths in Å.U.)

$m = 2$ 5895·930 D_1 5889·963 D_2	3 3302·94 3302·64	4 2853·038 2852·821	5 2680·443 2680·335	6 2593·927 2593·828	7 2543·875 2543·817
8 2512·210 2512·128	9 2490·733 —	10 2475·533 —	11 2464·397 —	12 2455·915 —	13 2449·393 —
14 2444·195	15 2440·046	16 2436·627	17 2433·824	18 2431·433	19 2429·428
20 2427·705	21 2426·217	22 2424·937	23 2423·838	24 2422·856	25 2421·997
26 2421·233	27 2420·520	28 2419·922	29 2419·380	30 2418·893	31 2418·454
32 2418·062	33 2417·695	34 2417·362	35 2417·058	36 2416·779	37 2416·518
38 2416·271	39 2416·046	40 2415·838	41 2415·651	42 2415·474	43 2415·305
44 2415·147	45 2415·006	46 2414·872	47 2414·746	48 2414·627	49 2414·518
50 2414·411	51 2414·313	52 2414·218	53 2414·131	54 2414·050	55 2413·971
56 2413·910	57 2413·873	58 2413·837	— —	— —	— —

In the case of the *elements of the second group*, the *arc spectra* are naturally considerably more complicated than those of the alkali metals. For, quite apart from the occurrence of triplet terms, we find in these spectra two different systems of terms between which spectroscopic transitions occur. The fundamental term, *i.e.*, the numerically highest one, belongs to the system of simple terms ; it is the $1s$-term, with the subordinate number which determines the eccentricity of the orbit equal to unity. In the emission spectrum it forms the final level for two principal series, one of which arises in transitions from P-levels, and the other in transitions from p_2-levels. (Transitions from a p_1- or p_3-level are excluded by the selection rule previously enunciated.) Both the $2P$-level and the $2p_2$-level are found to be lower than the $1s$-level. Hence, on the emission of monochromatic light, only the return to the fundamental state $1S$ is possible from the $2P$- and the $2p_2$-level. To these two transitions correspond the *two resonance lines* in the arc spectra of the elements of the second group (*viz.*, $2P \rightarrow 1s$ and $2p_2 \rightarrow 1S$).

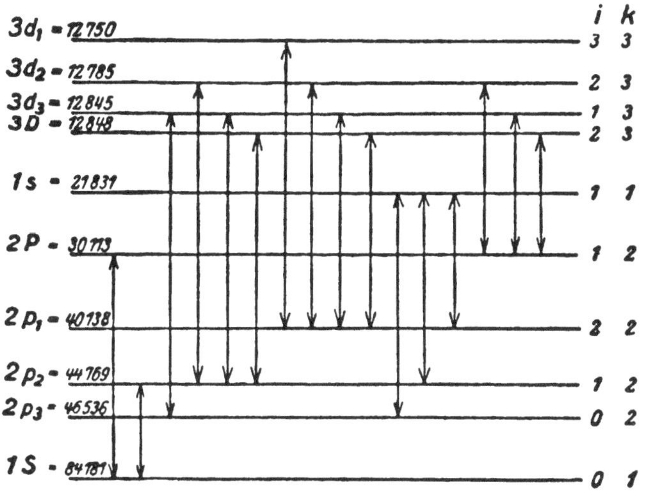

FIG. 102.—Energy-levels for mercury.

In Fig. 102 we see represented the most important terms and spectral lines in the arc spectrum of *mercury*, which belongs to the second group. The terms corresponding to the various levels are shown, so that the wave-length of a spectral line (in centimetres) is simply equal to the reciprocal of the difference between the terms of the levels between which the transition occurs. For example, the wave-lengths of the two resonance lines of mercury are the reciprocals of (84,181−44,769) and (84,181−30,113), or 1,849 and 2,537 Å.U.; both lines therefore lie in the ultra-violet. Fig. 102 also clearly shows the validity of Sommerfeld and Landé's selection rule; the values of the inner quantum number (i) and the subordinate quantum number (k), which determines the eccentricity of the orbit, are given on the right-hand side of the figure.

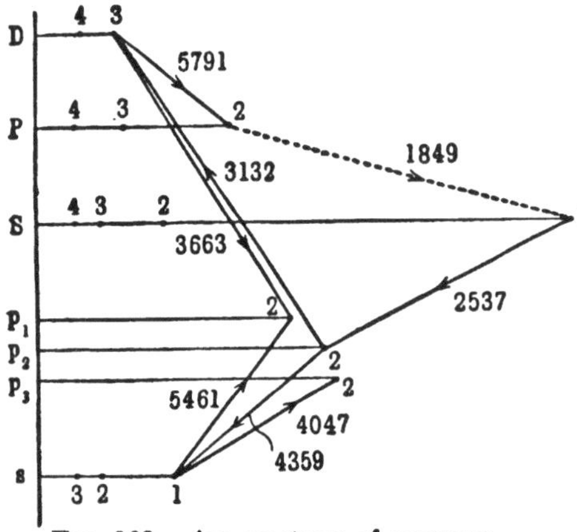

FIG. 103.—Arc spectrum of mercury.

Another method of representing the arc spectrum of mercury is that shown in Fig. 103. The numerical values of the terms are plotted to scale along horizontal lines drawn from the vertical zero-line (the numerals 1, 2, 3, 4 denote the sequence numbers). The connecting lines represent the transitions permitted by the selection rule, and the numbers written along these lines give the wave-lengths in Å.U. (The signification of the arrows will be discussed in a later section.)

§ 121. The Grouping of the Electrons in the Atoms.

It follows clearly from the previously discussed regularities in *Röntgen spectra* that the *electrons* surrounding the nucleus of an atom must be divided, in accordance with the *type* of their *normal orbits* [1] into *groups* and *sub-groups*. Thus, the groups are characterized by the *principal quantum number* (n) of the normal orbit, and the sub-groups by the *subordinate number* (k) which determines the *eccentricity of the orbit*. A sub-group may hence be denoted by the double numerical symbol n_k.

Atomic theory is now faced with a fundamental problem as to how the electrons, which are numerically equal to the chemical atomic number, are distributed amongst the groups and sub-groups in the atoms of the *various elements*.

A few clues as to this arrangement may first be derived from the *atomic numbers of the rare gases*, inasmuch as these are the elements which terminate the individual periods in the system of the elements, and in which, on account of their

[1] The expression " normal " is used here in contradistinction to the abnormal orbits of higher quantum number which give rise to the optical spectral lines.

chemical passivity, the grouping might be supposed to be particularly regular. The atomic numbers of the rare gases are as follows :—

$$(1)\quad \begin{cases} \text{He} & 2 = 2, \\ \text{Ne} & 10 = 2 + 8, \\ \text{A} & 18 = 2 + 8 + 8, \\ \text{Kr} & 36 = 2 + 8 + 8 + 18, \\ \text{Xe} & 54 = 2 + 8 + 8 + 18 + 18, \\ \text{RaEm} & 86 = 2 + 8 + 8 + 18 + 18 + 32. \end{cases}$$

As we have already mentioned in § 109, however, the numbers 2, 8, 18, and 32, which occur in this scheme, and also determine the *lengths of the individual periods*, may be represented in the form

$$(2)\qquad Z = 2r^2 \quad (r = 1, 2, 3, 4).$$

We shall now define a group of electrons as being developed to the rth stage when it consists of r sub-groups, each containing $2r$ electrons, and with subordinate numbers ranging from 1 to r. Moreover, according to the definition of the subordinate number, r can never be larger than the principal quantum number (n) of the group. The total number of electrons in a group completed to the rth stage of development is therefore $2r^2$.

Scheme (1) shows us that the *atoms of the rare gases* should be composed of *groups of electrons* in such *completed stages of development*. Moreover, the *Röntgen spectra* give us a means of determining the principal quantum numbers belonging to these groups. As we have seen in an earlier section (§ 118), the highest elements, from radium emanaiton onwards, are found to have 1 K-level, 3 L-, 5 M-, 7 N-, 5O-, and 3 P-levels. Since, as we saw in the same section, every type of orbit n_k is associated with two energy-levels (provided $k > 1$), the normal and the so-called anomalous level, the *emanation atom* must contain the following *six groups with fifteen sub-groups* :—

$$1_1;\ 2_1, 2_2;\ 3_1, 3_2, 3_3;\ 4_1, 4_2, 4_3, 4_4;\ 5_1, 5_2, 5_3;\ 6_1, 6_2.$$

Hence, in the emanation atom, only the orbits of the first four principal quantum numbers appear to be completely developed, whereas of the five possible types of the five-quantum orbits only three, and of the six possible types of the six-quantum orbits actually only two, are represented.

This fact at once suggests that, in the emanation atom, the one-, two-, three-, and four-quantum groups of electrons are completely developed, for in them $r = n$, whereas the five-quantum group is *only developed to the third stage*, and the six-quantum group *to the second stage*. The total number of electrons surrounding the nucleus would then be $2 + 8 + 18 + 32 + 18 + 8$, or 86, as in reality it is.

As we follow the Röntgen spectra down the natural series of the elements from radium emanation to xenon, the following types of orbit disappear in turn, as was shown in an earlier section (*cf.* Fig. 98): 6_1, 6_2, 5_3, and 4_4. This leads to the assumption that, in the *xenon atom*, there are five groups of electrons of which the one-, the two-, and the three-quantum groups are complete, whilst the four-quantum one is only developed to the third, and the five-quantum one to the second stage. The total number of electrons surrounding the nucleus would thus be $2 + 8 + 18 + 18 + 8$, or 54, as is actually the case.

We may well expect much the same to hold good for the other rare gases. If, therefore, we arrange the numbers in the various lines on the right-hand side of scheme (1) in such a way that, apart from the number two at the beginning of each line, the numbers first increase and then decrease symmetrically with respect to the centre, the numbers thus rearranged should at once show us the distribution of the electrons among the groups in order of increasing principal quantum number. In this way we obtain the scheme of the *electron grouping in the atoms of the rare gases*, which was developed by *Bohr* in 1921, and is represented in Table XIII.

TABLE XIII

Grouping of the Electrons in the Atoms of the Rare Gases

Number of electrons in the n_k-orbits.

	1_1	2_1	2_2	3_1	3_2	3_3	4_1	4_2	4_3	4_4	5_1	5_2	5_3	6_1	6_2
2 Helium . .	2														
10 Neon . . .	2	4	4												
18 Argon . .	2	4	4	4	4										
36 Krypton . .	2	4	4	6	6	6	4	4							
54 Xenon . .	2	4	4	6	6	6	6	6	6		4	4			
86 Emanation .	2	4	4	6	6	6	8	8	8	8	6	6	6	4	4

Let us now consider the elements in the natural sequence immediately following the rare gases, *i.e.*, the *alkali metals*, which commence the various periods. The idea at once suggests itself of ascribing to all the electrons in their atoms, apart from the *valency electron* which gives rise to the optical spectra, the *same grouping* as in the atoms of the *preceding rare gases*. Moreover, we know from the optical spectra that the fundamental term of the valency electron is an *s*-term, so that, in the normal state of the atom, the valency electron describes about the nucleus as focus an orbit whose *subordinate number* is *unity*. That is to say, its orbit is that one among all conceivable orbits with the same principal quantum number, which has the *greatest eccentricity* and *most elongated shape* (*cf*. the earlier Fig. 67 in § 99).

On the other hand, however, it is evident from Table XIII that, in lithium, the valency electron cannot possibly be situated in the completed one-quantum group of electrons, and in sodium there is no vacancy for it in the completed two-quantum group. On account of the great similarity between the alkali metals, we must hence quite generally conclude that, in all these metals, the valency electron describes a path whose *principal quantum number* is *larger by one* than the highest principal quantum number normally occurring in the preceding rare gas.

Thus, we obtain the *electron grouping in the alkali metals* by adding a single electron in each case in Table XIII, the types of orbit being as follows :—

$$\text{Li}, 2_1 ; \text{Na}, 3_1 ; \text{K}, 4_1 ; \text{Rb}, 5_1 ; \text{Cs}, 6_1.$$

Although these orbits of the valency electrons possess quite different principal quantum numbers, it follows nevertheless from the optical spectra that the *energy binding* the valency electron to the nucleus has *practically the same value* for all the alkali metals. This may be seen in Table XIV, which also shows that the binding in all the alkali metals is weaker than the binding of the electron in the normal one-quantum hydrogen atom, but stronger than that in the excited two-quantum hydrogen atom. The binding energy is represented in Table XIV by the *fundamental term* belonging to the principal series, for this, multiplied by the product $h\,c$, gives that energy.[2]

The nearly identical value of the binding energy in the different alkali metals is due to the *large eccentricity* of the orbits of the *valency electrons*. The very elongated orbit of the valency electron lies, indeed, for the most part far outside the electron groups with smaller principal quantum numbers :

[2] The hydrogen terms given in Table XIV are obtained on dividing the Rydberg term by the square of the principal quantum number.

TABLE XIV

Fundamental Terms of the Alkali Metals

Element.	Fundamental term.	Principal quantum number of the fundamental term.	Hydrogen term with the same principal quantum number.
—	—	1	109,678
3 Li	43,484	2	27,419
11 Na	41,445	3	12,186
19 K	35,006	4	6,855
37 Rb	33,685	5	4,387
55 Cs	31,407	6	3,047

thus, most of the orbit is located in regions from which the nuclear charge is almost completely screened off, except, as we have seen, for a single elementary quantum. At *perihelion*, however, owing to the great eccentricity of the orbit, the valency electron approaches *extremely close to the nucleus*, to within distances which are at any rate smaller than the radius of the 2_2-orbit, and most probably smaller even than the radius of the 1_1-orbit.[3]

The valency electron, therefore, penetrates during its revolution very deeply into the interior of the atom, and thus enters regions in which there is little or no screening of the nuclear charge, so that the full or almost full strength of the nucleus acts upon it. Hence, although, in the cæsium atom, for example, the valency electron normally describes a six-quantum orbit and consequently travels very far from the nucleus, and although, during the greater part of its revolution, only a single elementary quantum of the total nuclear charge acts upon it, yet, on the other hand, along a very small portion of its orbit the valency electron is attracted at extremely close range by a charge of 55 or 53 elementary quanta. The constant energy[4] along the orbit is therefore, in absolute amount, much larger than in a six-quantum hydrogen atom, although most of the orbit is that of a "hydrogen-analogue," inasmuch as only a single elementary quantum of the nuclear charge is effective.

Besides the rare gases and the alkali metals, a special

[3] *Cf.* the earlier Fig. 67 in § 99.

[4] In accordance with the conservation principle, this energy is, of course, constant only inasmuch as the mutual action between the valency electron and the other electrons may be neglected.

position is also occupied in the periodic system by the three elements which form group I B. These are *copper*, *silver*, and *gold*, which, both chemically and spectroscopically, exhibit far-reaching similarities to the alkali metals. The atomic numbers of these three metals are lower by eight than those of the alkali metals which commence the following periods.[5] This fact suggests that the grouping of the electrons in these three metals may be the same as in the next higher rare gases, except that, in each case, the highest-quantum group, which in the atoms of each of the rare gases contains eight electrons, will be represented in copper, silver, and gold by only a *single electron*. In the last eight places in each of the periods, the highest-quantum group would then be successively completed from one to eight electrons. The description of a very eccentric orbit of subordinate quantum number one by the valency electron in copper, silver, and gold follows both from their similarity to the alkali metals and from the fact of experimental spectroscopy that the optical fundamental term for copper, silver, and gold is an *s*-term. We assume, therefore, the grouping of the electrons in these three metals to be that represented in Table XV.

TABLE XV

Grouping of Electrons in the Atoms of Copper, Silver, and Gold

Number of electrons in orbit of type n_k.	1_1	2_1	2_2	3_1	3_2	3_3	4_1	4_2	4_3	4_4	5_1	5_2	5_3	6_1
29 Cu	2	4	4	6	6	6	1							
47 Ag	2	4	4	6	6	6	6	6	6		1			
79 Au	2	4	4	6	6	6	8	8	8	8	6	6	6	1

The way in which the highest-quantum groups are completed from one electron to eight electrons in the last eight places of the different periods is still an open question, to which only a partial answer can be obtained from the consideration of *optical spectra*. Since the *optical fundamental terms* of the *elements in the second group* of the periodic system are *S*-terms, or terms with the subordinate number unity, and since these elements appear divalent in their chemical

[5] Including the unknown element 87 with the alkali metals.

combinations, we must suppose that in each case *two electrons* are moving in the highest-quantum orbits with the *subordinate number unity*. This is the case not only with the elements succeeding Cu, Ag, Au, *viz.*, 30 (Zn), 48 (Cd), and 80 (Hg), but also with those which follow the alkali metals, *viz.*, 4 (Be), 12 (Mg), 20 (Ca), 38 (Sr), and 56 (Ba).

Again, we know from their optical spectra (as already mentioned) that the optical fundamental term for *aluminium, gallium, indium*, and *thallium* is a *p*-term, and hence that, in the atoms of these elements, the electron responsible for the optical spectra describes a *normal orbit with the subordinate number two*. Presumably, we have :

(3)
- in 13 (Al) two electrons in 3_1-orbits, one electron in a 3_2-orbit ;
- in 31 (Ga) two electrons in 4_1-orbits, one electron in a 4_2-orbit ;
- in 49 (In) two electrons in 5_1-orbits, one electron in a 5_2-orbit ;
- in 81 (Tl) two electrons in 6_1-orbits, one electron in a 6_2-orbit.

In other respects, the grouping of the electrons in the Al atom would obviously be the same as that in the Na atom, and the grouping in the atoms of Ga, In, and Tl the same as in those of Cu, Ag, and Au.[6]

§ 122. The Completion of the Inner Electron Groups.

As is clearly shown by the earlier Table XIII, an essential assumption of Bohr's modern theory of chemical periodicity is that the *development of the various quantum groups of electrons* from the principal quantum number three onwards takes place *in stages* and covers *several periods*. Thus, in the fourth period, the building up of the outermost four-quantum group of electrons must appear *interrupted* by the *completion* of the *inner* three-quantum group from 8 to 18 electrons. Similarly, in the fifth period, the development of the outermost five-quantum group is interrupted by the growth of the four-quantum group from 8 to 18 electrons, and in the sixth period the completion of the fifth group from 8 to 18 electrons interrupts the development of the outermost six-quantum group. Moreover, a further apparent delay in the

[6] The boron atom (the arc spectrum of which it has not yet been possible to investigate) probably also possesses an electron in a 2_2-orbit.

development of the peripheral group in the sixth period must be caused by the completion, still more deeply within the atom, of the four-quantum group from 18 to 32 electrons.

From the point of view of the *quantum theory*, a clue to this *development of the inner groups of* electrons may most readily be found from *optical spectra*. Let us consider three atoms, each of which possesses 18 electrons in the configuration of the argon atom and, in addition, a nineteenth, more loosely bound, electron; and let us suppose the nuclear charges of the atoms to be 19, 20, and 21 elementary quanta. These atoms would therefore be a *neutral potassium atom*, a *singly ionized atom of calcium* (Ca^+), and a *doubly ionized scandium atom* (Sc^{++}). In each of the three cases we shall suppose the loosely bound nineteenth electron to occupy one of *three different types of orbit, viz.*, a 4_1-, a 4_2-, and a 3_3-orbit; and we then ask ourselves how strong the *binding* of the electron will be.

As regards the neutral potassium atom, the *arc spectrum of potassium* gives us some help. We find that the highest *s*-term (corresponding to the lowest *s*-level) is 35,006, the highest *p*-term (a doublet) 22,021 and 21,963 respectively, and the highest *d*-term 13,471 and 13,468 respectively. From the considerations of the previous section (§ 121), we know that the highest *s*-term, as fundamental term, corresponds to a 4_1-orbit. We must likewise assume that the highest *p*-term (for which the subordinate number is equal to two) is associated with the principal quantum number four, seeing that the 3_2-sub-group was completed with argon, so that there is no place left in that sub-group for another electron. On the other hand, we may well ascribe the principal quantum number three to the highest *d*-term in potassium (the subordinate number being equal to three), for, in the argon atom, the 3_3-sub-group is quite unoccupied. This conception is also justified by the approximate agreement of the highest *d*-term for potassium with the three-quantum hydrogen term,[1] as might be expected from the circular form of the 3_3-orbit. In Fig. 104 the values of the three terms for potassium are shown, calculated, however, in volts according to the relation[2]

$$(1) \qquad \lambda V = 1{\cdot}234 \times 10^4 \, (\lambda \text{ in Å.U., } V \text{ in volts}).$$

The nature of the bindings in the singly ionized calcium atom is revealed by the *spark spectrum of calcium*, which is

[1] *Cf.* the last column in Table XIV.
[2] § 107 eqn. (3).

analogous to the arc spectrum of potassium with fourfold terms. Dividing the terms by four, we find for the highest *s*-term in the spark spectrum of calcium (*i.e.*, the 4_1-term) the value 23,930, for the highest *p*-term (*i.e.*, the 4_2-term) 17,632 and 17,576 respectively, and for the highest *d*-term (*i.e.*, the 3_3-term) 20,517 and 20,502 respectively. In Fig. 104 are represented the three terms for singly ionized calcium (Ca^+) calculated in volts.

Now Fig. 104 clearly indicates a considerable difference between the two spectra. In the arc spectrum of potassium, the 3_3-binding is weaker not only than the 4_1-, but also than the 4_2-binding. In the spark spectrum of calcium, however, the 3_3-binding is *stronger* than the 4_2-binding. Whereas the 4_1- and the 4_2-terms for Ca^+ are considerably *smaller* than in K, the 3_3-term in Ca^+ is considerably larger than in K. We do not, it is true, as yet know the spectrum of *doubly ionized scandium*. With the help of a very arbitrary *extrapolation*, however, as shown in Fig. 104, we may suppose that in doubly ionized scandium the 3_3-term is *larger* than the 4_1- and the 4_2-term.

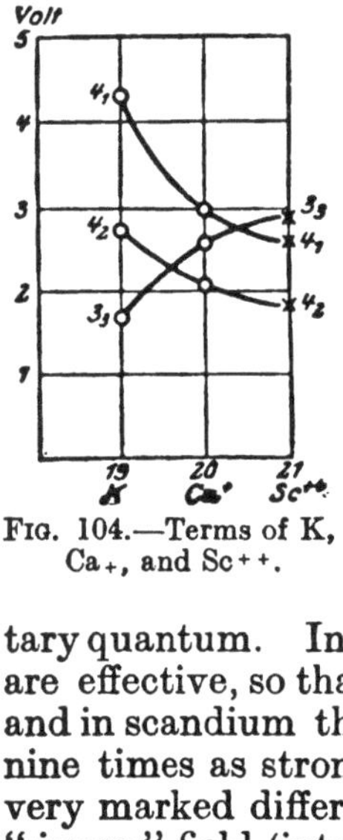

FIG. 104.—Terms of K, Ca_+, and Sc^{++}.

This rapid growth of the 3_3-term becomes intelligible to us when we remember that, *outside* the 18 electrons which are grouped as in the argon atom, the atomic nucleus of potassium has approximately only one effective elementary quantum. In calcium, however, two elementary quanta are effective, so that the binding energy is four times as large, and in scandium three elementary quanta, the binding being nine times as strong. Whereas, therefore, with potassium a very marked difference exists between the " outer " and the " inner " field (into which the eccentric orbits of subordinate numbers one and two penetrate [3]), this difference is not so large with calcium, and is still smaller with scandium. Hence, although with potassium the 4_1-term is about twice as large as the 3_3-term, the binding for scandium of a nineteenth electron in a 3_3-orbit is stronger than in an eccentric 4_1-orbit, which, although it penetrates in part into the interior of the atom, yet carries the electron much further from the nucleus than is the case for a circular 3_3-orbit.

[3] *Cf.* § 121.

Again, the grouping of the electrons in a neutral atom must evidently be exactly the same as if all the electrons surrounding the nucleus were torn away, and the atom were *reformed* by *successive binding of electrons to the nucleus.* In this process each electron would, in a sense, " fall as low as possible," *i.e.*, it would occupy the lowest free energy-level. We may therefore suppose that, in such a successive building up of a scandium atom, the nineteenth electron would take up a 3_3-orbit, and thus that scandium is the first element in which a 3_3-orbit is normally present. The *completion of the three-quantum group* from 8 to 18 electrons therefore evidently commences with *scandium* and, as we saw from the earlier Table XV, finishes with *copper.* The process of completion thus extends over *nine places* (Nos. 21–29); the *increase of ten electrons* in the inner electron group is explained by the fact that the outermost, four-quantum, electron group is represented in calcium by two electrons, but in copper by only one.

Since the 18 elements in the *fifth period* are complete homologues, in the same order, of the 18 elements in the fourth period, we must assume that in the fifth period the completion of the inner (*i.e.*, four-quantum) group of electrons also commences with the third element, *yttrium*, of atomic number 39, and ends with *silver*, of atomic number 47.

In the *sixth period*, the completion of the five-quantum group from 8 to 18 electrons evidently commences with *lanthanum* which, as third element in this period, is an exact homologue of scandium and yttrium, and ends with *gold.* In this case the completion extends over 23 instead of 9 places, *viz.*, from No. 57 to No. 79. Presumably, therefore, the four-quantum group also is completed between lanthanum and gold from 18 to 32 electrons.

In the elements of the sixth period, however, this four-quantum group lies fairly *deep within the atom.* Hence, during its completion, the *chemical* and also those *physical properties* such as valency and optical-spectral behaviour, which are conditioned by the *peripheral electrons*, must remain practically *unchanged.* In point of fact, the elements following after lanthanum, the so-called *rare earths*, closely resemble one another and lanthanum in chemical and in important physical properties, whereas lanthanum is very dissimilar from the immediately preceding element, barium. We must thus assume that the completion of the four-quantum group from 18 to 32 electrons *begins* with the element next after lanthanum, *cerium* (58). Corresponding to the difference between the numbers 32 and 18, there must naturally be a total of 14 elements resembling lanthanum, or, including this element, *fifteen rare earths* altogether.

The agreement of the peripheral atomic properties does, in fact, extend over 15 successive elements,[4] beginning with lanthanum and ending with element 71, *lutecium*. At the time when Bohr established his theory of the periodic system, element 72 was not yet known. Up till then it had commonly been supposed that it must also be a rare earth, and searches were therefore made for it in those minerals in which other rare earths had been found. It followed from the new theory of Bohr, however, that the unknown element could not possibly be a rare earth (because the theory does not permit of more than 15 rare earths), but rather that it must be a *homologue of zirconium*, which occupies the fourth place in the fifth period. Since homologous elements frequently occur together in nature, the idea suggested itself of looking for the unknown element in zirconium minerals. It constituted a brilliant triumph for Bohr's new theory when, in 1922, *Hevesy* and *Coster* actually succeeded in discovering the new element in zirconium minerals; in honour of the city of Copenhagen, where the discovery was made, they named it "*hafnium*." In point of fact, hafnium is by no means rare; there is, for example, only about five times more nickel than hafnium in the earth's crust. It can only be ascribed to its close resemblance to zirconium that hafnium was not discovered long ago, and also that it has not yet been possible to prepare it in a pure state.[5]

In the *seventh period* of the system of the elements, the six-quantum group must be completed from 8 to 18, and, in addition, the five-quantum group from 18 to 32 electrons. From the homology of actinium and lanthanum, of proto-actinium and tantalum, and of uranium and tungsten, we must conclude that the completion of the second outermost group of electrons (in this case, the six-quantum group) also begins in the seventh period with the third element, actinium, just as in the fourth, fifth, and sixth periods. On the other hand, up to and including uranium, the number of electrons in the five-quantum group further within the atom does not increase.

Fig. 105 represents the periodic system of the elements from the point of view of Bohr's new theory. The seven

[4] Including, it is true, the element 61, which is not yet known.

[5] The discovery was made by Röntgen spectroscopy. *Vide Nature*, 111 (1923), *passim*.

periods are arranged in vertical columns. Series of elements within which an inner group of electrons is understood to be completing itself are enclosed in rectangles; in the sixth period there are two such rectangles, one within the other.[6] Homology of elements in different periods is indicated by tie-lines, but lines are not drawn between elements which, despite the same valency, occupy different positions with respect to the rectangles; thus, there is none between Al and

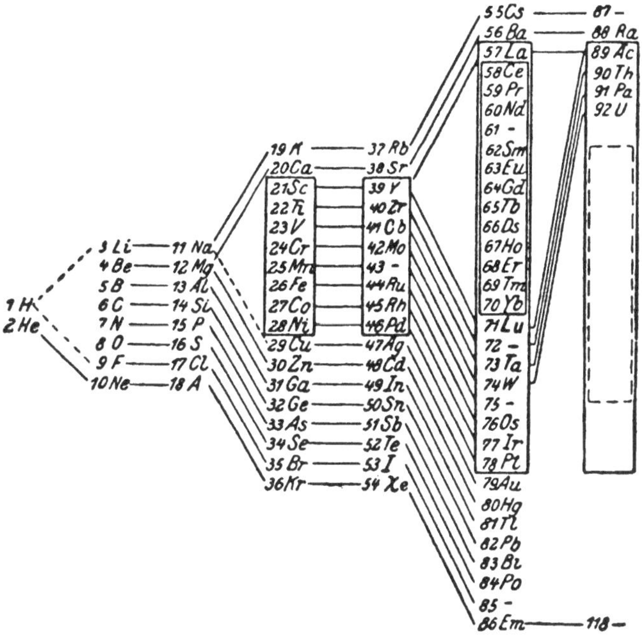

FIG. 105.—Periodic system, according to Bohr.

Sc, though there is between Sc and Y. Table XVI gives a survey of those elements (denoted by atomic number and chemical symbol) in which a definite type of orbit normally appears for the first time.

The arrangement reproduced in Fig. 105 clearly distinguishes from among the whole series of elements those in which an *inner group of electrons* occurs in an *incomplete stage of development*, *i.e.*, where its grouping is other than one

[6] The rectangles in Fig. 105 end before the elements in which the inner groups of electrons are completed. In the seventh period the inner rectangle is obviously quite arbitrarily drawn; we know only that the elements up to uranium inclusive lie outside it.

TABLE XVI

First Occurrence of the Different Types of Orbit in the Natural Sequence of the Elements

1_1	1 H	2_1	3 Li	3_1	11 Na	4_1	19 K	5_1	37 Rb	6_1	55 Cs	7_1	87 ?
		2_2	5 B	3_2	13 Al	4_2	31 Ga	5_2	49 In	6_2	81 Tl		
				3_3	21 Sc	4_3	39 Y	5_3	57 La	6_3	89 Ac		
						4_4	58 Ce						

of two, or twice four, or thrice six, or four times eight electrons. In point of fact, the elements enclosed in the rectangles in Fig. 105 exhibit, with their incomplete inner atomic structure, peculiarities lacking in the other elements. They are frequently *multivalent* ; they form *coloured salts* ; and, as a rule, they are *paramagnetic*.

The multivalency is evidently due to the possibility that electrons belonging to an incompletely developed inner group may be just as *loosely* bound as the peripheral valency electrons.[7] Hence, like the latter, they are also able to absorb quanta of *visible* light, so that their ions are coloured, despite the absence of valency electrons. Finally, as we shall discuss more fully in a subsequent section (§ 126), paramagnetism is ascribed to an incomplete balance in the magnetic moments of the electronic orbits in the individual atoms, and thus likewise appears as a consequence of asymmetry in the internal structure of the atom.[8]

A direct *experimental confirmation* of Bohr's attempts to explain chemical periodicity and homology from the point of view of the quantum theory is afforded by the *curves* representing the *dependency of the square roots of the Röntgen-spectroscopic terms of the elements on the atomic numbers* (*cf.* Fig. 100 in § 118).

In accordance with Moseley's Law, the form of these curves should be exactly *rectilinear*, provided that an increase in the atomic number by unity is accompanied by a like increase in the *effective fraction of the nuclear charge* for the term concerned, and a corresponding increase in the *binding*

[7] We have, moreover, to assume such a loose binding of the inner electrons in the elements which, though lying outside the rectangles in Fig. 105, yet border on them. Thus *copper* is both *univalent* in *cuprous* compounds and *divalent* in *cupric* compounds. Whereas cuprous ions are assumed to contain a complete three-quantum group of 18 electrons, the same group in cupric ions would only contain 17 electrons, and the inner atomic structure would thus be asymmetric. Actually, cuprous ions are colourless and cupric ions coloured.

[8] Ladenburg had, in 1920, already endeavoured before Bohr to ascribe the chemical and magnetic properties of the elements enclosed in the rectangles in Fig. 105 to the formation of an " intermediate shell " of loosely bound electrons

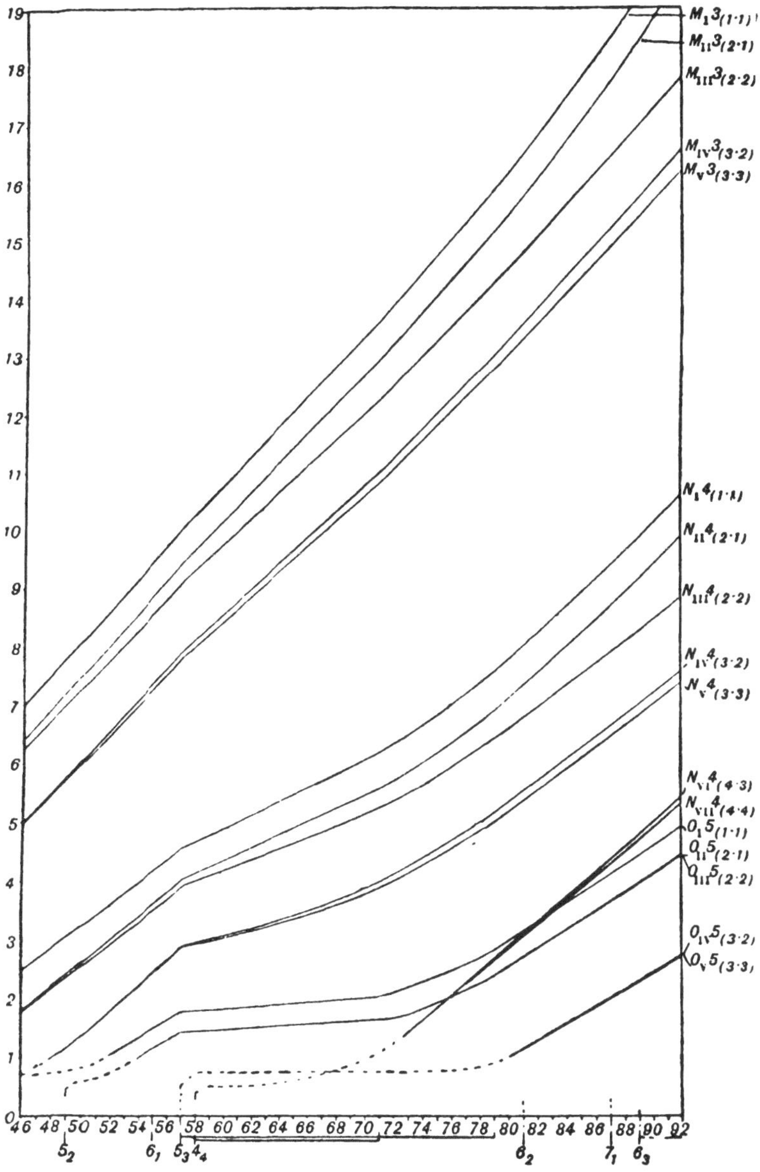

FIG. 106.—Röntgen terms of the elements.

energy. During the development of an inner electron group, however, the increase in the binding energy and hence of the term must at any rate be interrupted, or at the very least, retarded. This is due to the fact that the increase in the nuclear charge is entirely or partially counterbalanced by a similar increase of one in the number of *inner electrons screening* the nucleus. Hence, *during the development of an inner group of electrons* of given quantum number, the *curves* for the terms of higher quantum number should run *horizontally*, or at least at a *smaller angle of inclination* to the axis of abscissæ.

That such is the case is clearly shown by the earlier Fig. 100 and by Fig. 106, in which part of Fig. 100 is reproduced on a larger scale. Vertical lines underneath both figures indicate the elements in which (according to Table XVI) new types of orbit first occur, whilst the development of an inner group of electrons is shown by a horizontal line. In both figures the horizontal course of the five-quantum O-terms during the completion of the four-quantum group from 18 to 32 electrons is clearly visible. Moreover, during this completion, the curves for the N-terms run less steeply, for the addition of new electrons in the same group has, at least, a partial screening effect. The completion of the three- and four-quantum groups from 8 to 18 electrons is also rendered evident by the smaller inclination of the curves.

In Fig. 106 a further remarkable fact is manifest. The 4_4-term is at first smaller than the 5_1- and 5_2-terms, and at the very beginning it is even smaller than the 5_3-term. As soon as the development of the four-quantum group of electrons is completed [at 71 (Lu)], however, the binding energy of the 4_4-term rapidly increases, so that, between the 75th and the 83rd element, the 4_4-curves *cross* the 5_1- and 5_2-curves.

§ 123. Fluorescence.

The radiation emitted by a body whose atoms are excited by primary rays is called *fluorescent radiation.* In the special case in which the wave-lengths of the secondary and the primary radiation are identical, we speak of a *resonance radiation.* According to the discussion in a previous section (§ 120), we may expect to meet with such radiation when light of the wave-length of a *resonance line* of an element passes through the vapour of the element at a temperature below that at which any of its atoms are excited.

The phenomenon of resonance radiation was discovered and investigated long before the atomic theory was established, first by *Wood* in the case of *mercury*, and subsequently by *Dunoyer*, *Wood*, and *Strutt* in the case of alkali metal vapours, especially *sodium vapour.* At low pressures, the total energy of the re-emitted light does, in fact, agree with the energy absorbed from the primary radiation. It is only when the pressure is increased that a part of the absorbed primary radiation is changed into molecular energy of translation, with a resulting heating of the fluorescent vapour.

When a gas is irradiated with light of a frequency *greater than the resonance frequency* and corresponding to the difference in energy between the normal state and a *higher*, excited state, the *return* of the atom from this more highly excited state may take place *in steps.* In the fluorescent radiation, therefore, lines of greater wave-length than that of the exciting light may occur. Under no circumstances, however, can an atom which, before excitation, is in the normal state, emit fluorescent radiation of shorter wave-length than the primary radiation, and in this lies the explanation of *Stokes's rule of fluorescence*, which was mentioned in an earlier section (§ 91). Only when the irradiated atom has been previously excited can it happen, by way of exception (*e.g.*, on a return to the normal state), that the wave-length of the fluorescent radiation may be shorter than that of the exciting radiation.

An especially clear example of the phenomena of fluorescence is afforded in the case of *mercury vapour*, as has been shown by *Füchtbauer.*[1] In his researches, a quartz mercury vapour lamp served as the source of the primary rays. Of the characteristic rays of mercury produced within the lamp, only those wave-lengths as long as, or longer than, the resonance wave-length of 2,537 Å.U. can pass through the quartz walls, the remainder being absorbed. In the path of the rays from the lamp is placed a tube containing cold mercury vapour. Since the atoms in the cold vapour may be assumed to be initially in the normal state, only rays of the resonance wave-length of 2,537 Å.U. will be absorbed, because any rays of greater wave-length present are incapable of raising the atom from its normal state to a higher one.[2]

The atoms which absorb the rays of 2,537 Å.U. wave-length are raised to the excited $2p_2$-state. The atoms thus

[1] *Physikal. Zeitschr.*, 21 (1920), p. 635.

[2] The other resonance line of mercury, of 1,849 Å.U wave-length, which is due to a transition on absorption from the normal to the $2P$-state, is not found in the primary radiation because rays of this wave-length are absorbed by the quartz.

excited are then able to absorb rays of 4,359 and 3,132 Å.U. wave-length as well from the primary radiation, whereby they are raised to the still higher states $1s$ and $3D$ (*cf.* Fig. 103 in § 120). Now, in accordance with the selection principle (§ 120), the mercury atom can undergo transitions from the $1s$-state to the $2p_1$- or $2p_2$-states, with the emission of the lines 5,461 or 4,047. Again, the selection principle permits of transitions from the state $3D$ to $2P$ and $2p_1$, with the emission of the lines 5,791 and 3,663. All these lines could actually be observed by Füchtbauer in the fluorescence spectrum ; and it was evident that they were all produced by rays of 2,537 Å.U., because they all vanished when the latter were absorbed by a thin glass plate. At the same time, we see from Fig. 103 how the transition from the $2P$- to the normal $1S$-state (indicated by the dotted line) gives rise to an indirect fluorescent radiation of *shorter wave-length* than the primary radiation. Furthermore, Fig. 103 clearly shows how the primary light, which, being ultra-violet, is not visible to the human eye, gives rise as a result of fluorescence to a succession of rays whose wave-lengths lie between 4,000 and 8,000 Å.U., and are thus visible.

Just as atoms can be excited otherwise than by absorption of radiation, so also we must assume that an atom excited in any way can *return* to its normal state *without radiating*. In particular, we may expect the excess energy to be given off without radiation when the excited atom *collides* with another atom ; in that case, the excitation energy will be *transformed* into *translational energy*, with a consequent increase in the heat energy present.

Thus, of the excited atoms in a fluorescent gas, only a certain number would, in general, return to the normal state with the emission of radiation, the others returning without radiating. The ratio between the two numbers, however, evidently depends on the ratio between the *mean time between two collisions* (to be calculated from the theory of gases) and the *average duration of the excited state.* As long as the former interval is considerably greater than the latter, the total energy of the fluorescent radiation must be approximately as large as the total energy absorbed from the primary radiation by the fluorescing gas. If, however, the mean interval between two collisions be *shortened*, as by *increasing the pressure* by the *admixture of another gas*, the *fluorescence* should be gradually *extinguished* as the mean interval between two collisions becomes of the same order of magnitude as the average duration of the excited state.

The extinction of fluorescence was actually observed by *Wood*, when air was mixed with the fluorescent mercury vapour.[3] Both his observations and those of later investigators[4] lead to an approximate value of 10^{-8} sec. for the *mean duration of the excited state.* One-tenth as large (as we may remark in passing) is the time found by *Wien* for the *decay of the luminescence* of excited canal ray particles.[5]

Moreover, it has been found possible to demonstrate experimentally a *transference of excitation energy* from the atoms of one substance to those of *another*, the transference being obviously effected by collisions. This phenomenon, in which light is absorbed by atoms of one kind and re-emitted by atoms of another kind, is known as *sensitized fluorescence* It was first observed, in 1922, by *Franck* and *Cario*,[6] in a *mixture of mercury vapour and thallium vapour.* This mixture was irradiated at suitable pressure and low temperature with ultra-violet mercury light of the resonance wave-length of 2,537 Å.U. These rays are naturally absorbed by the mercury atoms, but not by the atoms of thallium, whose spectrum exhibits in the neighbourhood of 2,500 Å.U. no lines having the normal state as initial state for absorption. None the less, the spectrum of the fluorescent radiation contained all the lines of thallium whose excitation energy is smaller than a light-quantum of 2,537 Å.U. wave-length. Similar experiments were performed with mixtures of mercury and lead, mercury and bismuth, mercury and silver, and mercury and cadmium. At higher temperatures, indeed, sensitized fluorescence makes its appearance also with lines whose excitation energy is larger than a light-quantum of the primary radiation. In such cases the deficit in energy is evidently supplied during collisions by the translational energy of the mixture.

§ 124. Electron Collisions.

As we have seen in the previous considerations, Bohr's theory of spectra and atomic structure is based on the blending of two fundamental hypotheses. One of these is that

[3] *Phil. Mag.*, **21** (1911), p. 309.

[4] *Cf.* G. Cario, *Zeitschr. f. Phys.*, **10** (1922), p. 185.

[5] W. Wien, *Ann. der Phys.*, **60** (1919), p. 597, and **66** (1921), p. 229.—The previously discussed experiments of Füchtbauer also led to a probable value of 10^{-8} sec. for the average duration of excitation.

[6] *Cf.* the summary by J. Franck, "*Neuere Erfahrungen über quantenhaften Energieaustausch bei Zusammenstössen von Atomen und Molekülen,*" in the second volume of "*Ergebnisse der exakten Naturwissenschaften*" (Berlin: Springer, 1923).

the internal energy of atoms can assume only quite definite and discrete values; the other finds expression in Bohr's frequency condition. The theoretical explanation of *spectral* phenomena is to be found in a *combination* of these two fundamental assumptions.

It was, therefore, a great advance for atomic physics when, in 1913, *Franck* and *Hertz* succeeded in effecting a *direct experimental confirmation* of the first of the two fundamental hypotheses *quite independently* of the second. They were able to do this by investigating *collisions between atoms and free electrons* which, in the form of slow negative corpuscular rays, were made to pass through a gas or a vapour at a sufficiently low temperature. The phenomena of so-called *electron collisions*, as they are briefly termed, must for two reasons obviously afford much simpler and clearer experimental results than collisions between atoms. In the first place, the kinetic energy of a single electron can be exactly determined from a measurement of the electrostatic potential. On the other hand, the mass of an electron is negligible compared with that of an atom, and hence the exactly measurable kinetic energy lost by an electron can at once be equated to the increase in internal energy experienced by the atom involved in the collision.

Now, as long as the kinetic energy of an electron is smaller than the energy required to excite an atom, the kinetic energies of the electrons will not be diminished by the collisions. In this case, *only "elastic" impacts* are possible. *Inelastic impacts* (in which the colliding electron loses energy) will first be possible when a critical value of the potential is reached, such that, when multiplied by the elementary quantum of electricity, the product is equal to the excitation energy; this critical value is known as the *excitation potential.*[1] If we denote by λ^* the wave-length of the line emitted by the atom in its return from the excited state (in the narrower sense of the word) to the fundamental state, and by V^* the excitation potential, we have, by § 122 eqn. (1), the relation

$$(1) \qquad \lambda^* V^* = 1{\cdot}234 \times 10^4 \quad (\lambda^* \text{ in Å.U., } V^* \text{ in volts}).$$

For the elements in the first group of the periodic system, λ^* is to be given the value of the wave-length of the *resonance line*. With the *elements in the second group*, for which there

[1] Obviously, however, the kinetic energy of electrons can be increased before the excitation potential is reached, if they collide with previously excited atoms, which then give up their excess energy; in that case we speak of "electron collisions of the second kind."

are two resonance lines differing considerably in wave-length, a corresponding distinction is made between *two different excitation potentials.*

For many elements, it has been possible to make direct *experimental determinations* of the excitation potential by purely electrical methods. Using suitable apparatus, the electrons are made to pass through the gas or vapour, and the resulting current is measured with a galvanometer, while the applied potential difference is also observed. If the current be plotted on a graph against the potential, the curve exhibits a strongly marked discontinuity for the potential at which inelastic impacts first occur. With many elements, the *first flashing out of the resonance line* caused by the electronic collisions could be observed at this potential, and this in itself must be regarded as a brilliant experimental confirmation of Bohr's frequency condition.[2]

From the point of view of Bohr's theory, further discontinuities in the current are to be expected for values of the potential which, when multiplied by the elementary quantum of electricity, give the energy difference between the fundamental state and a state excited to a higher degree. The determination of these closely adjacent potential values, however, generally meets with experimental difficulties, and has hitherto only been successful in the case of mercury.[3] On the other hand, the potential at which the electron impacts first produce ionization of the atoms of the gas can be fairly sharply determined for many elements. This potential is called the *ionization potential.* It must be equivalent to the *fundamental term* of the atom, inasmuch as this corresponds to the work necessary for the removal of the electron concerned from the atom. The ionization potential must therefore satisfy eqn. (1) for a wave-length corresponding to the fundamental term.

It has been possible to *test* eqn. (1) *experimentally* for the *alkali metals* and the *metals of the second group* of the periodic system, both as regards the excitation potential and the ionization potential. Table XVII gives a summary of the calculated and the observed potential values. The calculations were carried out with the help of eqn. (1), taking, in the case of the alkali metals, the wave-length of the first line of the principal series ($2p \longrightarrow 1s$) and the $1s$-fundamental

[2] Thus, in mercury, Franck and Hertz observed the first inelastic impact at 4·9 volts, and at this same potential the first occurrence of the resonance line 2,537 Å.U.

[3] Franck and Einsporn, *Zeitschr. f. Phys.*, 2 (1920), p. 18.

term, and, for the metals of the second group, the wave-lengths of the two resonance lines and the $1S$-fundamental term. The observed values of the potential are those at which the first inelastic electron impacts and the first occurrence of ionization were actually observed. As Table XVII shows, a very good agreement exists between the calculated and the observed values.[4]

TABLE XVII

Excitation and Ionization Potentials (in Volts)

A. ALKALI METALS.

		Li	Na	K	Rb	Cs
Excitation potential	calc.	1·84	2·09	1·61	1·6	1·4
	obs.	—	2·12	1·55	1·6	1·48
Ionization potential	calc.	5·37	5·12	4·32	4·16	3·88
	obs.	—	5·13	4·1	4·1	3·9

B. METALS OF THE SECOND GROUP.

		Mg	Ca	Zn	Cd	Hg
Excitation potential (I)	calc.	2·7	1·88	4·01	3·78	4·87
	obs.	2·65	1·90	4·18	3·95	4·76 (4·9)
Excitation potential (II)	calc.	4·33	2·92	5·77	5·39	6·67
	obs.	4·42	2·85	5·65	5·35	6·45 (6·7)
Ionization potential	calc.	7·61	6·09	9·35	8·95	10·39
	obs.	7·75	6·01	9·3	8·92	10·2

The excitation and ionization potentials have been measured for numerous elements in the other groups, apart from the elements in the first two groups of the periodic system. Thus, for example, the excitation potentials for thallium and lead have been found to be 1·07 and 1·26 volts respectively, and the corresponding ionization potentials 7·3 and 7·93 volts. By means of the method of electron

[4] The observed values given are those of *Foote* and *Mohler*, who have developed Franck and Hertz's method. *Cf.* Foote and Mohler's "*The Origin of Spectra*" (New York, 1922). The bracketed values for Hg in Table XVII are based on measurements by Franck and Einsporn (*cf.* note 3).

collisions, it was in this way possible to derive two optical terms for lead, although a resolution of its spectrum into series has not yet been effected.

For the elements of the second group of the periodic system, and for some of the rare gases, the *ionization potentials of the second order* have also been observed, *i.e.*, the potentials at which a second electron is removed from the atom. Just as the ionization potential of the first order is equivalent to the fundamental term of the arc spectrum, so the ionization potential of the second order corresponds to the *fundamental term of the spark spectrum.* The ionization potential of the second order is naturally considerably higher than that of the first order; *e.g.*, for magnesium, the two potentials amount to 7·6 and 15 volts, and for helium to 24·5 and 78·6 volts.[5]

§ 125. Metastable Atomic States.

As we have seen in an earlier section, an atom generally persists in an excited state for an interval which is found to be only 10^{-8} sec. in order of magnitude; provided that the excited state has not already been destroyed, the atom then returns *spontaneously* to its normal state, with the *emission of monochromatic radiation.* It is, however, conceivable that an atom might be raised to a state of higher quantum number than the normal by some *other means* than the absorption of radiation, from which state a spontaneous return, with the emission of radiation, to the normal state would be *precluded by the spectroscopic selection principle.* In that case, the atom would have to persist in the excited state until this was *destroyed* by some *external influence*, such as collisions with other atoms. Thus, it appears theoretically possible for an abnormal state of this kind to continue for an arbitrary length of time.[1] Such an excited state, from which there is *no spontaneous return to the normal state*, is termed *metastable.*

Metastability has actually been detected in *atoms of*

[5] The frequency which, according to eqn. (1), corresponds to the ionization potential of the first order (24·5 volts) for helium must evidently represent the K-term of helium, since, quite generally, the K-term is a measure of the work required to liberate an electron from a normal one-quantum orbit. A comparison of the K-term of helium thus derived with the known K-terms of Be, B, C, N, and O (*cf.* § 117), and of the elements from Na onwards, clearly shows that, from helium to sodium the natural sequence of the elements completely agrees with the series formed in order of increasing atomic weights. The nuclear charge numbers of all the elements are thus unequivocally determined, for a comparison of the spectra of ionized helium and hydrogen immediately shows that the nuclear charge number of helium is two.

[1] Provided only that pressure and temperature be kept sufficiently low.

helium. In the arc spectrum of *neutral* helium [2] there are *two different systems of terms*, between which *no spectroscopic transitions* occur. The two systems of terms and series are distinguished by two names which are now only of historical significance; *viz.*, *parhelium* and *orthohelium.*[3] The normal orbit of parhelium, which corresponds to the normal state of the helium atom, must be assumed to be a one-quantum orbit, whereas the lowest orbit of orthohelium must be a two-quantum one. As we have just said, no spectroscopic transition between the normal para-term and the lowest (two-quantum) ortho-term is possible. By means of *electron collisions*, however, *Franck* has succeeded in transforming helium atoms from the normal para-state into the lowest ortho-state, for which process a so-called *transformation potential* of 19·8 volts is necessary. The actual metastability of the two-quantum ortho-state thus produced follows from the fact that this state is associated with a characteristic *resonance line.* Its term representation is $2_2 \rightarrow 2_1$ (both terms, of course, being ortho-terms), and its wave-length is 10,830 Å.U., or approximately 1μ[4]. For, if helium in which a sufficient number of metastable atoms has been produced by electron collisions be irradiated with infra-red light of about 1μ wave-length, perfect resonance ensues. This clearly shows that, in spectroscopic respects, the 2_1-ortho-state represents a normal state from which there is no spontaneous return, with the emission of radiation, to the true normal state (*viz.*, the normal para-state).[5]

The metastable state is, of course, associated with a *smaller ionization energy* than the normal state. The difference between the ionization potentials of the normal and the metastable atom must, indeed, be equal to the transformation potential. Since the ionization potential of normal

[2] We must, of course, distinguish between neutral helium and the hydrogen-analogue, ionized helium.

[3] The series of orthohelium consist of very narrow doublets, whereas the lines of the parhelium spectrum are singlets. The well-known strong yellow helium line (the Fraunhofer line D_3) belongs to orthohelium; its term representation is $2p - 3d$.

[4] As was mentioned in § 86, a micron (μ) is one-thousandth of a millimetre.

[5] If, on the other hand, helium be irradiated with infra-red light of about 2μ wave-length, only incomplete resonance results, for in the spectrum of parhelium there is a line which has approximately this wave-length (to be more exact, 20,582 Å.U.); its term representation is $2S - 2P$ (the para-terms being distinguished by capital letters). This line is absorbed by such atoms as have been *previously* excited from the normal state $1S$ into the para-state $2S$. After absorption of the 2μ line, the atoms thus still further excited can now return to the normal state with the emission of radiation, so that only some of these atoms emit a pure resonance radiation

helium is 24·5 volts, and the transformation potential 19·8 volts, the ionization potential of metastable helium must amount to 4·7 volts. The metastable helium atom must therefore lose an electron more easily than lithium and sodium,[6] and hence must be still *more electropositive* than these alkali metals, and, in contrast to the rare gases, it should exhibit considerable *chemical activity*. This also follows from the fact that, at sufficiently high densities, a band spectrum of helium appears, and this is obviously due to the *formation of molecules* by metastable atoms.[7]

§ 126. Atomic Magnetism.

Since every revolving electron in an atom represents a convection current whose strength is given by the product of the elementary quantum of electricity and the revolution frequency (ω), every *electron orbit* in an atom is associated [according to § 60 eqn. (1)] with a *magnetic moment* of magnitude

$$(1) \qquad M = \frac{e\,\omega\,F}{c}$$

where F denotes the area of the orbit.

Let us consider, in particular, the *normal hydrogen atom*, for whose circular orbit (of radius a) the angular momentum (U) is equal to $h/2\pi$; then

$$(2) \qquad U = \frac{h}{2\pi} = m\,a\,v = m\,a\,.\,2\pi\,a\,\omega = 2\,m\,\omega\,F\,.$$

Hence, denoting the specific charge of an electron by γ, we have

$$(3) \qquad M^* = \frac{U\gamma}{2c} = \frac{h\gamma}{4\pi c}\,.$$

Introducing the known values of h and γ, we find

$$(4) \qquad M^* = 9{\cdot}21 \times 10^{-21} \text{ abs. units.}$$

The quantity M^* is known as *Bohr's magneton*.

Although every electron orbit is associated with a magnetic moment, yet an atom as a whole may be non-magnetic, owing to a mutual *compensation* of the magnetic moments of the electronic orbits. This will be the case when

[6] *Cf.* Table XVII in § 124.

[7] According to two recent notes in "*Nature*" [114 (1924), p. 861, and 115 (1925), p. 337] J. J. Manley has succeeded in combining mercury and helium under the influence of a glow discharge to form a compound HgHe. Details of the experiment are still lacking, but, in view of the existence of the metastable state of helium, the formation of such a compound does not appear improbable.

two parallel orbits, of equal area and with the same revolution frequency, are traversed in opposite senses. A compensation of this kind must, however, become imperfect when the previously non-magnetic atom is exposed to the action of an external *magnetic field*. For, in every circuit, an *induction current* will then be produced, and the sense of this current will depend (according to § 63) only on whether the magnetic flux of force traversing the circuit area increases or decreases. Thus, although the original, mutually compensating, currents may be opposed in sense, yet the superimposed induced currents will be *similar in sense*, so that the original compensation will be partially destroyed. Since no resistance is offered to the induced currents, they must persist until, on removal of the magnetic field, they disappear as a result of a second induction effect opposed to the first one. In a magnetic field, a non-magnetic atom therefore exhibits *induced magnetism*. According to Lenz's rule (§ 63), however, a *force of repulsion* must exist between the induced currents, when they are generated by the approach of a magnet, and the inducing magnet. Consequently, the substance in which the induced currents are flowing must appear *diamagnetic*.

Atomic theory regards as *paramagnetic* those substances in which, because of an imperfect compensation (*cf.* § 122) in the atoms (or molecules), the vectorial sum of the magnetic moments differs from zero from the very beginning, whilst an external magnetic field exerts *directing forces* on the atoms. In *ferromagnetic* bodies, which exhibit a permanent magnetism, there appear to be, in addition, *molecular forces* acting between the molecules, these forces ceasing to act, however, above a certain temperature, known as the *Curie point*. Above this temperature the body in question exhibits only the properties of paramagnetism (in the narrower sense) ; *e.g.*, for iron, the Curie point lies at 750°, and for nickel at 375° C.[1]

In 1911, *Pierre Weiss* came to the conclusion that, in a state of magnetic *saturation* (only attainable at low temperatures, because at higher temperatures the continuous molecular collisions do not admit of it),[2] the *magnetic moment* referred to a gram-molecule is, in ferromagnetic substances, an *integral multiple* of 1,123 abs. units. For example, below the Curie point it is three times that amount in nickel, nine times in cobalt, and ten or eleven times in iron.

[1] At the Curie point a sudden change in the specific heat also occurs.

[2] In the state of saturation the vectors of the magnetic moment are presumably similarly directed in all molecules.

The magnetic moment of a single molecule would thus be an integral multiple of the amount obtained on dividing 1,123 by the Loschmidt number, *i.e.*, of

$$(5) \qquad M' = 1{\cdot}85 \times 10^{-21} \text{ abs. units.}$$

This quantity, which is known as *Weiss's magneton*, is one-fifth as large as Bohr's magneton. More recent investigations (which we cannot discuss here), however, have explained the apparent discrepancy in favour of the quantum theory.

The concept of atomic magnetism dates back for more than a century. It is based principally on the well-known experimental fact that, even in the smallest fragments of a magnet, the algebraical sum of the quantities of magnetism contained therein vanishes. On the analogy of the equivalence between closed currents and magnetic shells which he had discovered, *Ampère* (1822) referred atomic magnetism to electric currents circulating in the individual molecules. A further important support for this concept was provided by *Faraday's* discovery (1845) that magnetism is a universal property of all matter.[3]

The *reality* of the so-called *molecular currents* conjectured in theory was first demonstrated in 1915 by *direct experiment* by *Einstein* and *de Haas*, who succeeded in causing a soft iron rod to *rotate* by *rapid reversals of magnetism*.[4] An effect of this kind could, indeed, be predicted on the basis of eqn. (3), according to which every change in the magnetic moment of an atom connotes a change in its angular momentum, so that, in virtue of the Principle of the *Conservation* of the Total Angular Momentum, a jerky rotation of the whole body must result as a *reaction effect*. In 1917, *Barnett* succeeded in producing the converse of the effect discovered by Einstein and de Haas, *viz.*, in *magnetizing a rod of soft iron by rapid rotation*.[5] It may be remarked incidentally that the results of observations on both effects give a value for the specific charge of an electron which is exactly half that obtained from deflection experiments on cathode rays.[6]

An important supplement to Einstein's and Barnett's experiments was furnished by *Gerlach* and *Stern* in 1922,

[3] The above-given explanation of diamagnetism from the point of view of the molecular current hypothesis is essentially that of *Wilhelm Weber* (1852). The development of Weber's ideas on an electron-theoretical basis is principally due to *Gans* and *Langevin*.

[4] *Verhandl. der Deutsch. Physikal. Gesellsch.*, **17** (1915), p. 152.

[5] *Phys. Rev.*, **10** (1917), p. 7.

[6] The theoretical explanation of the factor 1/2 is connected with similar difficulties in the interpretation of the so-called anomalous Zeeman effect, which are not yet solved.

when they passed a beam of rapidly moving *atoms of silver* through a *non-homogeneous magnetic field.* (A non-homogeneous magnetic field exerts upon a magnet placed in it not only directing, but also deflecting, forces.) This experiment enabled Gerlach and Stern to show that, just as was demanded by the *quantum theory* in opposition to the classical theory, only *quite definite orientations* of the atoms relative to the magnetic field are possible.[7] The value of the magnetic moment of a single silver atom was found from the experiment to be actually that of Bohr's magneton.[8]

The greatest difficulty with which the older theory of atomic magnetism had to contend was the question as to how *resistance-free circuits* could possibly exist.[9] The modern theory, it is true, is also not in a position to solve this difficulty, but it recognizes in it only a special case of a much more general difficulty, which arises from a *contradiction* between *Bohr's atomic theory* and *Maxwell's theory* of the electromagnetic field. It follows from *Poynting's theorem,* which is a necessary consequence of Maxwell's equations, and from the concept of convection currents, that every *electric charge* when *accelerated* in any way must *radiate electromagnetic energy.* According to Poynting's theorem, the same must obviously also hold for every electron revolving within an atom, in which case it would no longer be possible for the electrons to describe stationary orbits as postulated by Bohr's atomic theory. A key to the removal of this contradiction, however, already appears to be provided in the so-called correspondence principle of Bohr.[10]

§ 127. The Formation of Molecules.

Our previous considerations have shown that an atom, on *losing* any of its loosely bound electrons, becomes a

[7] As Debye and Sommerfeld have shown, it follows from the quantum theory that an n-quantum atom can only assume $2n$ positions in a magnetic field. The angle (ϕ) included between the normal to the orbital plane and the magnetic lines of force is given by the relation

$$\cos \phi = \pm \frac{n'}{n} \; (n' = 1, 2 \ldots n).$$

[8] For further details of Gerlach and Stern's researches, as well as of the problems discussed in this section, *cf.* W. Gerlach's report, "*Magnetismus und Atombau*" in the second volume of "*Ergebnisse der exakten Naturwissenschaften.*"

[9] At extremely low temperatures, moreover, a number of metals become what are termed *supraconductors*, and in them *Kamerlingh Onnes* and his collaborators have been able to produce currents practically without resistance. *Cf.* the report by Crommelin in the *Jahrb. f. Radioakt. u. Elektronik*, **19** (1922), p. 38.

[10] *Cf.* § 99.

positive ion; on the other hand, it is transformed into a *negative ion* when the number of electrons surrounding the nucleus is *raised* above the nuclear charge number. The actual existence of such negative ions is rendered evident by the phenomena of *electrolysis;* halogens are electrolytically separated as univalent negative ions, elements of the sixth group of the periodic system as divalent negative ions, and so on.

The formation of negative ions is presumably to be explained by ascribing to *neutral atoms* of certain elements a definite *affinity for electrons, i.e.*, an ability to *combine* with free electrons, with the *liberation of energy.* Just as energy is liberated when a positive ion recombines with an electron previously removed from the atom,[1] so also must energy be liberated when an atom which is neutral, but has an affinity for electrons, takes up an electron.

An empirical confirmation of this conception is apparently afforded, as *Franck* has shown,[2] by *spectroscopic* experiments which permit of a quantitative determination of the electron affinity by optical means, *i.e.*, a determination of the energy liberated in the formation of negative ions. For, under certain circumstances, this energy can be transformed into electromagnetic radiation energy, and a light-quantum thereby generated must obviously be equal to the sum of the energy liberated on the formation of an ion and the kinetic energy possessed by the electron before its binding. Since the kinetic energy may have any arbitrary value, which, however, must always be positive, we may expect the appearance of a *continuous spectral band with a limit on the side of longer wave-lengths*, and the frequency of this limit multiplied by h must be a measure of the electron affinity.

Since *oppositely charged* ions exert an *electrostatic force of attraction* upon each other, the combination of ions may lead to the formation of a *molecule of a chemical compound.* Observations of the *evolution of heat* accompanying chemical processes give us an idea, correct at least in order of magnitude, of the amount of *energy liberated* in the formation of such a compound. It must not be forgotten, however, that alterations in the state of aggregation frequently find expression in the observed evolutions of heat.[3] In the most usual chemical reactions, the evolution of heat referred to

[1] This energy is, of course, identical with the ionization energy.
[2] *Ann. der. Phys.*, 64 (1921), p. 673.
[3] Therein also lies the explanation of the observation of apparently negative evolutions of heat.

the gram-molecule amounts to between 10,000 and 100,000 calories,[4] *i.e.*, to between 4×10^{11} and 4×10^{12} ergs. On dividing by the Loschmidt number, we obtain the amount associated with a single molecule—approximately 7×10^{-13} to 7×10^{-12} ergs. By way of comparison, it may be remembered that the excitation energies of the alkali metals lie between 2×10^{-12} ergs and one and a half times that amount.

Up to now we have limited our considerations to molecules formed by the union of oppositely charged ions; such compounds are termed *heteropolar*. It is, however, possible for two neutral atoms to exert a force of attraction on each other, provided that the electrons in the one atom come sufficiently near the protons in the other; in that case, *homopolar* compounds are formed.

The most important examples of homopolar compounds are represented by those formed *from atoms of the same element*. It follows from measurements of the gas density (and from Avogadro's Law [5]) that, in the gaseous state, the molecules of *hydrogen*, *nitrogen*, *oxygen*, and the *halogens* are all *diatomic*, as are those of selenium and tellurium. Besides diatomic molecules, molecules of *eight atoms* have been found in *sulphur*, and of *four atoms* in *phosphorus* and *arsenic*, below certain limits of temperature.[6] The modification of oxygen known as *ozone* has been proved to be *triatomic*. Determinations of the vapour densities of the electropositive metals have invariably shown that these do not form molecules, but are, chemically speaking, *monatomic*. That such is also the case with the rare gases in general is evident from their chemical inertia.[7] Homopolar combinations between whole molecules are also obviously possible; and in this way arise the so-called *complex compounds* in which whole molecules appear to group together.[8]

[4] The heat evolved, for example, in the formation of one gram-molecule of water from H_2 and O amounts to 69,000 cal., and in the formation of CO_2 from C and O_2 to 96,000 cal.

[5] *Cf.* § 148.

[6] At higher temperatures, both phosphorus and arsenic vapours are apparently diatomic.

[7] The assumption of a formation of molecules in helium under certain circumstances has already been discussed in § 125.

[8] For example, both the ammonia molecule (NH_3) and the hydrochloric acid molecule (HCl) are, as a whole, neutral. Nevertheless, in consequence of the attraction exerted by the nitrogen ion, the NH_3 molecule is able to combine with the hydrogen ion in the HCl molecule, thereby forming the complex compound ammonium chloride (NH_4Cl). Indeed, the binding between the two molecules is so strong that, even in electrolysis, ammonium chloride behaves like a compound of a negative chlorine ion and a univalent cation NH_4, known as "ammonium," which in its combinations plays a precisely similar part to

The concept of the *electrostatic nature* of the *chemical forces* acting between atoms was suggested by *Berzelius* in 1812. Since, however, both homopolar compounds and complex compounds, as they gradually became known, appeared to contradict Berzelius' theory, it dropped into the background after ruling chemistry for several decades. It was not until the existence of both positive and negative charges in every atom was recently established that *Kossel*, in 1916, was able to revive Berzelius' theory by explaining the contradiction.

By the *absorption of light-quanta*, molecules can be brought into a state in which, because of their *increased energy*, they are enabled to *take part in chemical reactions* not possible for them in their normal state. The phenomena based on this absorption are termed *photochemical*. In 1912, *Einstein* formulated his *law of photochemical equivalence*, which states that the number of chemically reacting molecules must be equal to the number of absorbed light-quanta. Experimental measurements seem to confirm this principle.[9] Among the most important and best-known photochemical reactions are the formation of hydrochloric acid (HCl) from H_2 and Cl_2, the assimilation of carbon dioxide by plants, and the decomposition of the silver halides upon which photography is based.

Like the phenomena of fluorescence, photochemical reactions can also be "*sensitized*." That is to say, under certain circumstances the reactions can also be produced by light of wave-lengths which are not absorbed by the reacting molecules, provided only that a sensitizer which can absorb the light be added. Owing, presumably, to molecular collisions, the surplus energy absorbed is transferred from this sensitizer to the molecules of the substances concerned, which can then react. This appears especially clearly from an experiment of *Franck* and *Cario*, in which *mercury vapour* served as sensitizer for the *dissociation of hydrogen*.[10] It is known from thermochemical measurements that the energy required for the resolution of a molecule of hydrogen into

the likewise univalent ions of the alkali metals. The chlorine atom in the HCl molecule, like the hydrogen atom, can also join on to other molecules. The number of atoms contained in the molecules of organic compounds can be very large, *e.g.*, a molecule of cane-sugar consists of 45 atoms (corresponding to the formula $C_{12}H_{22}O_{11}$). A molecule of hæmoglobin, the red colouring matter in blood, is probably composed of 2,378 atoms, corresponding to the hypothetical formula $C_{758}H_{1203}N_{195}O_{218}FeS_3$.

[9] As a result of the frequent occurrences of intermediate reactions, most photochemical reactions appear very complicated.

[10] Cario and Franck, *Zeitschr. f. Phys.*, **11** (1922), p. 161.

its two atoms amounts to about 6×10^{-12} ergs. Dividing this by the elementary quantum of action, we obtain a frequency of about 9×10^{14} sec.$^{-1}$, or a wave-length of approximately 3,000 Å.U.[11] A non-excited hydrogen atom, however, cannot absorb any radiation of greater wave-length than its resonance line, which (as the first line of the Lyman series) lies at 1,216 Å.U., far in the ultra-violet. Franck and Cario introduced mercury vapour into the hydrogen, and submitted the mixture at sufficiently low temperature to the action of the rays of 2,537 Å.U. wave-length emitted by a quartz mercury lamp. Although, in accordance with what we have just said, these rays could not be absorbed by the hydrogen, yet a dissociation of the hydrogen took place. This was clearly proved by a reduction of metallic oxides which, as is known from experience, can only be effected at the temperature in question by monatomic, but not by diatomic, hydrogen. For the rest, the best-known instance of a photochemical sensitization is afforded by the *sensitization of a photographic plate* with suitable dyes which possess in the red the absorption lines lacking, in that region, in the silver halides.[12]

The converse of photochemical processes is represented by *chemico-luminescence*, due to a transformation of chemical energy into light-quanta. A well-known example of this is the oxidation of aldehydes at low temperatures, which is accompanied by an intensive luminescence. *Haber* and *Zisch* have shown that, on leading a stream of chlorine through sodium vapour, chemico-luminescence ensues; it can be recognized by the appearance of the characteristic *D*-lines of sodium at a temperature at which a luminescence due to heat is obviously excluded.[13]

Chemical reactions may be produced not only by the absorption of light-quanta, but also by *electron collisions*. The experimental investigation of these processes promises

[11] The corresponding potential is 3·5 volts. This "dissociation potential" actually represents the excess potential necessary for the generation of definite hydrogen lines over and above the values equivalent to the initial levels on emission. From this we must conclude that a hydrogen molecule must be split up into atoms before excitation can take place.

[12] Radiation absorbed by molecules can, however, be re-emitted by them in the form of fluorescent radiation. Hence it in no way follows that a molecule which absorbs an amount of energy sufficient for dissociation should necessarily be dissociated. At low pressures this hardly ever occurs. For example, diatomic iodine vapour, when excited by the green mercury line, radiates fluorescent light, although the energy of a green light-quantum is half as large again as the dissociation energy of the iodine molecule.

[13] Haber and Zisch, *Zeitschr. f. Phys.*, **9** (1922), p. 267.

to be of great importance, because it permits of an exact determination of the *minimum energy* requisite for the initiation of a chemical reaction.[14]

§ 128. The Structure of Crystals.

A special case of the formation of molecules is represented by the *formation of crystals*. The example of the rock-salt crystal, discussed in an earlier section, showed that crystals represent single *giant molecules*. On account of the absolutely regular arrangement in space-lattices, however, the different species of atoms are intercombined in very simple numerical proportions, so that the formation of crystals may be associated with a simple chemical formula and, correspondingly, a definite " molecular weight."

As was explained in an earlier section (§ 106), the possibility of crystal analysis is based on the fact that atoms of different mass diffract Röntgen rays differently. Diffracting power depends, in fact, on the number of electrons surrounding the atomic nucleus, and this number increases in general with the atomic weight. Instead of the original Laue photographs, line spectra of known wave-lengths are now used for the investigation of crystal structure, whence, by means of Bragg's equation, the constants characteristic of the structure can be derived. Moreover, the ground available for research in crystal analysis has been greatly extended through the application of the crystal powder method (already mentioned in § 106).

Most *elements* crystallize in *face-centred cubic lattices*, such as is obtained by replacing the Na and Cl atoms in the rock-salt crystal (*cf.* the earlier Fig. 86) by atoms of the same kind. Crystal structures of this nature have, for example, been found in copper, silver, gold, calcium, aluminium, lead, cobalt, nickel, rhodium, palladium, iridium, platinum, and also in nickel-steel and manganese-steel. The lattice constants of these crystals lie between $3{\cdot}5$ and $5{\cdot}5 \times 10^{-8}$ cm. The very great *stability* of the face-centred cubic arrangement is proved by its persistence, with unchanged lattice constant, even when the metals are sprayed ; thus, particles consisting of only a few hundred atoms probably still have the same crystal structure. *Space-centred* cubic crystal lattices have been discovered in some metals, such as lithium, sodium, tantalum, chromium, molybdenum, tungsten, and iron.

[14] Buch-Andersen has investigated the formation of ammonia (NH_3) from nitrogen and hydrogen on these lines : *Zeitschr. f. Phys.*, **10** (1922), p. 54.

Lattices of the rock-salt type are also found in numerous compounds, *e.g.*, in KCl, KBr, KI, LiF, and in others. Crystals of cæsium chloride represent a related type to that of rock-salt ; the Cs atoms by themselves form a simple cubic

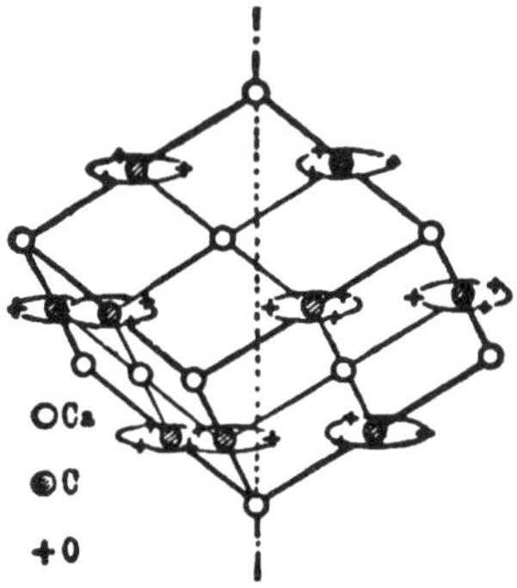

Fig. 107.—Structure of calcite.

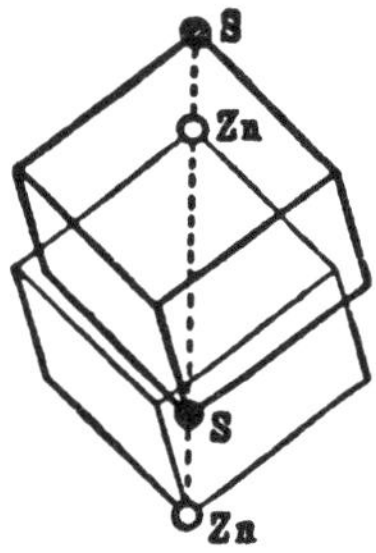

Fig. 108.—Structure of zinc blende.

lattice, as do the Cl atoms, and the points in one lattice coincide with the centres of the cubes in the other.

Besides the rock-salt type previously discussed, the calcite ($CaCO_3$) type is also of great importance (Fig. 107). We can

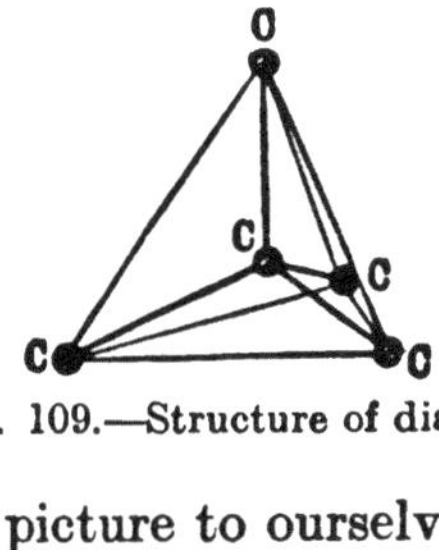

Fig. 109.—Structure of diamond.

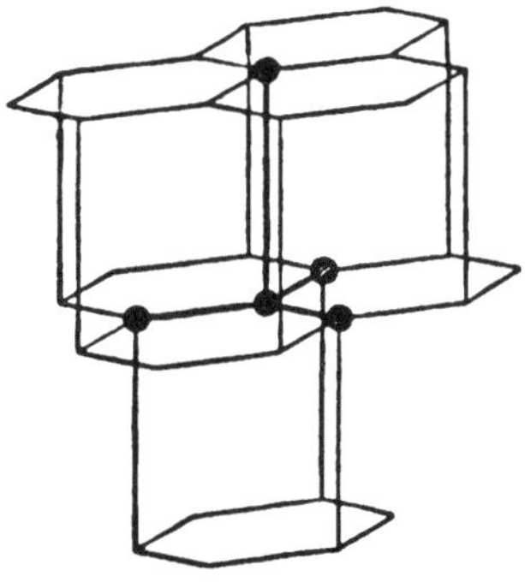

Fig. 110.—Structure of graphite.

best picture to ourselves the transition from one type to the other by imagining the rock-salt lattice to be placed with a diagonal of the cube vertical, and the Na and Cl atoms to be replaced by Ca and C atoms. Then we suppose three O atoms, to be added round each of the C atoms in a horizontal plane, *i.e.*, at right angles to the vertical cube diagonal. Finally, we must imagine a deformation of the rock-salt cube into a rhombohedron.[1]

A peculiar type of crystal is represented by *zinc blende* (ZnS), consisting of two interpenetrating, face-centred, cubic lattices of zinc and sulphur atoms (Fig. 108). Carborundum (SiC) and fluorspar (CaF_2) crystallize in the same way. On substituting carbon atoms for both the zinc and the sulphur atoms in zinc blende, we obtain the *diamond lattice* (Fig. 109). In this each atom represents the centre of a tetrahedron whose corners are occupied by four other carbon atoms. Silicon crystallizes like diamond.

Graphite, another modification of carbon, crystallizes *rhombohedrally* as shown in Fig. 110; and the crystals of arsenic, antimony, and bismuth are similar in structure. Numerous elements, finally, have been found to possess *hexagonal* lattices, *e.g.*, beryllium, magnesium, zinc, cadmium, titanium, zirconium, ruthenium, and osmium.[2]

§ 129. Molecular Spectra.

Although, as yet, we know nothing of the way in which the molecules of a polyatomic gas are constructed from the atoms, yet it seems at least plausible to assume that the atoms in the molecules can *oscillate* with respect to each other, and further that the molecules can execute *rotations* (apart from translatory motions). On these assumptions, the *internal energy of a molecule* would consist of three components: the internal energy of the atoms (E_{I}), the energy of the oscillatory motion (E_{II}), and the energy of the rotatory motion (E_{III}). Every *variation* in the internal energy of a molecule could therefore be represented in the form

$$(1) \qquad \Delta E = \Delta E_{\text{I}} + \Delta E_{\text{II}} + \Delta E_{\text{III}}\,.$$

If we wish to apply *Bohr's frequency condition* also to the variations of intra-molecular energy, we must assume that a variation of the energy by ΔE is connected with the emission or absorption of a light-quantum whose frequency is equal to $\Delta E/h$. From the point of view of the quantum theory, we must moreover suppose that the quantities E_{II} and E_{III} can only assume *quite definite values determined by quantum relations*, as we know already from previous considerations

[1] $MnCO_3$, $FeCO_3$, and $NaNO_3$ crystallize like calcite.

[2] A review of crystal structures hitherto investigated, with references to the literature, is to be found in "*X-Rays and Crystal Structure*," by Sir W. H. Bragg and Prof. W. L. Bragg, and also in Ewald's book, "*Kristalle und Röntgenstrahlen*" (Berlin: Springer, 1923).

to be the case for the quantity $E_{\rm I}$. Granted that this is so, however, the possible values of $\Delta E_{\rm II}$ divided by h must then also represent a manifold of quite definite frequency values ($\nu_{\rm II}$) ; and, likewise, the possible values of $\Delta E_{\rm III}$ divided by h will form another manifold of quite definite frequency values ($\nu_{\rm III}$). We have, in addition, the manifold of optical spectral lines ($\nu_{\rm I}$), due to those variations which only affect the internal atomic energy ($E_{\rm I}$). Hence, on adding the frequency values of these three manifolds, we obtain the manifold of frequency values of the *molecular spectrum* in the form

$$\nu = \nu_{\rm I} + \nu_{\rm II} + \nu_{\rm III} \,. \tag{2}$$

When, in particular, both $E_{\rm I}$ and $E_{\rm II}$ remain unchanged during a variation of the molecular energy, *i.e.*, when ν is equal to $\nu_{\rm III}$, we speak of a *pure rotation spectrum ;* the question then arises as to what are the quantum relations through which the manifold of possible values of the *rotational energy* can be determined. The simplest and most obvious assumption to make is that a molecule consists of *rigidly inter-connected atoms*, so that a diatomic molecule might be compared approximately with a dumbbell.

On this assumption, the molecular *moment of inertia* (I) may be regarded as *constant*. Again, it follows from the elementary formulæ of the mechanics of rigid bodies [§ 28 eqn. (14) and § 29 eqn. (1)] that the angular momentum (U) and the rotational energy ($E_{\rm III}$) are given by

$$U = w\,I\,, \quad E_{\rm III} = \frac{1}{2} I\,w^2\,, \tag{3}$$

where w denotes the angular velocity. If we postulate that (in accordance with § 92) the *angular momentum* multiplied by $2\,\pi$ shall be an *integral multiple of the elementary quantum of action*, we obtain from eqn. (3) the relations

$$w = \frac{n\,h}{2\,\pi\,I}\,, \tag{4}$$

and

$$E_{\rm III} = \frac{n^2\,h^2}{8\,\pi^2\,I}\,, \tag{5}$$

where n invariably denotes a whole number—a quantum number. Hence, whereas the energy of an electron in an atom is inversely proportional to the square of its quantum number, the rotational energy of the molecules is *directly* proportional to the square of the quantum number.

If n and s be two quantum numbers, we thus have

$$\nu_{III} = \frac{h}{8\pi^2 I}(n^2 - s^2)\,. \tag{6}$$

Now, it may also be shown by the same methods of reasoning as led to the selection principle of Rubinowicz (*vide* § 99) that the quantum number of the rotational energy can only vary by *unity*.[1] We have therefore

$$n^2 - s^2 = \pm 2n - 1\,. \tag{7}$$

Putting for brevity

$$\frac{h}{8\pi^2 I} = \nu^*\,, \tag{8}$$

we thus obtain

$$\nu_{III} = \nu^* (\pm 2n - 1)\,. \tag{9}$$

The frequencies of the individual lines in a pure rotation spectrum must therefore be *odd integral multiples* of the fundamental frequency defined by eqn. (8). An absorption spectrum possessing this property has actually been detected in some elements *far in the infra-red*, notably by *Rubens* and *Eva von Bahr* in *water vapour*.[2]

A pure rotation spectrum can, however, be *superimposed* upon one of frequency ν_{II} arising from a variation of the oscillatory energy. In that case, provided ν_{II} be considerably greater than ν^*, there appears a so-called *rotation-vibration spectrum* which, according to eqn. (9), is described by the formula

$$\nu = \nu_0 + \nu^* (\pm 2n - 1)\,. \tag{10}$$

A spectrum of this kind must thus consist of approximately *equidistant lines*,[3] the constant frequency difference between which must be the same as that in the pure rotation spectrum. The scheme of such a spectrum is represented in Fig. 111; underneath the individual lines are given the values of the numbers n and s corresponding to the transition $n \rightarrow s$, whereby, in accordance with the above-mentioned selection principle, s can only differ from n by unity.

Series of *absorption lines* not far in the infra-red (at wave-

[1] The numbers n and s must, of course, *not* be identified with the *principal* quantum number of earlier considerations.

[2] Rubens, *Berlin. Ber.*, 1913, p. 513; and Bahr, *Verh. der Deutsch. Physikal. Gesellsch.*, **15** (1913), pp. 731 and 1150.

[3] The equidistance is only approximate, because the intervals between the lines depend not directly on the frequency differences, but on differences between angles of deflection.

lengths of a few μ), corresponding to this scheme, have actually been detected and investigated in many substances, especially in the hydrogen compounds of the halogens (HCl, HF, HBr).[4] Several series of lines occur in the various substances, which is due to the fact that ν_0 may also be a

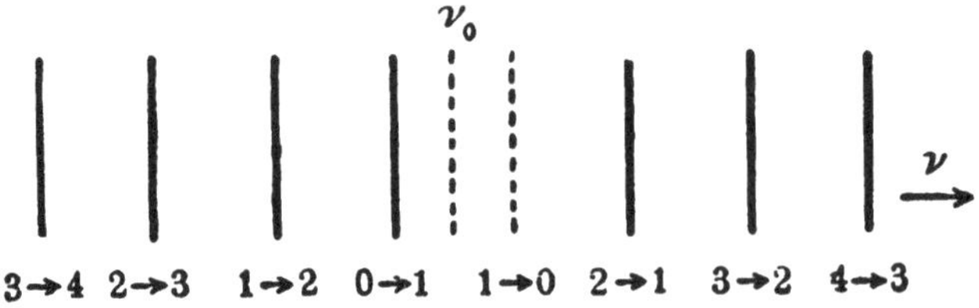

FIG. 111.—Scheme of the rotation-vibration spectrum.

so-called harmonic. In Fig. 112 is reproduced a photograph by *Imes* [5] of a series of absorption lines, a so-called *absorption band* belonging to hydrochloric acid and lying between 3 and 4 μ (30,000 and 40,000 Å.U.). The ordinates give the absorption expressed as a percentage, and the abscissæ angles of deflection measured in connection with the photograph,

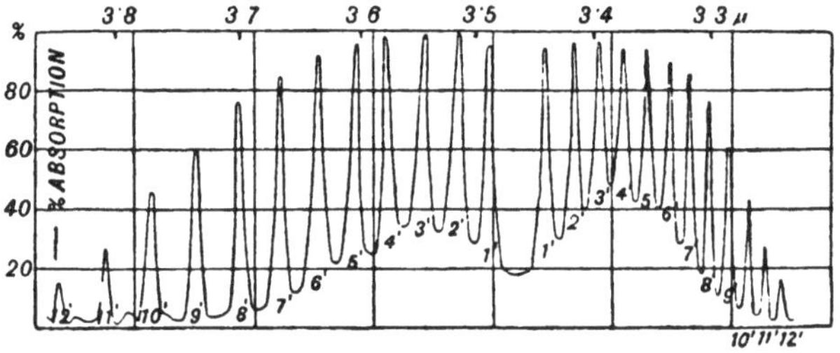

FIG. 112.—Absorption bands in HCl.

which was taken with a reflection grating. The maxima along the curve represent the absorption lines, and the corresponding wave-lengths (in μ) are given above them; below the maxima are inserted the relevant values of the number n following from eqn. (10). The form of the curve and a comparison with Fig. 111 (in which the transitions are given for the case of emission) show that the state for which $n = 0$ is not represented as an *initial state* in the *absorption spectrum*. We must therefore assume that a non-excited

[4] The experimental investigation of rotation-vibration spectra is principally due to Rubens, Eva von Bahr, Sleator, and Imes.
[5] *Astrophys. Journ.*, 50 (1919), p. 251.

molecule *never*, or only *very rarely*, exists in a *rotationless* state.

The value of ν_0, the so-called " vibration frequency," evidently depends on the *mass* of the oscillating atom,[6] and it is thus to be expected that *isotope doublets* (or multiple lines) will occur in the rotation-vibration spectra of com-

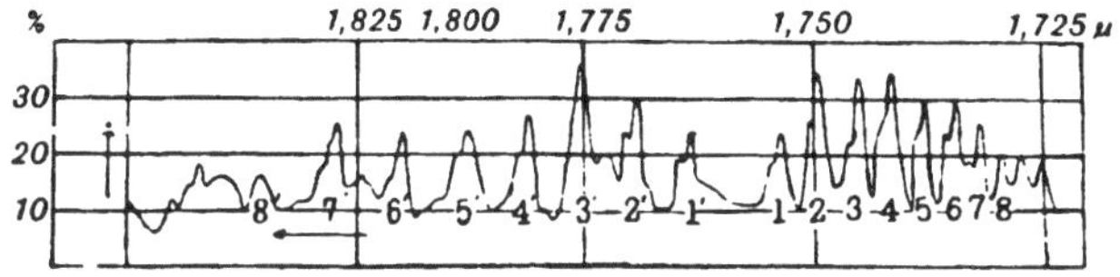

FIG. 113.—Isotope effect in the rotation-vibration spectrum of HCl.

pounds of mixed elements.[7] Doublets of this kind, due to the isotopes of *chlorine*, are in fact clearly recognizable in Fig. 113, in which is reproduced an absorption band of HCl at 1·76 μ, according to Imes.[8] In the neighbourhood of the principal maxima due to the Cl_{35} atoms, the curve exhibits

3884 3872 3862 3855 3850

FIG. 114.—The so-called cyanogen bands in the nitrogen spectrum.

weaker subordinate maxima of smaller frequency, occasioned by the Cl_{37} atoms.[9]

Inasmuch as the values both of ν_{II} and of ν_{III} lie in the infra-red, they are relatively small as compared with optical frequencies. If, therefore, changes in the molecular energy

[6] For ν_0 corresponds to the variation in oscillatory energy of the atoms combined in the molecule.

[7] The existence of isotopes also finds expression in a difference in the molecular moments of inertia.

[8] This band corresponds to a " harmonic," for the fundamental frequency in Fig. 113 is double that in Fig. 112.

[9] The difference in frequency between the two lines of an isotope double of Cl is theoretically given by

$$\frac{\delta\nu}{\nu} = \frac{1}{2}\left(\frac{1}{35} - \frac{1}{37}\right).$$

It follows that the wave-length difference of the doublet in the band reproduced in Fig. 113 is 13 Å.U.

be accompanied by a *variation in the intra-atomic energy* (E_i) which, when divided by h, gives an *optical* frequency, a series of closely packed lines will be produced in the optical region of the spectrum, forming a so-called *band*. The positions at which the lines appear particularly closely packed are called the *heads of the bands*. The whole assembly of bands forms the so-called *band spectrum* of the molecule concerned. By way of illustration, a portion of the band spectrum of *nitrogen* (the so-called cyanogen bands) is reproduced in Fig. 114 ; the numbers underneath the figure give the wave-lengths in Å.U.[10]

The theory of band spectra (into which we cannot enter here)[11] is very complex, since the electron transitions generating the optical frequencies (ν_i) will obviously also produce variations in the molecular moment of inertia. The most important laws of band spectra were discovered empirically by *Deslandres* in 1887.[12] An explanation of these spectra from the point of view of the quantum theory was first attempted by *Bjerrum* in 1912, or before the birth of Bohr's theory of the atom.[13] In 1916 *Schwarzschild* applied Bohr's frequency condition and the principle of quantizing the angular momentum to molecular spectra, and since then the theory has been developed principally by *Heurlinger*, *Lenz*, and *Kratzer*, although no real finality has as yet been reached.

§ 130. Dispersion.

In consequence of the revolutions of the electrons, the oscillations of the atoms, and the rotations of the molecules, numerous "*characteristic frequencies*" are to be observed in the molecules of all substances : these frequencies lie in the

[10] The *many-line spectrum of the hydrogen molecule* is to be included among band spectra. The absence of band heads in it can be explained by the exceptional smallness of the moment of inertia of the hydrogen molecule. Helium also shows a many-line spectrum, which must be referred to a transitory formation of molecules : *cf.* the remarks in § 125 on metastable helium atoms.

[11] *Cf.* Sommerfeld's "*Atomic Structure and Spectral Lines*," Ch. vii, §§ 3–5 ; and A. Kratzer's "*Der heutige Stand der Theorie der Bandenspektren*," in "*Ergebnisse der exakten Naturwissenschaften*," Vol. I.

[12] The most important regularity discovered by Deslandres is to the effect that, with very fair approximation, the individual lines in a band may be represented in the form

$$\nu = \nu_1 + an^2,$$

where ν_1 is the frequency of the head of the band, a a constant characteristic of the band in question, and n successively assumes all integral values.

[13] Bjerrum, it is true, quantized the rotational energy instead of the angular momentum, and, also incorrectly, equated the frequencies of the spectral lines simply to the revolution numbers of the rotation.

infra-red, in the optical region of the spectrum, and in the ultra-violet. Hence, when a molecule is penetrated by a light-wave of a frequency which does not differ greatly from a characteristic frequency, a kind of *resonance* ensues, and to this is to be ascribed the well-known phenomenon of *dispersion*, discovered by *Newton* in 1672.

It has not yet been possible to provide a satisfactory explanation of dispersion from the point of view of the quantum theory.[1] The older theory, which assumed quasi-elastic forces between the electrons in an atom, led to a dispersion formula that agrees well with experimental results. This formula [2] gives the following expression for the square of the *refractive index* :—

$$n^2 = 1 + \sum \frac{N_i e^2 z_i}{\pi m_i (\nu_{0i}^2 - \nu^2)} . \quad (1)$$

In this equation, different kinds of electrical charges are assumed to exist and they are distinguished by the subscript i ; $z_i e$ is the quantity of electricity in such a charge, m_i its mass, ν_{0i} the frequency of its characteristic oscillation and N_i the number of these charges in a unit-volume.

Eqn. (1) shows in the first place that, with increasing values of ν, the individual summands likewise increase, both when $\nu > \nu_{0i}$ and when $\nu < \nu_{0i}$. (For, in the first case, the summand is positive and its denominator decreases, and, in the second case, the summand is negative and its denominator increases.) This is actually confirmed by experience, for, in almost all substances, the refractive index in the visible region of the spectrum *increases from the red towards the violet end.*

On the other hand, a very different behaviour must be expected when the frequency of the light differs but little from a characteristic frequency of the substance. In accordance with eqn. (1), as the increasing value of ν approaches the value ν_{0i}, the summand in question becomes larger and larger, and would finally, if eqn. (1) were actually obeyed exactly, become infinite. As ν increases beyond ν_{0i}, it follows from eqn. (1) that n^2 must first become negative, *i.e.*, the refractive index would be imaginary. It is not until ν

[1] *Cf.* Ladenburg and Reiche, "*Absorption, Zerstreuung und Dispersion in der Bohrschen Atomtheorie,*" *Naturwissenschaften*, **11** (1923), p. 584.

[2] If, in particular, we equate to zero the frequency ν in eqn. (1), we obtain the *dielectric constant* in the narrower sense, characterizing the electro*static* field. The formula thus specialized gives a description of the *deviations from Maxwell's relation.* This relation is obeyed exactly only in such wave-length regions as contain no characteristic oscillations of the substance in question.

becomes larger than ν_{0i} by a definite amount that n will again become real, and, starting from zero, again increase.

In the region of a continuous spectrum near the frequency of a characteristic oscillation, light with a frequency larger than ν_{0i} will therefore be less refracted than light with a frequency smaller than ν_{0i}. This is, of course, a contradiction to the rule, generally confirmed by experience, that the refractive index increases with the frequency. If, therefore, a characteristic oscillation of a substance falls within the *visible region* of the spectrum, the successive order of colours in the latter may differ from the usual one. An *anomalous dispersion* of this nature is actually observed in some substances, notably in *iodine vapour* and *fuchsine*. Unlike the older experimental physics, however, the electron theory does not discriminate between normal and anomalously dispersing substances. Indeed, according to the electron theory, *every* substance should exhibit the phenomena both of normal and of anomalous dispersion. If the dispersion in the visible part of the spectrum be " normal " for most substances, the electron theory only draws the conclusion that the characteristic oscillations responsible for anomalous colour dispersion lie *outside* the visible region of the spectrum in the majority of substances.

Waves whose frequencies differ very little from a characteristic frequency are very strongly reflected and absorbed by the substance in question. This phenomenon, which has been observed in many substances, is known as *selective reflection* and *absorption*. Observations on the phenomenon render possible a determination of the values of the characteristic frequencies, and, with the help of the dispersion formula, they also give clues as to the nature and the number of the charges responsible for the dispersion. Indeed, as might be expected, observations have shown that characteristic frequencies in the ultra-violet arise from valency electrons, whereas those in the infra-red, which in many substances extend into the region of short electric discharge waves, are to be referred to atomic oscillations.

PART IV

THEORY OF HEAT

Chapter XVI

STATISTICS

§ 131. Statistical Probability.

By resolving the object of physical processes, so-called *matter*, into an *assembly of individual particles*, it becomes possible for theoretical physics to investigate all physical phenomena and states by means of methods which are known quite apart from the province of physics as *statistical methods*. Among the manifold kinds of problems known to general statistics, problems of the following nature are of particular importance to physics. Let us imagine an assembly of individuals to be given, and let us consider one definite property of the individuals ; [1] we suppose that this property can be described in terms of one or more definite quantities which can be *graduated in intervals* in some way or another. In this way we obtain a simple or manifold plurality of states to be distinguished from each other. The question then arises as to how, under any given conditions, the *individuals* will be *distributed among the different states*.

Thus, for instance, the children born in a year in a large town represent an assembly of individuals. One of their many properties which we may consider is their weight at birth. This property can be described by a single quantity, the weight, which may be graduated in various ways, *viz.*, in tenths of kilograms or in decagrams, *etc.* A definite state is then represented, for example, by a weight of between 3·2 and 3·3 kg. Another property of the children, for the description of which, however, two quantities are necessary, is the age of their parents at the time of birth. In this case we have to deal with a double plurality of states, the age both of the father and of the mother being graduated, for instance, in years or months. A definite state would then be that in which the father's age lies in the interval between thirty-three and thirty-four years, and the mother's age

[1] *I.e.*, one among many ; *e.g.*, in human beings the height of the body, or the age, or the financial status, *etc.*

between twenty-one and twenty-two years. A further example of a distribution of individuals among states is represented by the distribution of the population of a country according to their yearly income. The distribution is then bound by the condition that, if we multiply the number of people belonging to each interval of income by the value of the income for that interval, and then take the sum over all the intervals, the result must give the total annual income of the population.[2]

Just as general in character as the problem of individual distribution are some concepts to which the treatment of this problem leads. Let us consider an assembly of N individuals, and suppose the different intervals of state to be distinguished by indices ; thus N_i will be the number of individuals in the state characterized by the index i for a certain condition of distribution of states. A definite *distribution of states* is then described by the numbers N_1, N_2, ... N_i, . . . , *etc.*, with the qualification, however, that the sum of all the numbers N_i over all the states must give the total number of all the individuals. We must therefore have

$$\sum N_i = N \,. \tag{1}$$

Should we include states which contain no individuals at all,[3] this naturally would not at all affect the consideration, because for such states the corresponding number N_i vanishes.

A distribution of states is therefore completely determined as soon as we know *how many* individuals the different states contain. *Which* the individuals are is a matter of indifference. Hence, if two individuals in different states are interchanged, *no new* distribution arises, for the numbers N_1, N_2, . . . N_i, . . . , *etc.*, are not altered. A new "*complexion*" does, however, arise ; and thus, in general, a definite distribution of states is associated with many complexions. The number of complexions is termed the *statistical probability* of the distribution of states in question.

The statistical probability, which must always be a positive integer, is obviously given by the number of *permutations* of N individuals taken altogether, omitting, however, those arising from an interchange of two individuals in the same state. Now the total number of all possible permutations

[2] Naturally not with absolute accuracy, for the intervals have been chosen as finite.

[3] This would be the case, for instance, if we took into consideration weights below 1 kg. or above 8 kg. in discussing the weights of newly born infants.

of x elements is $x\,!$ (" factorial x "), where, by definition of the symbol,

(2) $$x\,! = 1\,.\,2\,.\,3\,\ldots\,(x-2)\,.\,(x-1)\,.\,x\,.$$

The total number of permutations of N individuals taken altogether is therefore $N\,!$. The number of permutations resulting from an interchange of the individuals in the first state is $N_1\,!$, *etc.* Hence, we obtain the statistical probability of a given distribution of states on dividing $N\,!$ in the first place by $N_1\,!$, because all the combinations arising from interchanges between individuals in the first state represent, as a whole, only one complexion (and not $N_1\,!$). Then we have to divide by $N_2\,!$, and so on. If W be the statistical probability, we therefore have [4]

(3) $$W = \frac{N\,!}{N_1\,!\;N_2\,!\ldots N_i\,!\ldots}\,.$$

We now introduce new quantities, defined by the equation

(4) $$w_i = \frac{N_i}{N}\,;$$

i.e., the quantity w_i gives the fraction of the individuals occurring in the state characterized by the index i. According to eqn. (1), we must have

(5) $$\sum w_i = 1\,.$$

The expression for the statistical probability can be considerably simplified with the help of an easily derivable mathematical *formula of approximation.* This formula concerns the logarithm of $n\,!$. We suppose the curve

$$y = \log_e x$$

to be constructed in an x-y-coordinate system, and at all points whose x-coordinate is *integral,* but not greater than a whole number n, the corresponding ordinates to be drawn. We then obtain (Fig. 115) a figure $A\ B\ C$, the coordinates of the point A being $x = 1$, $y = 0$; those of the point B,

[4] *E.g.*, in an assembly of four individuals, *a b c c*, the two *c*'s form a subgroup of two individuals, because an interchange between them does not lead to any new complexion. The possible complexions are

a b c c	*a c b c*	*a c c b*
b a c c	*b c a c*	*b c c a*
c a b c	*c a c b*	*c b a c*
c b c a	*c c a b*	*c c b a*

The number of these complexions is, in fact, equal to $4\,!\,/\,2\,!$, or 12.

$x = n$, $y = 0$; and those of the point C, $x = n$, $y = \log_e n$. The area of this figure is given by

$$F = \int_1^n \log_e x \, dx \, .$$

We carry out the integration, in the usual way, with the help of the identity [5]

$$(6) \qquad \log_e x \, dx = d\,(x \log_e x) - dx \, ,$$

and find

$$(7) \qquad F = n \log n - n + 1 \, .$$

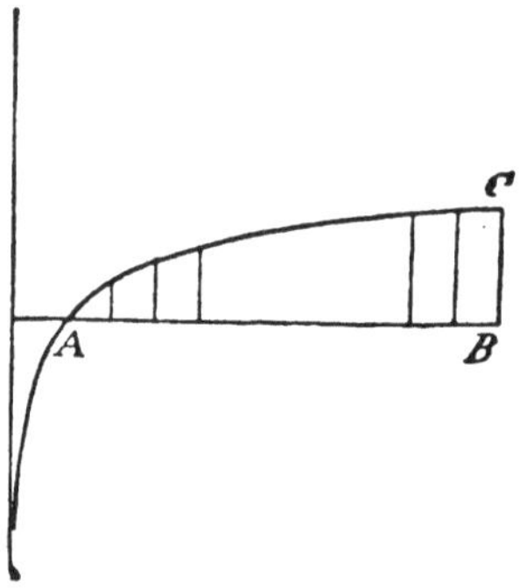

FIG. 115.

On the other hand, the area of the figure $A\ B\ C$ is divided into a number of *strips* by the ordinates from the n points with whole number abscissæ. Apart from the lower values of x, these strips can be regarded as small *rectangles*, for then the logarithmic curve runs approximately parallel to the x-axis. The width of these rectangles is always unity, and their height $\log_e x$ ($x = 2$ to n). On this assumption, the area of the figure appears as the sum

$$(8) \qquad \begin{aligned} &\log_e 2 + \log_e 3 + \ldots + \log_e (n-1) + \log_e n \\ &\quad = \log_e [1 \,.\, 2 \,.\, 3 \ldots (n-1) \,.\, n] = \log_e (n\,!). \end{aligned}$$

If n be large, the error due to our regarding the strips as rectangles even for small values of x vanishes. Hence, for large values of n, expressions (7) and (8) may be taken as identical. Since the number 1 may then be disregarded in comparison with n, for large values of n we find the approximate formula [6]

[5] We have

$$dx = x \frac{1}{x}\, dx = x\, d\,(\log_e x)\, .$$

[6] This formula represents a special case of a more exact approximation formula, known as *Stirling's formula* (*vide* textbooks on probability). According to Stirling's formula,

$$(9b) \qquad \log_e (n\,!) = n \log_e n - n + \frac{1}{2} \log_e (2\, n\, \pi)\, .$$

For large values of n, this formula is extraordinarily accurate; even for $n = 10$ the error is less than 1%. Eqn. (9) is, of course, less accurate than Stirling's formula, the last term in the latter being omitted. For large values of n, however, it is easy to see that this term is quite unimportant.

(9) $$\log_e (n\,!) = n \log_e n - n\,.$$

This approximate formula may be applied to eqn. (3), if we assume that the large majority of individuals occur in states each of which contains a large number of them. For, even if there be only a few individuals in any state k, but many individuals in a state h, $\log_e (N_k\,!)$ may simply be neglected as compared with $\log_e (N_h\,!)$, inasmuch as, with increasing x, $x\,!$ increases more rapidly than e^x.[7] Applying, therefore, the approximation formula to eqn. (3), we find

$$\log_e W = N \log_e N - N - \sum N_i \log_e N_i + \sum N_i\,,$$

or, bearing in mind eqns. (1) and (4),

$$\begin{aligned}\log_e W &= N \log_e N - \sum N w_i (\log_e N + \log_e w_i) \\ &= N \log_e N - N \log_e N \sum w_i - N \sum w_i \log_e w_i\,.\end{aligned}$$

Finally, taking eqn. (5) into account, we obtain the simple relation

(10) $$\log_e W = - N \sum w_i \log_e w_i\,.$$

This formula determines the statistical probability for a given distribution of states.

Now, the larger the statistical probability of any distribution, the larger also is the probability that any given distribution will change into it. If A, B, C are three distributions, and their statistical probabilities are W_1, W_2, W_3, where $W_2 < W_1 < W_3$, the transition from A to C must be more probable than that from A to B. Since, however, these distributions are quite arbitrary, the probability is that, on a change of distribution, the *statistical probability* of a system will *increase*.

On the other hand, it is evident that there must be an *upper limit* to the statistical probability of a system, for, by definition, this cannot possibly be larger than the *total number of all complexions*. This number is given by the Nth power of z, when N individuals are distributed over z states.[8]

[7] We have

$$\frac{x\,!}{e^x} = \frac{1}{e} \cdot \frac{2}{e} \cdot \frac{3}{e} \cdots \frac{x-1}{e} \cdot \frac{x}{e}.$$

From the third term onwards, however, all the factors are larger than unity and increase throughout.

[8] For, if there were only two individuals, there would be z possible states for the first of them. To each single possible state, however, there correspond z further complexions, because, even when the state of the first individual is fixed, there are still z possible states for the second. With two individuals, therefore, the total number of all complexions is given by z^2; and hence, in general, with N individuals by z^N.

The number of *possible* complexions is, of course, smaller when the distribution is subject to conditions which can be satisfied by only some of the complexions. In every case, however, there is an upper limit, and this limit determines the *most probable distribution*, towards which the system *apparently strives*, and which, on account of its particularly *large number of complexions, occurs most frequently.*[9]

§ 132. The Modulus of Distribution.

The most important condition governing a distribution of individuals over different states is as follows : Let the system possess a certain *amount* B of any quantity, this amount being equal to the *sum of the component amounts* associated with the separate individuals. Let the component amount associated with the individual depend, on the one hand, on the state in which the individual occurs, and on the other hand, on quantities which, at any moment, are the same for all individuals in the system, and which may be termed the *parameters of the system*. Denoting these parameters, r in number, by $u_1, u_2, \ldots u_r$, and by β_i the component amount associated with an individual in the definite state characterized by the index i, then β_i is a function only of the parameters, so that

$$\beta_i = f(u_1, u_2, \ldots u_r) . \qquad (1)$$

Again, if N_i be the number of individuals in the state characterized by the index i, we must have

$$B = \sum N_i \beta_i . \qquad (2)$$

We meet, now, with a *special* case of great importance when the distribution is such that its *statistical probability* represents an *extreme value* for those *variations of the distribution* in which the total amount of the quantity and the number of individuals remain constant. A distribution which

[9] The correctness of this theorem is, indeed, confirmed by experiences of everyday life. If we place layers of blue and yellow powder in a glass tube, the contents of the tube after a little shaking will appear uniformly green. But, however much we shake, we shall never be able to make the lower layer once more appear blue and the upper yellow. Order changes to disorder by itself, but the reverse process, the fortuitous appearance of order, is only to be observed with a very small number of individuals. The chance that four men who die successively in a town depart from this life in the alphabetical order of their names is, on an average, 1 in 24 ($24 = 4!$). In a series of ten deaths, the probability of such an alphabetical arrangement is less than a millionth, in a series of twenty deaths less than 1 in 10^{18}, and in a series of thirty deaths less than 1 in 10^{30}.

satisfies both this main condition and the subordinate conditions is termed a *canonical distribution*. Since it is more convenient to calculate with the logarithm of the probability rather than with the probability itself, the main condition may be expressed by the equation

$$(3) \qquad \delta\,(\log_e W) = 0\,;$$

and the subordinate conditions by

$$(4) \qquad \delta N = 0, \qquad \delta B = 0\,.$$

In accordance with § 131 eqns. (4) and (5), the constancy of the number of individuals may be expressed by the formula

$$(5) \qquad \sum \delta w_i = 0\,.$$

Since the variation refers only to the quantities w_i, we have

$$(6) \qquad \delta B = \sum \beta_i\, \delta N_i\,.$$

On account of the constancy of N, the subordinate condition of the vanishing of the variation of B may be expressed by the equation

$$(7) \qquad \sum \beta_i\, \delta w_i = 0\,.$$

It follows from § 131 eqn. (10) that, since N is constant, the variation of the probability logarithm is given by

$$(8) \qquad \delta\,(\log_e W) = -\,N \sum \log_e w_i\, \delta w_i - N \sum \delta w_i\,.$$

Since the last sum vanishes [by eqn. (5)], the principal condition of a canonical distribution is expressed by the equation

$$(9) \qquad \sum \log_e w_i\, \delta w_i = 0\,.$$

Now, in accordance with the rules of the calculus of variations, we can combine this equation, expressing the principal condition, with eqns. (7) and (5), expressing the subordinate conditions, to form a single equation. This we do by multiplying each of eqns. (7) and (5) by a factor (at first left undetermined), and then adding them to eqn. (9). (The process is the same as in Lagrange's method of undetermined multipliers.[1]) Denoting the two undetermined factors by λ_1 and λ_2, we thus obtain

$$\sum (\log_e w_i + \lambda_1\, \beta_i + \lambda_2)\, \delta w_i = 0.$$

As the variations are quite arbitrary, this equation can only be satisfied when the expression in brackets vanishes.[2] We must therefore have

[1] *Vide* § 20.
[2] *Cf.* the analogous considerations in § 20.

$$\log_e w_i = -\lambda_2 - \lambda_1 \beta_i$$

or,

(10) $$w_i = e^{-(\lambda_2 + \beta_i \lambda_1)},$$

We shall now introduce two new quantities, ψ and Θ, defined by the relations

(11) $$\lambda_1 = \frac{1}{\Theta}, \qquad \lambda_2 = -\frac{\psi}{\Theta}.$$

Then

(12) $$w_i = e^{\frac{\psi - \beta_i}{\Theta}}.$$

The two quantities ψ and Θ are determined in accordance with the subordinate conditions in such a way that

(13) $$N \sum w_i \beta_i = B$$

and

(14) $$\sum w_i = 1.$$

From eqns. (14) and (12) we obtain the relation

(15) $$e^{-\frac{\psi}{\Theta}} = \sum e^{-\frac{\beta_i}{\Theta}},$$

while we find from eqns. (13) and (12) that

$$B = N \sum \beta_i \cdot e^{\frac{\psi - \beta_i}{\Theta}} = \frac{N \sum \beta_i e^{-\frac{\beta_i}{\Theta}}}{e^{-\frac{\psi}{\Theta}}},$$

or, by eqn. (15),

(16) $$\frac{\sum \beta_i e^{-\frac{\beta_i}{\Theta}}}{\sum e^{-\frac{\beta_i}{\Theta}}} = \frac{B}{N}.$$

This equation determines the quantity Θ, which is known as the *modulus of distribution*. Once that this is given, ψ can be calculated from eqn. (15). Introducing, for the simplification of subsequent calculations,

(17) $$N\psi = \Psi,$$

we find by eqn. (15)

(18) $$\Psi = -N\Theta \log_e \sum e^{-\frac{\beta_i}{\Theta}}.$$

The value of the probability logarithm for the case of canonical distribution follows from eqn. (12), and from § 131 eqn. (10):—

$$\log_e W = -N \sum \frac{\psi - \beta_i}{\Theta} e^{\frac{\psi - \beta_i}{\Theta}},$$

or,

$$\log_e W = -\frac{N\psi}{\Theta} \sum e^{\frac{-\beta_i}{\Theta}} + \frac{N}{\Theta} \sum \beta_i e^{\frac{\psi - \beta_i}{\Theta}}.$$

According to eqns. (12) and (14), the first sum is equal to unity, and the second sum, by eqns. (12) and (13), is equal to B/N. Hence, it follows from eqn. (17) that, for a canonical distribution,

(19) $$\log_e W = \frac{B - \Psi}{\Theta}.$$

Since β_i depends only on the parameters,[3] we see from eqn. (18) that Ψ, and hence also ψ, can be a *function only of the parameters and the modulus of distribution.* In accordance with eqn. (16), the same thing applies also to the quantity B/N (the average value of the component amount), and, by eqn. (19), to the probability logarithm.

Now, in virtue of its definition, a canonical distribution is characterized by the constancy of the probability logarithm for small variations of the distribution. Obviously, therefore, an *especially large number of complexions* will have a value of the probability logarithm corresponding to a canonical distribution. Since, however, the number of complexions by means of which a distribution can be realized is simply the statistical probability, the extreme value of the *logarithm of the probability* for a canonical distribution must be a *maximum.*

A canonical distribution thus represents at the same time the most probable and the most frequent distribution. Moreover, it is distinguished by definition by the maintenance of its canonical character for infinitesimal variations. Otherwise, in accordance with what we have said above, variations in distribution are accompanied by a movement in the direction of increasing statistical probability rather than in the reverse direction. *Infinitesimal canonical variations,* on the other hand, occur without any change in the

[3] *Vide* eqn. (1).

statistical probability, and thus appear to be equally probable in either directions ; they are apparently *reversible.*[4]

§ 133. The Canonical Distribution of Energy.

In the following discussion, we shall apply our previous statistical considerations by identifying the quantity B with *energy,* for which the *Law of Conservation* holds good. The canonical distribution of energy is therefore determined by the formula

$$(1) \qquad w_i = e^{\frac{\psi - \epsilon_i}{\Theta}},$$

where ψ and Θ must satisfy the two conditions

$$(2) \qquad \sum e^{\frac{\psi - \epsilon_i}{\Theta}} = 1$$

and

$$(3) \qquad N \sum \epsilon_i e^{\frac{\psi - \epsilon_i}{\Theta}} = E\,.$$

If, now, the parameters u_1, u_2, *etc.*, are varied by the infinitesimal amounts du_1, du_2, *etc.*, and if, as we have assumed, the quantity ϵ_i depends only on the parameters, then the variation in the energy of a single individual is given by

$$\frac{\partial \epsilon_i}{\partial u_1} du_1 + \frac{\partial \epsilon_i}{\partial u_2} du_2 + \ldots .$$

For an infinitesimal variation of the parameters, the total variation in energy of a system in a state of canonical distribution is given by a summation over all the individuals. Hence, the total variation of the energy is represented by the expression

$$\sum \left\{ N_i \frac{\partial \epsilon_i}{\partial u_1} du_1 + N_i \frac{\partial \epsilon_i}{\partial u_2} du_2 + \ldots \right\}.$$

In accordance with the axiomatic *Principle of the Conservation of Energy,* this quantity must be equal and of opposite sign to the *work* performed by the system as a result of the variation of the parameters. We thus obtain the relations

$$(4) \quad dA = - N\, du_1\, w_i \sum \frac{\partial \epsilon_i}{\partial u_1} - N\, du_2 \sum w_i \frac{\partial \epsilon_i}{\partial u_2} - \ldots .$$

[4] The variations must be regarded as being infinitesimal ; *cf.* the analogous considerations on the Principle of Virtual Displacements (§ 18).—For an exact proof of the assertion that the canonical distribution is also the most probable, *cf.* in particular the article by P. and T. Ehrenfest in the *Mathematische Enzyklopädie* IV., 2, II., part 6.

The summands in this equation may, however, be greatly simplified with the help of eqn. (2). In doing this, we must remember that ϵ_i is a function only of the u values, whilst ψ is a function only of the u values and Θ. We may therefore regard the summand expressions in eqn. (2) as functions of the u values and Θ, and hence differentiate them partially with respect to the parameters. In this way we obtain

$$\frac{1}{\Theta}\sum \frac{\partial(\psi - \epsilon_i)}{\partial u_1} e^{\frac{\psi - \epsilon_i}{\Theta}} = 0 ,$$

or,

$$(5) \qquad \sum \frac{\partial \epsilon_i}{\partial u_1} e^{\frac{\psi - \epsilon_i}{\Theta}} = \frac{\partial \psi}{\partial u_1} \sum e^{\frac{\psi - \epsilon_i}{\Theta}} ,$$

or, taking eqns. (1) and (2) into account,

$$(6) \qquad \sum \frac{\partial \epsilon_i}{\partial u_1} w_i = \frac{\partial \psi}{\partial u_1} .$$

Eqn. (4) thus becomes, on replacing $N\psi$ by Ψ,

$$(7) \qquad dA = -\frac{\partial \Psi}{\partial u_1} du_1 - \frac{\partial \Psi}{\partial u_2} du_2 - \ldots$$

Since the quantity Ψ thus plays the same part for the system as a whole as potential energy would, it is called the *free energy* of the system. [It follows, in fact, from § 132 eqn. (19) that Ψ and Θ must always have the same dimensions as B, for $\log_e W$ must, of course, always be a pure number.]

We can write eqn. (7) in another important form, when we remember that, in accordance with what was said above, Ψ must be determined by the parameters and the modulus of distribution. Thus

$$(8) \qquad d\Psi = \frac{\partial \Psi}{\partial \Theta} d\Theta + \frac{\partial \Psi}{\partial u_1} du_1 + \frac{\partial \Psi}{\partial u_2} du_2 + \ldots$$

Hence, by eqn. (7),

$$(9) \qquad dA = -\,d\Psi + \frac{\partial \Psi}{\partial \Theta} d\Theta .$$

But, by § 132 eqn. (18),

$$(10) \qquad \Psi = -\,N\Theta \log_e \sum e^{-\frac{\epsilon_i}{\Theta}} ,$$

and therefore

$$\frac{\partial\Psi}{\partial\Theta} = -N\log_e \sum e^{-\frac{\epsilon_i}{\Theta}} - \frac{N\Theta\sum \epsilon_i e^{-\frac{\epsilon_i}{\Theta}}}{\Theta^2 \sum e^{-\frac{\epsilon_i}{\Theta}}},$$

or, bearing in mind eqn. (10) and § 132 eqn. (16),

(11) $$\frac{\partial\Psi}{\partial\Theta} = \frac{\Psi}{\Theta} - \frac{E}{\Theta}.$$

Hence, by eqn. (9),

(12) $$dA = -d\Psi - \frac{E-\Psi}{\Theta}d\Theta.$$

If we suppose an *infinitesimally small quantity of energy* (dQ) to be introduced into the system considered,[1] then in general both the "*internal energy*" of the system will be thereby altered, and also *work* will be done by the variation of the parameters. According to the Principle of the *Conservation of Energy*, we must have

(13) $$dQ = dE + dA.$$

In this expression dE is a complete differential, because, by eqns. (3) and (8), E is completely determined by the modulus of distribution and the parameters, and consequently depends only on the state of the system. On replacing dA in eqn. (13) by its value as given in eqn. (12), we obtain the relation

(14) $$dQ = d(E-\Psi) - \frac{E-\Psi}{\Theta}d\Theta.$$

We can easily see that the right-hand side of this equation becomes a *complete differential* when the equation is divided by Θ. We then have

$$\frac{dQ}{\Theta} = \frac{d(E-\Psi)}{\Theta} - \frac{E-\Psi}{\Theta^2}d\Theta,$$

or

(15) $$\frac{dQ}{\Theta} = d\left(\frac{E-\Psi}{\Theta}\right),$$

or, finally, taking § 132 eqn. (19) into account,

(16) $$\frac{dQ}{\Theta} = d(\log_e W).$$

[1] dQ is, of course, no more a complete differential than is dA.

From a mathematical point of view, the *modulus of distribution* thus represents an *integrating divisor*, by virtue of which the infinitesimal energy introduced becomes a complete differential. The function of which the total differential is the fraction dQ/Θ, however, has (as we have shown in §§ 131 and 132) the property of appearing to alter in one sense only, and of remaining constant for reversible variations in distribution.

For a system the *energy* of which is *canonically distributed*, the following two important theorems therefore hold good:

I. If an infinitesimal quantity of energy be introduced into the system, while the latter performs work, the *difference between the energy introduced and the work performed* represents the *complete differential of a function* which depends only on the *instantaneous state of the system*, and is completely determined by the modulus of distribution and the parameters.

II. The *quotient of the infinitesimal quantity of energy introduced and the modulus of distribution* represents the *differential of a second function* which likewise depends only on the instantaneous state of the system, and is likewise completely determined by the modulus of distribution and the parameters. This function has the property of remaining *constant* in *reversible* processes, but otherwise it generally *increases*.

§ 134. Equilibrium and Modular Equalization.

When *several systems together* form a *separate whole*, this again may be regarded as a single system and associated with a definite *statistical probability*. The relation of the latter to the probabilities of the component systems is very simple. For the sake of simplicity, we shall first assume that we are dealing with only two component systems of given distributions, the statistical probabilities being W_1 and W_2. The given distribution of the first system can then be realized by W_1 complexions, and that of the second system by W_2 complexions. Since, however, each complexion of the first system can be combined with each of the W_2 complexions of the second, the total number of complexions through which the given distribution can be realized in the composite system made up of the two component systems is W_1 times W_2.[1]

[1] We might quite easily draw the following false conclusion. If the number of individuals in the two component systems be N' and N'', and if the first state contain N_1' and N_1'' individuals respectively, we might suppose the statistical

Since this consideration may be directly extended to any arbitrary number of component systems, the statistical probability of the whole is equal quite generally to the *product of the statistical probabilities of the component systems.* The logarithm of the probability of the composite system is therefore equal to the *sum of the probability logarithms of the component systems.* We have

$$(1) \qquad \log_e W = \log_e W_1 + \log_e W_2 + \ldots$$

Like the component probabilities, the total probability must obviously have an upper limit which cannot be greater than the product of the upper limits of the component probabilities.[2] Hence, there is also a *maximum* for the probability logarithm of the composite system, and this will evidently be attained in a distribution, the probability logarithm of which does not vary when *infinitesimal energy interchanges* occur between the *component systems.*

If we consider two canonical component systems, which may be distinguished simply by the indices 1 and 2, and if a quantity of energy (dQ) be transferred from the first to the second system, then, by § 133 eqn. (16),

$$(2) \qquad -\frac{dQ}{\Theta_1} + \frac{dQ}{\Theta_2} = d(\log_e W_1) + d(\log_e W_2)\,.$$

When

$$(3) \qquad \Theta_1 = \Theta_2\,,$$

the right-hand side of the equation vanishes, and the sum of the probability logarithms of the two systems then remains constant. If, therefore, the composite system be made up of component systems, each of which is canonically distributed and has the *same value of the modulus of distribution,* the probability logarithm of the whole is not altered by infinitesimal interchanges of energy between the component systems. In that case, the component systems are said to be in a state of *statistical equilibrium.*

In statistical equilibrium, the probability logarithm represents a *maximum.* Since, under other circumstances, it seems to tend towards this, there is an apparent endeavour

probability of the composite system to be obtained by dividing $(N' + N'')!$, first by $(N_1' + N_1'')!$, then by $(N_2' + N_2'')!$, *etc.* This conclusion, however, would be quite wrong. For, if an individual in any state in the first component system be interchanged with an individual in the second, this gives rise not only to a new complexion, but also, in general, to a new distribution in the two component systems.

[2] Should certain conditions exist, it will, of course, be smaller than this product.

on the part of every composite system to produce statistical equilibrium between its component systems. When systems with different moduli of distribution are connected in such a way that a transfer of energy is possible between them, an *equalization in the values of the moduli of distribution* will, in general, occur.

Inasmuch as an equalization of this kind is always accompanied by an increase of the total probability logarithm, a change in state of a system is only *reversible* on two conditions. First, the distribution in the system itself must remain canonical; and secondly, during the change, any combinations with other systems into which the system enters must be associated with only *infinitesimal deviations from statistical equilibrium.* Moreover, in consequence of the existence of an equation of state (to be discussed later), a change in the modulus of distribution must entail a change in at least one of the other quantities of state. If, therefore, no change is apparent in the original conditions of two systems when they are brought in contact, it may be concluded that they both have the *same value* of the modulus of distribution.

Since, on an equalization of moduli, the total probability logarithm increases, it follows from eqn. (2) (Θ_1 being not equal to Θ_2) that

$$(4) \qquad -\frac{dQ}{\Theta_1}+\frac{dQ}{\Theta_2}>0\,.$$

That is to say, that component system which *loses* energy must have the larger modulus. Rare as is a process in which the statistical probability diminishes, the *transfer of energy of its own accord from a system with a smaller to one with a larger modulus* must be just as *rare.*

If, in an equalization of moduli, the first system has a far larger number of individuals and a far greater energy than the second, the common value of the modulus after the equalization cannot differ appreciably from the original value (Θ^*) of the modulus in the first system. In this special case, therefore, the process of equalization simply consists in an assumption by the second system of the modulus of distribution of the first. Compared with the second, the first system thus acts as a "*reservoir*" with a modulus value of Θ^*.

§ 135. Homogeneous Systems.

As we have seen, the free energy of a system is completely determined by the modulus of distribution and the parameters, and the same must therefore hold for the partial derivatives of the free energy with respect to the parameters, *i.e.*, for the quantities

$$(1) \qquad U_1 = -\frac{\partial \Psi}{\partial u_1}, \qquad U_2 = -\frac{\partial \Psi}{\partial u_2}, \ldots$$

In accordance with § 133 eqn. (7), the dependency of the quantities U_1, U_2, *etc.*, defined by these equations, on the work performed for an infinitesimal variation of the parameters is given by the relation

$$(2) \qquad dA = U_1\, du_1 + U_2\, du_2 + \ldots$$

The quantities U_1, U_2, *etc.*, may therefore be termed generalized *statistical forces*, although their dimensions are by no means necessarily those of a force.

Since the statistical forces are thus completely determined by the modulus of distribution and the parameters, we have, if the number of parameters be r, r equations of the form

$$(3) \qquad \begin{cases} U_1 = f_1(\Theta, u_1, u_2, \ldots u_r), \\ U_2 = f_2(\Theta, u_1, u_2, \ldots u_r), \text{ etc.} \end{cases}$$

The number r may be termed the number of *statistical degrees of freedom*. The quantities Θ, U_1, $U_2, \ldots U_r$, $u_1, u_2, \ldots u_r$ are called the *quantities of state of the system*, and processes in which one or more of these quantities change are defined as *changes of state*. Owing to the existence of the r "*equations of state*" (3), which can be solved with respect to the parameters, only $(r + 1)$ out of the $(2r + 1)$ quantities of state are mutually independent. The state of a system with r degrees of freedom can therefore always be represented by $(r + 1)$ *independent variables of state*, to be chosen at will from among the quantities of state.

An important special case is presented by a system which has only *one degree of freedom*, the *single parameter* being defined by the *quotient of the volume* occupied by the system *and the number of individuals.*[1] A system of this nature is said to be *statistically homogeneous*. We then have for the single parameter

$$(4) \qquad u = \frac{V}{N}.$$

[1] V denotes the space within which the individuals are contained: it must not be confused with the sum of the volumes occupied by the separate individuals.

When the parameter u is increased, the system performs work against the *pressure* (p) acting upon its surface. This work may readily be calculated.

Let us consider an element (df) of the surface of the system. A force of magnitude $p \cdot df$ will thus act upon the element. Let us suppose that, for an infinitesimal variation of the volume, the surface-element is displaced from the position represented by the interval AB to that indicated by the interval $A'B'$ (Fig. 116), the mid-point of the element being thereby moved from O to O'. The *work* performed in the displacement of the surface-element is then equal to $p \times df \times OO' \times \cos\gamma$, where γ is the angle included between the interval OO' and the normal to the surface. (For an infinitesimal displacement, the normals to the surfaces AB and $A'B'$ will evidently differ in direction by only an infinitesimal amount).

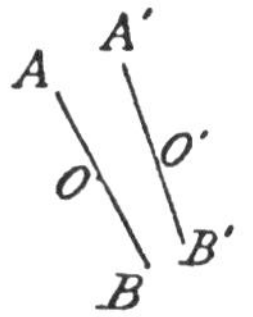

FIG. 116.

But $df \times OO' \times \cos\gamma$ is the volume of the infinitesimally small cylinder whose ends are formed by the two equally large surface-elements AB and $A'B'$, and whose height is equal to $OO' \times \cos\gamma$. The work performed in the displacement of the surface-element is thus equal to the product of the pressure and the volume of the cylinder formed in the displacement. If we take the sum of this volume over the whole surface of the system, we obtain the total increase in volume, which may be denoted by dV. Hence, the total work (dA) performed by the system for an increase in its volume by the amount dV is given by

$$dA = p\,dV = Np\,du. \tag{5}$$

By eqn. (2), therefore,

$$U = Np, \tag{6}$$

and hence, according to eqn. (1),

$$p = -\frac{1}{N}\frac{\partial\Psi}{\partial u}, \tag{7}$$

or, by eqn. (4),

$$p = -\frac{\partial\Psi}{\partial V}. \tag{8}$$

When the number of individuals is constant, the volume is proportional to the parameter u, and therefore, in a homogeneous system, instead of the parameter the *volume* itself can be regarded as a *quantity of state*. In a canonical distribution, pressure, volume, and modulus of distribution are thus always connected by an *equation of state*. It also follows that none of these three quantities can vary without at least one of the other two varying. Hence if, when two homogeneous systems are brought into contact in such a way that a transfer of energy is possible, neither volume nor pressure alter, we must conclude that both systems have the *same* modulus of distribution.

§ 136. Cyclic Processes.

As has been shown in § 133, the application of the principle of energy to canonical systems leads to the two principal equations

$$dQ = dE + dA\,, \tag{1}$$

and

$$\frac{dQ}{\Theta} = d(\log_e W)\,. \tag{2}$$

We can at once see that, in two special cases, the equations must assume an especially simple form.

In the first place, if the reversible change of state of the system occurs without any transfer of energy to or from outside sources, *i.e.*, if it occurs *adiabatically* (because no energy then passes through the surface of the system [1]), dQ is to be equated to zero in the integration of eqn. (1). Since, however, the internal energy depends only on the instantaneous state, integration of eqn. (1) in this case gives

$$\sum dA = E_1 - E_2\,, \tag{3}$$

where E_1 and E_2 are the values of the internal energy in the initial and the final states.

Secondly, if the reversible change of state occurs in such a way that the value of the modulus of distribution remains constant, eqn. (2) then assumes a simple form. For, if we denote the statistical probabilities of the initial and the final states by W_1 and W_2, it follows from eqn. (2) that, since Θ is constant,

$$\sum dQ = \Theta \log_e \frac{W_2}{W_1}\,. \tag{4}$$

Like the quantity E, the quantity W (according to § 133) also depends solely on the instantaneous state of the system. For the moment, we shall define a reversible process in which the modulus of distribution remains constant as *isomodular*

[1] The Greek for "to penetrate" is "diabainein."

—a term which is not in general use. We may realize a process of this kind by supposing the system to be connected to a *reservoir* of the modulus in question throughout the duration of the change of state.

Since the quantities E and W depend only on the instantaneous state of the system, especially simple relations result for the case in which, at the conclusion of the changes of state, a system finally returns to the same definite state whence it started. A process of this kind, wherein the initial and final states are identical, is called a *cyclic process*. From eqns. (1) and (2) we obtain for such a process the simple but none the less important relations

$$\sum dQ = \sum dA\,, \tag{5}$$

and

$$\sum \frac{dQ}{\Theta} = 0\,. \tag{6}$$

We shall now consider a special cyclic process of great importance, composed of two isomodular and two adiabatic component processes. The nature of such a process can best be illustrated with the help of a *graphical representation.* Owing to the existence of the equation of state, each state of the system considered can be described by two variables. We can therefore represent each state graphically by a point in a plane coordinate system, the abscissa of the point being equal to the value of the volume, and the ordinate to the value of the pressure. Finite *changes of state* are then represented by *curves*, the successively arranged points on which indicate the consecutive states through which the system passes during the process. Curves representing isomodular or adiabatic changes of state will, for the sake of brevity, be called isomodular or adiabatic curves.

Let us now consider a definite state, graphically represented by the point A (Fig. 117). We allow the system to expand isomodularly (the volume-abscissæ therefore increasing) to a second state, represented by the point B. To this change of state will correspond the segment AB of an isomodular curve. Let the second component process to which the system is subjected be an adiabatic expansion. This will lead to a third state, represented by the point C (with a

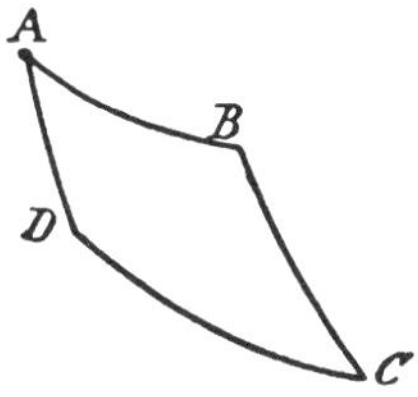

FIG. 117.

still larger abscissa), the interval BC being a segment of an adiabatic curve. We now draw an isomodular curve through the point C, on the one hand, and on the other an adiabatic curve through the point A. Let the two curves intersect at a point D,[2] representing a fourth state. The third component process will then consist in an isomodular passage of the system from the third to the fourth state, and the fourth component process in an adiabatic return from the fourth to the original first state.

Let Q_1 and Q_2 be the quantities of energy taken up during the first and third isomodular processes (it is hardly necessary to point out that Q_2 will obviously be negative). The quantity of energy Q_1 will be taken up at a modulus value of Θ_1, and the quantity Q_2 at a modulus value of Θ_2. Since the second and fourth component processes are adiabatic, the total quantity of energy introduced into the system from without during the cycle is equal to $(Q_1 + Q_2)$. Applying eqns. (5) and (6) to this special case of a cyclic process, we therefore obtain the two relations

$$(7) \qquad Q_1 + Q_2 = \sum dA\,,$$

and

$$(8) \qquad \frac{Q_1}{\Theta_1} + \frac{Q_2}{\Theta_2} = 0\,.$$

If we denote the sum of dA briefly by A, meaning thereby the total work performed by the system during the cycle,[3] we have, by eqns. (7) and (8),

$$(9) \qquad A = Q_1 + Q_2 = Q_1 - Q_1\frac{\Theta_2}{\Theta_1} = Q_1\frac{\Theta_1 - \Theta_2}{\Theta_1}\,.$$

Thus, of the quantity of energy taken up by the system during the first component process, *only a fraction* ($\eta\, Q_1$) *is transformed into work.* The quantity η is called the *efficiency of the cycle*, and is given by the relation

$$(10) \qquad \eta = \frac{\Theta_1 - \Theta_2}{\Theta_1}\,.$$

The efficiency of the cycle considered is therefore *quite independent of all properties of the system* subjected to the cyclic

[2] It is admittedly taken for granted that, starting from the state A, the modulus *can* be brought, by an adiabatic change of state, to the value corresponding to the point C.

[3] Work performed by the system is counted as positive, and work expended on the system as negative. For the rest, dA is obviously not the differential of A; *cf.* § 133, note 1.

process. It depends simply and solely on the *values of the moduli* associated with the two reservoirs in the cycle.

This last fact, now, enables us to construct a *scale of modulus values* without our requiring to know anything about the nature of the internal energy of the systems. It is sufficient for us to be able to determine the quantities of energy transferred to or from the systems with respect to external sources, their volume and their pressure, and the work performed when the volume is changed. Apart from this, we require to have only two systems, which we may call S_1 and S_2, in quite definite states (easily identifiable and reproducible). These states, which we shall regard as *fundamental*, must evidently be associated with two definite values of the modulus of distribution (Θ_1 and Θ_2).

We may, quite *arbitrarily*, take the modulus of distribution of the system S_2 as the *zero-point of a scale*, and, likewise arbitrarily, *define a scale-unit* by making the difference between the moduli of distribution of the systems S_1 and S_2 equal to a scale-units. Then we carry out a cyclic process between the systems S_1 and S_2 with the aid of an auxiliary system with respect to which those systems act as reservoirs; let the measured efficiency of this process be η^*. This gives us at once, however, the number of scale-units that the *absolute zero* of modulus values lies beneath the arbitrary zero. For, if this number be ζ, we have

$$(\Theta_1 - \Theta_2) : \Theta_1 = a : (a + \zeta),$$

and hence, by eqn. (10),

$$\zeta = a\left(\frac{1}{\eta^*} - 1\right). \tag{11}$$

If, now, any system (S') be given, we can find the exact value (z) on the scale corresponding to its modulus value (Θ') by carrying out a cycle between the systems S' and S_2. Let the measured efficiency be η'; then

$$\eta' = \frac{\Theta' - \Theta_2}{\Theta'},$$

or,

$$\eta' = \frac{z}{z + \zeta}; \tag{12}$$

and, since ζ is known already, z can be calculated.

Thus, it is possible for an observer to prepare an exact scale of modulus values even when the interior of the system is completely hidden from him, and he can only follow the behaviour of the system from outside.

§ 137. Theorem of Equipartition.

Apart from the quantities which describe the state of a system as a whole (modulus, parameters, statistical forces), we can also consider those quantities which describe the state of the individual and are therefore known as *individual quantities of state.*[1] In our derivation of the previous results, a consideration of these individual quantities of state would, indeed, have been superfluous, but, in investigating special cases, the necessity for such a consideration may well arise.

Let us suppose the individual state to be determined by *n mutually independent quantities*, denoted by $q_1, q_2, \ldots q_n$. If these quantities be graduated in definite *intervals*, any given individual state will be defined by the quantities lying in the intervals between q_1' and $(q_1' + \Delta q_1')$, between q_2' and $(q_2' + \Delta q_2')$, and so on.

We shall now consider a *special case*, in which the *individual energy* (ϵ_i) can be represented as a pure *quadratic function* of s out of these n individual quantities of state, its form being

$$(1) \qquad \epsilon_i = \alpha_1 q_1^2 + \alpha_2 q_2^2 + \ldots + \alpha_s q_s^2 .$$

We suppose, therefore, the influence both of the parameters and also of the remaining individual quantities of state, q_{s+1} to q_n, to be negligible. We shall further assume these remaining individual quantities (q_{s+1} to q_n) to lie between definite limits. The multiple integral

$$(2) \qquad \iint \ldots \int dq_{s+1} \ldots dq_n = \Omega$$

is then a *finite* quantity.

According to eqn. (1),

$$(3) \qquad e^{-\frac{\epsilon_i}{\Theta}} = e^{-\frac{\alpha_1 q_1^2}{\Theta}} . e^{-\frac{\alpha_2 q_2^2}{\Theta}} \ldots e^{-\frac{\alpha_s q_s^2}{\Theta}} .$$

We shall denote by $\Delta\tau$ the product of the intervals of all the quantities of state, so that

$$(4) \qquad \Delta\tau = \Delta q_1 . \Delta q_2 \ldots . \Delta q_n .$$

By multiplying the right-hand side of eqn. (3) by $dq_1, dq_2, \ldots . dq_n$, and integrating over all values of $q_1, q_2, \ldots . q_n$, we shall evidently obtain the sum of $e^{-\epsilon_i/\Theta}$ over all individual states, multiplied by $\Delta\tau$. Since, however, the separate individual quantities of state are mutually *inde-*

[1] Planck suitably discriminates between *macro-states* and *micro-states.*

pendent, the integral can be regarded as the product of single integrals. Hence we find

$$\sum e^{-\frac{\epsilon_i}{\Theta}} = \frac{\Omega}{\Delta\tau}\int e^{-\frac{\alpha_1 q_1^2}{\Theta}} dq_1 \int e^{-\frac{\alpha_2 q_2^2}{\Theta}} dq_2 \dots \int e^{-\frac{\alpha_s q_s^2}{\Theta}} dq_s .$$

Now we know that, for any quite arbitrary variable ξ, the definite integral [2]

$$(5) \qquad \int_{-\infty}^{+\infty} e^{-\xi^2}\, d\xi = \sqrt{\pi}.$$

Thus, on integrating with respect to the quantities q_1 to q_s over all possible individual states, *i.e.*, with $-\infty$ and $+\infty$ as integration limits, we obtain

$$\int e^{-\frac{\alpha_1 q_1^2}{\Theta}} dq_1 = \sqrt{\frac{\Theta_1}{\alpha_1}}\int_{-\infty}^{+\infty} e^{-\frac{\alpha_1 q_1^2}{\Theta}} d\left(q_1\sqrt{\frac{\alpha_1}{\Theta}}\right) = \frac{\Theta^{\frac{1}{2}}.\,\pi^{\frac{1}{2}}}{\sqrt{\alpha_1}}.$$

Hence

$$(6) \qquad \sum e^{-\frac{\epsilon_i}{\Theta}} = \frac{\Omega}{\Delta\tau}\cdot\frac{\Theta^{\frac{s}{2}}.\,\pi^{\frac{s}{2}}}{\sqrt{\alpha_1\alpha_2\dots\alpha_s}}.$$

Differentiating this equation partially with respect to Θ, we have

$$(7) \qquad \frac{1}{\Theta^2}\sum \epsilon_i e^{-\frac{\epsilon_i}{\Theta}} = \frac{\Omega}{\Delta\tau}\cdot\frac{s}{2}\cdot\frac{\Theta^{\frac{s}{2}-1}.\,\pi^{\frac{s}{2}}}{\sqrt{\alpha_1\alpha_2\dots\alpha_s}}.$$

On multiplying now eqn. (7) by Θ^2, and dividing by eqn. (6), we obtain on the left-hand side [in accordance with

[2] Eqn. (5) is most simply derived as follows :—Let the (as yet unknown) value of the definite integral be J. If we suppose an arbitrary plane coordinate system to be constructed with a ξ- and an η-axis, then we must have

$$\int_{-\infty}^{+\infty}\int_{-\infty}^{+\infty} e^{-(\xi^2+\eta^2)}\,d\xi\,d\eta = J^2.$$

On introducing polar instead of rectangular coordinates, we have

$$\xi^2 + \eta^2 = r^2,$$

and the surface-element $d\xi.\,d\eta$, with the integration limits $-\infty$ and $+\infty$, is replaced by the surface-element $r\,d\phi\,.\,dr$ with the integration limits 0 and ∞ for r, and 0 and 2π for ϕ. Hence

$$J^2 = \int_0^\infty\int_0^{2\pi} e^{-r^2} r\,dr\,d\phi = 2\pi\int_0^\infty e^{-r^2} r\,dr = \pi\int_0^\infty e^{-r^2}\,d(r^2) = \pi,$$

and therefore

$$J = \sqrt{\pi}.$$

§ 132 eqn. (16)], the *mean value of the individual energy*, which may be denoted by $\bar{\epsilon}$. This division gives the simple relation

$$\bar{\epsilon} = s\frac{\Theta}{2}. \tag{8}$$

In the special case considered, therefore, the mean value of the individual energy is equal to half the modulus of distribution multiplied by the number of terms between which and the individual energy a quadratic relation exists. Each of these terms is associated with the same amount of energy, and, consequently, the law expressed by eqn. (8) is known as the *theorem of equipartition*.

We shall now still further specialize our considerations by assuming the system to be statistically *homogeneous*, and the *positional coordinates* of the individuals (regarded as points) to be among the individual quantities of state upon which the energy does not depend. Then, since the individual quantities of state are mutually independent, and the integral over the positional coordinates represents the *volume*,

$$\Omega = \omega V, \tag{9}$$

where ω must, of course, be independent both of the volume (V) and Θ. The free energy of the system can be calculated from the formula previously derived [§ 132 eqn. (18)]:

$$\Psi = -N\Theta \log_e \sum e^{-\frac{\epsilon_i}{\Theta}}. \tag{10}$$

We have, by eqn. (6),

$$\Psi = -N\Theta \log_e V - N\Theta \log_e\left(\frac{\omega\,\Theta^{\frac{s}{2}}\pi^{\frac{s}{2}}}{\Delta\tau\,\sqrt{\alpha_1\alpha_2\ldots\alpha_s}}\right). \tag{11}$$

Since, however, the free energy of a statistically homogeneous system is completely determined by the volume and the modulus, and all the quantities in the last term of eqn. (11) are independent of the volume, it follows that

$$\left(\frac{\partial\Psi}{\partial V}\right)_\Theta = -\frac{N\Theta}{V}. \tag{12}$$

In accordance with § 135 eqn. (8), the left-hand side of this equation simply represents the *pressure*, taken with the opposite sign. In the special case under consideration, therefore, the *equation of state* assumes the extremely simple form

$$pV = N\Theta. \tag{13}$$

The validity of this equation is evidently subject only to the conditions that the individual energy can be represented as a pure quadratic function of some of the individual quantities of state, and that the positional coordinates of the individuals (regarded as points) shall be among those quantities of state upon which the individual energy does not depend.

§ 138. Rayleigh and Jeans's Principle.

In addition to the *special case of a distribution of energy* considered in the previous section, there is another of the

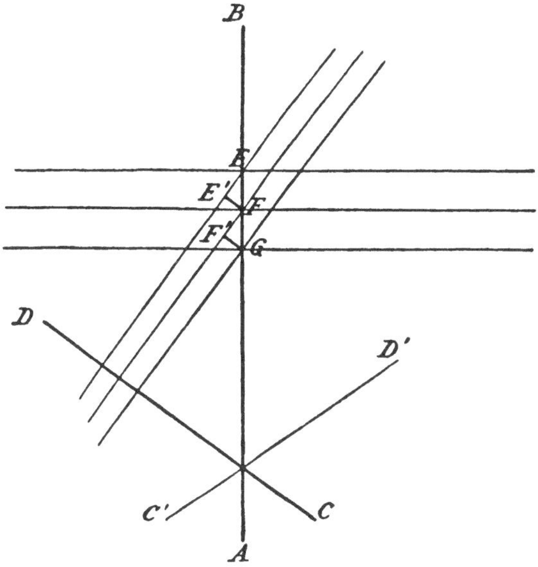

Fig. 118.

greatest importance for modern theoretical physics. This concerns the *canonical distribution of quanta of oscillatory energy.* If we consider an enclosed space, which, for simplicity's sake, we shall assume to be a *cube*, it may contain *stationary vibrations*, so-called *characteristic vibrations*, of the most varied kinds. Any investigation of the energy distribution must first be preceded by a solution of the question as to how these vibrations are distributed over the various frequency values. If we graduate the frequency values in intervals, we must first find out how large is the *number of vibrations* whose *frequencies lie within a definite interval*, for a given volume of cube.

In answering this question, we begin with the well-known fact that, along a stretched string of length l, only such stationary vibrations are possible as have wave-lengths (λ), which, for integral values of the *number of loops* (n), satisfy the equation

$$\frac{\lambda}{2} = \frac{l}{n}. \tag{1}$$

In a cube the nodal points are replaced by *nodal planes* parallel to the faces of the cube, and the values of the segmental numbers may differ for the three mutually perpendicular directions parallel to the edges. We shall imagine a coordinate system to be constructed with a corner of the cube as origin, and the three edges meeting at this corner as coordinate axes. Let the number of segments in the x-, the y-, and the z-direction be denoted by n_1, n_2, n_3, and the angles included between an arbitrary ray and the three axes by α, β, γ. If, now, the straight line AB be parallel to the z-axis (Fig. 118), and if CD be the direction of a ray, then, when the nodal planes parallel to the x-y-plane intersect AB at E, F, G at intervals of a/n, the wave-planes passing through E, F, G must also exhibit differences in phase of half a period. We must therefore have

$$E'F = F'G = \lambda/2\,.$$

But

$$E'F = EF\cos\gamma = \frac{a}{n_3}\cos\gamma\,,$$

and hence

$$\frac{\lambda}{2} = \frac{a}{n_3}\cos\gamma\,.$$

Since analogous relations must hold for the x- and the y-axis, we obtain a triplet of equations:

$$\left\{\begin{aligned} \frac{2a}{\lambda}\cos\alpha &= n_1\,, \\ \frac{2a}{\lambda}\cos\beta &= n_2\,, \\ \frac{2a}{\lambda}\cos\gamma &= n_3\,. \end{aligned}\right. \tag{2}$$

Again, according to a well-known formula,

$$\cos^2\alpha + \cos^2\beta + \cos^2\gamma = 1. \tag{3}$$

Hence, on squaring the three eqns. (2) and adding, we find

$$(4) \qquad \frac{2a}{\lambda} = \pm \sqrt{n_1^2 + n_2^2 + n_3^2}.$$

Inserting this value in eqns. (2), we obtain the relations

$$(5) \qquad \begin{cases} \cos\alpha = \pm \dfrac{n_1}{\sqrt{n_1^2 + n_2^2 + n_3^2}}, \\ \cos\beta = \pm \dfrac{n_2}{\sqrt{n_1^2 + n_2^2 + n_3^2}}, \\ \cos\gamma = \pm \dfrac{n_3}{\sqrt{n_1^2 + n_2^2 + n_3^2}}. \end{cases}$$

(In Fig. 118, therefore, in addition to the direction CD, the symmetrical direction $C'D'$ would be possible.)

Thus we see that to *every numerical triplet*, n_1, n_2, n_3, there corresponds a *characteristic vibration of quite definite wave-length and quite definite direction*. Conversely, *only* those vibrations are possible for which the values of λ, α, β, γ satisfy eqns. (4) and (5) for integral values of n_1, n_2, n_3.

If we replace the wave-length in eqn. (4) by the frequency (ν) and the wave-velocity (v), according to the well-known formula

$$\frac{1}{\lambda} = \frac{\nu}{v}$$

[§ 42 eqn. (9)], we may also write eqn. (4) in the form

$$(6) \qquad \frac{2a}{v}\nu = \sqrt{n_1^2 + n_2^2 + n_3^2}.$$

Let us now consider a quite arbitrary *vibration interval*, extending from a frequency ν' to a frequency ν'' (differing but slightly from ν'). We shall denote the difference $(\nu'' - \nu')$ by $\Delta\nu$. The question then arises as to how large is the number of frequencies possible, in view of eqn. (6), which lie within the selected interval.

To answer this question, we may conveniently make use of a simple *geometrical construction*. We imagine a rectangular coordinate system to be constructed, using, however, only the positive halves of the axes, so that an *octant* is cut out of the whole space. We then suppose a *spatial point-lattice* to be constructed in this octant, with the unit of length as lattice-constant. The individual points in the lattice will thus everywhere possess integral coordinates; and, just as each point is associated with a numerical triplet (n_1, n_2, n_3) as coordinate values, so, conversely, each numerical triplet

(n_1, n_2, n_3) unequivocally determines a point in the lattice. Consequently, each lattice-point determines a stationary vibration of given frequency and direction, while, conversely, every such vibration has a representative in a quite definite point in the lattice.

Now the distance between a lattice-point and the origin of the octant is given by

$$r = \sqrt{n_1^2 + n_2^2 + n_3^2}, \tag{7}$$

or, by eqn. (6),

$$r = \frac{2a}{v}\nu, \tag{8}$$

where ν is taken to mean the frequency of the characteristic vibration corresponding to the lattice-point concerned.

The *number of characteristic vibrations,* with frequencies between ν' and ν'', which are possible within the cube is therefore *equal simply to the number of lattice-points* lying within the space bounded by two spherical surfaces whose radii are r' and r'', where

$$r' = \frac{2a}{v}\nu', \qquad r'' = \frac{2a}{v}\nu''. \tag{9}$$

Since, on an average, each unit of volume contains one lattice-point (because the distance between two neighbouring points is equal to the unit of length), the number of points within the spatial octant bounded by the two spherical surfaces is approximately equal to one-eighth of the volume of the whole spherical shell enclosed by the two concentric surfaces. We shall assume—as will always be the case in subsequent considerations—that ν' and ν'' are extremely large numbers [1]; then, even when $\Delta\nu$ is small compared with ν', the corresponding difference $(r'' - r')$ can be regarded as large in comparison with the unit of length. The volume of an eighth of the spherical shell will thus be approximately given by

$$V' = \frac{1}{8}\, 4\pi r'^2 (r'' - r'). \tag{10}$$

The number of possible characteristic vibrations in the interval between ν' and ν'' is equal in magnitude to this volume. Denoting this number by ΔZ, we have [in virtue of eqn. (9)]

$$\Delta Z = \frac{\pi}{2}\left(\frac{2a}{v}\nu'\right)^2 . \frac{2a}{v} . \Delta\nu ,$$

[1] In optics the frequency is of the order of magnitude of hundreds of billions.

or, replacing ν' simply by ν,

$$\Delta Z = \frac{4\pi a^3 \nu^2}{v^3} \Delta\nu. \tag{11}$$

Since a^3 represents the volume of the cube, the quantity

$$z_\nu = \frac{4\pi \nu^2}{v^3}, \tag{12}$$

referred to the unit of volume, may be defined as the *density of the vibrations in the frequency interval between* ν *and* $(\nu + d\nu)$. On multiplying z_ν by the magnitude of the interval and the volume, we obtain the number of vibrations whose frequencies lie within the interval. Eqn. (12), which shows that the *vibration density is proportional to the square of the frequency*, was first derived by *Rayleigh* (1900), and more accurately established by *Jeans* in 1905.[2]

§ 139. The Statistical Equilibrium of a Vibrating System.

Let us again consider a cube filled with stationary vibrations, and fix our attention upon those whose frequencies lie in the very small interval between ν and $(\nu + d\nu)$. These vibrations of *one and the same frequency* ν may be regarded as an assembly of individuals whose energy is assumed to be canonically distributed. According to the *quantum principle* (§ 91), every characteristic vibration can then occur in a series of *discrete states* in which the associated values of the energy will be 0, $h\nu$, $2h\nu$, $3h\nu$, *etc.* If Θ be the modulus of distribution, we have thus

$$\sum e^{-\frac{\epsilon}{\Theta}} = \sum_{n=0}^{n=\infty} e^{-\frac{nh\nu}{\Theta}}. \tag{1}$$

In accordance with the well-known summation formula for a geometrical progression,

$$1 + y + y^2 + y^3 + \ldots = \frac{1}{1-y}, \tag{2}$$

when $y < 1$. Hence

$$\sum e^{-\frac{\epsilon}{\Theta}} = \frac{1}{1 - e^{-\frac{h\nu}{\Theta}}}. \tag{3}$$

[2] Admittedly, in this section, only the special case of a cube has been considered. But it will be evident from the considerations of the next section, and even more so from subsequent applications, that eqn. (12) must hold quite independently of all geometrical shape.

On differentiating partially with respect to Θ, and then multiplying both sides of the equation by Θ^2, we find

(4) $$\sum \epsilon_i e^{-\frac{\epsilon_i}{\Theta}} = \frac{h\nu \, . \, e^{-\frac{h\nu}{\Theta}}}{\left(1 - e^{-\frac{h\nu}{\Theta}}\right)^2}.$$

If we now divide eqn. (4) by eqn. (3), it follows from §132 eqn. (16) that the left-hand side must represent the *mean value of the energy* associated with a single vibration of frequency ν. Thus

(5) $$\bar{\epsilon} = \frac{h\nu}{e^{\frac{h\nu}{\Theta}} - 1}.$$

On multiplying this value by the density of the vibrations of frequency ν [§ 138 eqn. (12)], we obtain the *energy density* for the frequency considered, and this may be denoted by u_ν.[1] Hence, we have

(6) $$u_\nu = \frac{4\pi h \nu^3}{v^3} \cdot \frac{1}{e^{\frac{h\nu}{\Theta}} - 1},$$

where v again denotes the velocity of propagation of the vibrations.

The whole system of stationary vibrations can be resolved into *component systems* of definite frequency values if we graduate these values into intervals. We have hitherto considered such a component system, composed of vibrations of the same frequency. If, however, the whole system of vibrations is to be *in statistical equilibrium*, it follows from § 134 that this is only possible when *all the component systems, despite the different frequencies*, have the *same value of modulus of distribution*.

When such is the case, the true *energy density*, denoted by η, may be calculated from the formula

(7) $$\eta = \int_0^\infty u_\nu \, d\nu,$$

Θ being regarded as constant during the integration. Hence

(8) $$\eta = \frac{4\pi h}{v^3} \int_0^\infty \frac{\nu^3 \, d\nu}{e^{\frac{h\nu}{\Theta}} - 1}.$$

[1] It must be remembered, however, that the physical dimensions of u_ν are not those of an energy density. That is true only for the product of u_ν and $d\nu$.

To simplify this expression, let us put

(9) $$\frac{h\nu}{\Theta} = x, \qquad \text{or } \nu = \frac{x\Theta}{h}:$$

then

(10) $$\eta = \frac{4\pi\,\Theta^4}{h^3\,v^3}\int_0^\infty \frac{x^3\,dx}{e^x - 1}.$$

We can evaluate the definite integral in this equation by starting off with eqn. (2), in which we replace y by e^{-x}: then

$$\frac{1}{1 - e^{-x}} = \frac{e^x}{e^x - 1} = 1 + e^{-x} + e^{-2x} + e^{-3x} + \ldots.,$$

or,

$$\frac{1}{e^x - 1} = e^{-x} + e^{-2x} + e^{-3x} + \ldots.$$

Hence

$$\int \frac{x^3\,dx}{e^x - 1} = \int e^{-x}\,x^3\,dx + \frac{1}{2^4}\int e^{-2x}\,(2x)^3\,d(2x)$$
$$+ \frac{1}{3^4}\int e^{-3x}\,(3x)^3\,d(3x) + \ldots.$$

Now, according to a well-known formula [2] for an arbitrary variable z,

(11) $$\int_0^\infty z^3 e^{-z}\,dz = 3\,!.$$

Since 3 ! is equal to 6, we have

(12) $$\int_0^\infty \frac{x^3\,dx}{e^x - 1} = 6\alpha,$$

where α denotes the sum of the convergent power series [3]

(13) $$\frac{1}{1^4} + \frac{1}{2^4} + \frac{1}{3^4} + \frac{1}{4^4} + \ldots. = 1{\cdot}0823.$$

[2] This formula may be derived as follows :—For an arbitrary parameter

$$\int_0^\infty e^{-tz}\,dz = \left|\frac{1}{t}\cdot e^{-tz}\right|_\infty^0 = \frac{1}{t}.$$

On successive partial differentiations with respect to t, we find

(11b) $$\int_0^\infty z\,e^{-tz}\,dz = \frac{1}{t^2};\quad \int_0^\infty z^2\,e^{-tz}dz = \frac{1.2}{t^3};\quad \int_0^\infty z^3\,e^{-tz}dz = \frac{1.2.3}{t^4}.$$

When t is put equal to 1, we obtain eqn. (11).

[3] Although it does not concern the subsequent considerations at all, we may remark in passing that α is nothing else than $\pi^4/90$.

Hence

(14) $$\eta = \frac{24\pi\alpha\Theta^4}{h^3 v^3}.$$

In statistical equilibrium, therefore, the *energy density is proportional to the fourth power of the modulus of distribution.*

In order next to calculate, in like manner, the *free energy* of the quantized vibrating system, we shall first consider the component system of those vibrations whose frequencies lie in the interval between ν and $(\nu + d\nu)$. It follows from § 132 eqn. (18) that the amount of free energy associated with a single individual is given by

$$\psi = -\Theta \log_e \sum_{n=0}^{n=\infty} e^{-\frac{nh\nu}{\Theta}},$$

or, by eqn. (3),

(15) $$\psi = \Theta \log_e \left(1 - e^{-\frac{h\nu}{\Theta}}\right).$$

We now introduce the well-known logarithmic series $(y < 1)$

(16) $$\log_e (1 - y) = -y - \frac{y^2}{2} - \frac{y^3}{3} - \frac{y^4}{4} - \dots.$$

Then, making use of the abbreviation given in eqn. (9), we find

(17) $$\psi = -\Theta\left[e^{-x} + \frac{e^{-2x}}{2} + \frac{e^{-3x}}{3} + \frac{e^{-4x}}{4} + \dots\right].$$

On multiplying this quantity by the density of the vibrations in the frequency interval between ν and $(\nu + d\nu)$, and then integrating with respect to ν from 0 to ∞, we obtain the *density of the free energy,* which may be denoted by ζ. Thus,

$$\zeta = -\frac{4\pi\Theta}{v^3}\int_0^\infty \left[e^{-x} + \frac{e^{-2x}}{2} + \dots\right]\nu^2\, d\nu,$$

or, by eqn. (9),

(18) $$\begin{cases} \zeta = -\dfrac{4\pi\Theta^4}{h^3 v^3}\Bigg[\displaystyle\int_0^\infty e^{-x}x^2 dx + \frac{1}{2^4}\int_0^\infty e^{-2x}(2x)^2\, d(2x) \\ \qquad\qquad + \dfrac{1}{3^4}\displaystyle\int_0^\infty e^{-3x}(3x)^2\, d(3x) + \dots\Bigg]. \end{cases}$$

But, for any arbitrary variable (z),

(19) $$\int_0^\infty e^{-z} z^2\, dz = 2$$

[according to eqn. (11b)]. Hence, introducing the abbreviation used in eqn. (13),

$$\zeta = -\frac{8\pi\alpha\Theta^4}{h^3 v^3}. \tag{20}$$

As we see by a comparison with eqn. (14), the *free energy in a quantized vibrating system* is therefore *equal*, but opposite in sign, *to one-third of the total energy*. We have

$$\Psi = -\frac{E}{3}. \tag{21}$$

This equation leads directly to a further conclusion. In accordance with eqn. (14), the energy density depends only on one variable, the modulus of distribution (for the wave velocity may be regarded as constant). The energy itself thus depends, apart from on the modulus of distribution, on only a single parameter, the volume. In the sense of the earlier considerations of § 135, a quantized vibrating system thus appears to be *statistically homogeneous*, so that its *pressure* is given [according to § 135 eqn. (8)] by the simple relation

$$p = -\left(\frac{\partial\Psi}{\partial V}\right)_{\Theta = \text{const.}} \tag{22}$$

or, by eqn. (21),

$$p = \frac{1}{3}\left[\frac{\partial(\eta V)}{\partial V}\right]_{\Theta = \text{const.}}. \tag{23}$$

Since, however, we have shown that η depends only on the modulus of distribution, we have simply

$$p = \frac{\eta}{3}. \tag{24}$$

In a *quantized vibrating system*, the *pressure* is *equal to one-third of the energy density.*

Chapter XVII

THERMODYNAMICS

§ 140. Temperature Scales.

A THEORETICAL explanation is afforded by statistics for those phenomena which are known as heat phenomena, if, in every case in which *energy* is *distributed among a very large number of individuals* in a manner not directly perceptible to human senses, this energy be *interpreted as heat.* The individuals among which heat energy is distributed *throughout bodies* are identified by theoretical physics with *molecules.*[1]

All quantities which, in a statistical sense, are proportional to the number of individuals must therefore, in a thermal sense, be *proportional to the number of molecules,* or, what amounts to the same thing in chemically homogeneous substances, to the *mass.* Thus all these quantities may be *referred* to a *gram-molecule,* or to the *unit of mass.*

It follows from the electron theory that the forces acting between the particles of matter must be regarded as central forces and depend only on the distances between the particles. According to § 17, however, a system of this kind is governed by the Principle of the Conservation of Energy, which was one of our hypotheses in our previous statistical considerations of energy. If, now, the internal forces only depend on the distances separating the individual molecules, then the internal energy of a homogeneous body will evidently depend only on one parameter, *viz.,* on the volume of the unit of mass, or the so-called *specific volume.* It seems justifiable, therefore, to regard bodies of a *physically uniform nature* as being also *statistically homogeneous.*

Experimental physics is acquainted with various phenomena in which *energy* arising from the *motion of bodies,* or *electromagnetic energy, apparently disappears with evolution of heat.* This is the case when energy of motion is destroyed by any obstacles such as *friction* or the resistance of the

[1] In elements which do not form molecules, the atoms are correspondingly regarded as "monatomic molecules."

medium ; when an electric current is transmitted through a conductor possessing a *galvanic resistance ;* when electromagnetic waves strike a body which does *not completely reflect them.* In all these instances, theoretical physics assumes that the energy which has apparently disappeared is not really lost, but is distributed among a large number of individuals. The *heat* produced in such cases is therefore *measurable in units of energy,* if the apparent loss of energy be determined, as may easily be done with frictional heat and electrical heat.

According to this concept, the energy introduced into or abstracted from a system of molecules appears as *heat absorbed* or *emitted.* Now if, on connecting two bodies in such a way that a heat transfer becomes possible, no change in their initial state occurs, it must be concluded from the existence of the equation of state that both bodies have the *same modulus of heat.* A body whose state is readily variable may thus, to a certain extent, serve as a thermoscope, *viz.,* as an *instrument for the detection of the equality of heat moduli.* For, if the volume at constant pressure, or the pressure at constant volume, indicated by this instrument be the same when it is connected with two bodies A and B, we may conclude that, at the moment when they were connected with the instrument, the bodies A and B had the same heat modulus. Furthermore, if the change of state of a body K occurs in such a way that the state of a thermoscope connected with the body does not vary throughout the change, we know that the change of state of K is an *isomodular* one. On the other hand, we can effect an *adiabatic* change of state in a body by surrounding it completely with some substance which is known from experience to be impermeable to heat (as is very approximately the case with, for instance, wool).

Measurements made on *cyclic processes* based on these facts permit of the establishment of an *exact scale of values of heat moduli.* Let us suppose a homogeneous body (preferably a gas, because of its ready variability of state) to be in a state I, and a thermoscope connected with it to give a reading a. Before commencing the cyclic process, let the volume be adiabatically increased until the thermoscope gives any reading b, the state of the body, as defined by its volume and pressure, being IV. Then we have to carry out a cyclic process in the following manner :—

We transfer the body from state I to an arbitrary state (II), by means of an isomodular expansion, and then we let it expand adiabatically to a state III in which the thermoscope gives the above-mentioned reading b. Next, the body is brought to state IV, by means of an isomodular compression, and finally adiabatically transferred to state I. We measure in units of energy, on the one hand, the quantity of heat introduced, for example, electrically by means of a suitable regulation of current during the isomodular transference from I to II, and, on the other hand, the total work performed during the cycle. This may most simply be done by representing the process graphically on a p-V-plane, and measuring the area of the figure bounded by the four component curves (the so-called indicator diagram), for this area is equal to $\int p\,dV$, or the work performed. On dividing this work by the above quantity of heat (in energy units), we obtain the efficiency of the cycle.

We introduce, now, as *fundamental values* of the heat modulus the values possessed by *melting ice and boiling water at normal atmospheric pressure*,[2] it being empirically established (from reasons which will appear later [3]) that the reading of the thermoscope does not change during either the melting or the boiling process. In conformity with experimental physics, we choose the ice-point as the zero-point on an arbitrary scale, and we divide the difference between the two fundamental values of the modulus arbitrarily into *one hundred* scale-units or *degrees*. Then we define the number on this arbitrary scale corresponding to the heat modulus as the *temperature*.

If we carry out a cycle in such a way that the isomodular component processes are performed at 100° and 0°, we can determine [according to § 136 eqn. (11)] how many degrees below the ice-point the *absolute zero of temperature* lies. As is found from other observations which are easier to make and will be discussed later,[4] this number amounts to 273·1°. The number of degrees obtained by adding the number 273·1 to that reckoned from the ice-point is known as the *absolute temperature*; it is generally denoted by the letter T. Modulus of heat and absolute temperature can thus only differ by a universal proportionality factor.

Finally, by carrying out the cycle in such a way that, in the first component process, the thermoscope gives any

[2] Normal atmospheric pressure is equal to the force exerted, in virtue of its weight, by a column of mercury (in thermal equilibrium with melting ice) 76 cm. in height and 1 cm.^{2} in cross-section, at sea-level in the geographical latitude of 45°. In absolute measure, this pressure is equal to 1,013,250 gm. cm.$^{-1}$ sec.$^{-2}$.

[3] *Vide* § 147.

[4] *Vide* § 148.

arbitrary reading, while the third component process is performed at the temperature of melting ice, we can derive the correct value of the temperature for any given reading of the thermoscope, with the help of § 136 eqn. (12). Used in this manner, the thermoscope becomes an exact measurer of temperature—a *thermometer*.

The development of the construction of temperature scales in practical physics has followed along much simpler lines than the method given above, which must be regarded as the most perfect from the point of view of theoretical physics. If V_0 be the volume of a thermometric substance (such as mercury) at the ice-point, V_{100} the volume at the boiling-point, and V the volume at an unknown temperature θ, θ is simply equated to $(V - V_0)/(V_{100} - V_0)$. The quantity θ thus appears as the temperature in the scale of the thermometric substance used. The practical utility of this simple process is supported by the fact that the scales of various suitable thermometric substances (mercury, alcohol, air, *etc.*) differ but little from each other. For reasons which will appear subsequently,[5] *hydrogen* and *helium* are regarded as the most perfect thermometric substances, and the scales of other thermometric substances are therefore corrected to the hydrogen or the helium scale.[6] The idea of basing a truly accurate temperature scale upon measurements on cyclic processes is due to *William Thomson* (*Lord Kelvin*) (1855). (In § 146 we shall discuss a suitable method for the establishment of the Kelvin temperature scale.)

§ 141. Calorimetry.

The construction of a temperature scale permits of a *simple measurement of quantities of heat*, without it ever being necessary to embark upon the complicated process of a direct measurement in energy units. In accordance with the Law of the Conservation of Energy, the quantity of heat required for a finite change of state is given by the expression

$$(1) \qquad \sum dQ = E_2 - E_1 + \int_{V_1}^{V_2} p\, dV,$$

where the subscripts 1 and 2 refer to the initial and the final

[5] *Vide* § 149.

[6] It is well known from experimental physics that the scale on which the ice-point and the boiling-point are chosen as fundamental points, the fundamental interval being made equal to one hundred degrees, is due to *Celsius* (1742).

state. When, however, the change of state occurs at *constant pressure*, the integral is simply equal to $p\,(V_2 - V_1)$, and, in that case, the total amount of heat introduced depends simply and solely on the initial and final states. It follows, therefore, from the equation of state, that the quantity of heat required to raise a given body from an initial temperature T_1 to a final temperature T_2, at constant normal atmospheric pressure, is always the *same*, and, because of the molecular structure of the heated body, this quantity of heat must also be proportional to the mass of the body.

Thus, a *unit of heat* may be arbitrarily created by defining as a *calorie* the quantity of heat required to raise 1 gram of water at constant normal atmospheric pressure from 15° to 16° (measured from the ice-point).[1] Using the same quantities of heat (which may, for instance, be generated by sending equally strong electric currents through the same resistance for the same length of time), we can at one time raise x gm. of water from 15° to 16°, and at another time y gm. of water from $\theta_1°$ to $\theta_2°$. We know then that x/y calories are necessary to heat 1 gm. of water from $\theta_1°$ to $\theta_2°$. In this way we can find out how many calories are required to raise 1 gm. of water through one degree at any particular temperature (*e.g.*, to heat 1 gm. of water from 30° to 31°, 0·9968 cal. are required). Any arbitrary quantity of heat can thus be measured in calories, as soon as we know the rise in temperature produced by it in a given quantity of water.

The number of energy units equivalent to one calorie is known as the *mechanical equivalent of heat.*[2] It can be directly determined by heating a quantity of water by frictional or electrical means, and expressing the quantity of heat generated, on the one hand, in calories, and on the other hand (as determined by the apparent loss in energy) in energy units. The value found by experiment for the mechanical equivalent of heat is [3]

$$(2) \qquad J = 4{\cdot}19 \times 10^7 \text{ ergs} = 427 \text{ gm.-m.}$$

[1] The quantity of heat required to raise 1 gm. of water from 0° to 1° is called the zero-point calorie; it is approximately 1·008 times as large as the calorie above defined. Sometimes, however, the latter is also defined as being the quantity of heat necessary to raise 1 gm. of water from 14½° to 15½°; the difference, however, is very slight. The term "mean calorie" is given to the one-hundredth part of the quantity of heat used in heating 1 gm. of water from 0° to 100°; it is very nearly equal to the true calorie (of 15°). The large calorie is understood to be equal to one thousand "small" calories, the quantity of heat then being referred to 1 kgm. instead of 1 gm.

[2] Its reciprocal value may be called the calorimetric equivalent of work.

[3] Referred to the zero-point calorie, $J = 4{\cdot}22 \times 10^7$ ergs $= 430$ gm.-m.

§ 142. The Two Principal Laws of Thermodynamics.

Since temperature, measured in degrees, and modulus of heat only differ by a universal proportionality factor, the theorems obtained in general statistics must be equally valid if modulus of distribution be replaced by the product of absolute temperature and a proportionality factor, and heat be substituted for energy. As applied to the theory of heat, the theorems stated at the end of § 133 form the fundamental relations of that theory, and are known as the *two Principal Laws of Thermodynamics*.

The *First Principal Law* may be expressed as follows: If an infinitesimal quantity of heat be introduced into a system, while work is simultaneously performed, the *difference between heat introduced and work performed* represents the *complete differential* of a function, the *internal energy*, which depends only on the instantaneous state of the system and, when r parameters exist, is therefore determined by $(r + 1)$ *independent variables of state.*

Moreover, this statement of the First Principal Law embodies a fact which follows directly from the statistical-energetic concept of heat, *viz.*, that in all cases where heat is developed with apparent loss in energy, a *constant transformation ratio* exists between the heat measured in calories and the energy loss measured in energy units.

The nucleus of the *Second Principal Law* is contained in the conclusion derived from general statistics, that the *quotient of the infinitesimal quantity of heat introduced and the temperature* must represent the *complete differential* of a *second function*, which likewise can depend only on the instantaneous state of the system, and must therefore also be determined by $(r + 1)$ independent variables of state when the number of parameters is r. This function is known in thermodynamics as the *entropy*, and is thus defined by the equation

$$(1) \qquad dS = \frac{dQ}{T}.$$

Entropy must thus possess the same properties as the probability logarithm, to which it is proportional. *Entropy* likewise *appears to tend to a maximum*, and to remain *constant only in reversible processes.*

According to § 134 eqn. (1), the entropy of a system of bodies is equal to the *sum of the entropies* of the individual bodies, so that the entropy of a single body is proportional to its mass.[1] If we speak of *thermodynamic* instead of statistical *equilibrium*, we see that such is only possible *between bodies at the same temperature*. When the temperatures differ, there is a tendency to *equalization*, and that in such a way that *heat is not of itself transferred from a colder to a warmer body*.

In accordance with eqn. (1), every addition of heat involves an increase in the entropy of a body. Since this applies to every process in which the energy (other than that of heat) of a closed system is transformed into heat by distribution amongst the constituent individuals, the energy contained in a closed system must apparently tend to be *gradually transformed completely into heat*. This apparently progressive transformation is known as the *dissipation*, scattering or *degradation of energy*.[2]

Denoting the universal proportionality factor between modulus of heat and absolute temperature by k, so that

(2) $$\Theta = kT\,,$$

we have [by eqn. (1) and § 133 eqn. (16)]

(3) $$S = k \log_e W\,.$$

The definition of entropy is thereby tacitly supplemented by the postulate that it shall not differ by any additive constant from the probability logarithm. The quantity k is called *Boltzmann's constant* after Boltzmann, who founded the statistical concept of entropy in 1866 ; and the relation which states that the entropy is proportional to the logarithm of the statistical probability of the individual distribution of heat energy is known as *Boltzmann's entropy law*.

Thermodynamics first developed into a branch of theoretical physics in the middle of the 19th century. *Thermometry*,[3] however, was established as early as the beginning of the 18th century, and *calorimetry*[4] towards the end of that century. In 1822, moreover, *Fourier* created an exact theory of *heat conductivity* in his famous "Analytical Theory

[1] This follows, moreover, directly from the proportionality of the probability logarithm to the number of individuals.

[2] Energy is "degraded" because heat cannot be changed back again into other forms of which use might once again be made, for, in such a transformation back, without some kind of compensation, the entropy would be diminished.

[3] The real founder of thermometry was *Fahrenheit*.

[4] The differences in the specific heats of various bodies were first observed by *Wilke* (1772). Conspicuous in the further development of calorimetry were Black, Dulong and Petit, Lavoisier and Laplace, and Regnault.

of Heat." In 1842 *Robert Mayer* established the important principle of the existence of a *constant transformation ratio* between the amount of heat developed when an apparent loss of energy takes place and the energy which has apparently disappeared. The same principle was simultaneously established by *Joule*, though quite independently of Mayer. The establishment of this principle upon an accurate theoretical basis was chiefly due to *Helmholtz* (1847). A determination of the mechanical equivalent of heat was effected by Mayer along theoretical lines,[5] and by Joule by measurements of electrical and frictional heat.

The Second Principal Law of Thermodynamics was derived by *Clausius* (1850) and by *William Thomson* (1851) from purely phenomenological considerations. The concept of entropy is due to Clausius and that of the degradation of energy to Thomson. In the derivation of the Second Principal Law, it was possible to utilize the important results obtained as early as 1824 by *Sadi Carnot* in his researches on *thermal cyclic processes*.[6]

The middle of the 19th century also saw the establishment of the *mechanical theory of heat* which regards heat as individually distributed *energy of motion*.[7] This assumption afforded an immediate explanation of the First Principal Law, and the kinetic concept was applied with great success by *Krönig*, *Joule*, *Clausius*, and *Maxwell* to explain the properties of *gases* (*vide* § 148). To *Boltzmann* we owe, as already mentioned, the statistical interpretation of the Second Principal Law, the *relegation of irreversibility to atomistics*. His ideas have been further developed and broadened, especially by *Smoluchowski* (at the beginning of the 20th century); whilst the development of statistical methods through the treatment of *statistical mechanical problems* is principally due to *Gibbs*.

§ 143. Irreversibility.

The law of the *increase of entropy* is based on probability considerations in general statistics. It is, therefore, not a law in the strict sense of the word, but rather a *rule* which, in physical processes, is *more frequently obeyed than disobeyed*. If a certain process leads to a z-fold increase in the statistical

[5] *Vide* § 148.

[6] An exact mathematical form was given to Carnot's ideas by *Clapeyron* (1833).

[7] This concept was first stated in a scientific form by *Rumford* (about 1800).

probability of a system, and if we assume that this process can only take place in the one direction, being thus *irreversible*, then, for every z instances in which the *assumption* is found to be *correct*, there will be only *one single* case in which it will be *contradicted by an experimental fact*. The number z may thus be regarded as a *measure of the irreversibility*.

If, now, in any process the entropy increases by the amount ΔS, the logarithm of the statistical probability will be increased by the amount $\Delta S/k$. The number z, which shows how many times the original value of the statistical probability has been raised, is then determined by the relation

$$z = e^{\frac{\Delta S}{k}}. \tag{1}$$

In order to be able to calculate z for definite cases, we must, however, know the value of Boltzmann's constant (k); this is, as we shall find in subsequent considerations (§ 148), which we may anticipate, equal to $1{\cdot}34 \times 10^{-16}$ erg . deg.$^{-1}$.

As a very simple example of a process associated with an increase in entropy, we shall consider an *equalization of temperature*. We suppose two previously isolated bodies, differing in temperature by 1° (say, 15° and 16°), to be brought into contact with each other. Let the very small quantity of heat equivalent to 1 erg be transferred from the body which is warmer by one degree (*i.e.*, an amount not even sufficient to raise 1 mgm. of water 1/1000°).[1] The increase in the probability logarithm thereby produced is, by § 134 eqn. (2),

$$\frac{\Delta S}{k} = \frac{1\,\text{erg}}{k}\left(\frac{1}{288^\circ} - \frac{1}{289^\circ}\right) = 9 \times 10^{10}.$$

Hence, in this case, according to eqn. (1)

$$z > 10^{10^{10}}.$$

In the insignificant equalization of temperature considered, the irreversibility as measured by the number z is therefore large beyond all human imagination, for, in order to write down the number z in figures, we should require a strip of paper as long as the circumference of the earth. For z cases in which the erg is transferred from the warmer to the colder body, there is but one instance in which the reverse process takes place.

Thus, if we make the assumption that heat is never

[1] If we suppose the two bodies to be sufficiently large, the heat quantity of one erg can be regarded as "infinitesimally small" in comparison with the heat content of either body.

spontaneously transferred from a colder to a warmer body to an observable extent (without some kind of compensation), the probability of refuting this assumption by experiment is inconceivably small.[2] Even if millions of physicists were to carry out afresh every minute throughout millions of years the experiment of bringing into contact at room temperature two bodies differing by a degree, the prospect of any one of them observing even only once a loss of heat by the colder instead of by the warmer body, to the extent of at least an erg, would be incredibly small. Inconceivably rare as is such an instance, however, its probability is still greater than zero; and, within *sufficiently long intervals of time*, which would certainly be long beyond all human conception, an observer would undoubtedly meet *at least once* with this *unimaginably rare exception.*

In the case of a system whose individual energy distribution is constantly changing, it may be said that *thermodynamic equilibrium* represents the *normal state;* the deviations therefrom, which are actually always present, would almost always completely escape notice. Only on occasions of the very greatest rarity would the deviations come within the limits of human perceptibility, and the further, within these limits, the more seldom the occasions. Even *macroscopically* recognizable mechanical and electrical processes can occur spontaneously in a system through variations of distribution. A period of macroscopical processes of this kind, which of itself arises from a state of equilibrium and tends to return to it, would certainly represent only an *inconceivably rare and exceptional episode* in the history of the system. Moreover, however long this period might last, it would yet be unimaginably brief as compared with the time which must elapse before such a period of macroscopical processes of itself *returned* once again.

§ 144. Free Energy and Thermodynamic Potential.

In accordance with § 132 eqn. (19), *free energy* (for which, in thermodynamics, the usual symbol is F instead of Ψ) is connected with the probability logarithm and the modulus of heat by the relation

$$E - F = \Theta \log_e W,$$

[2] The probability of the exceptional case becomes larger, however, when a smaller quantity of heat than an erg is considered, and when the temperature difference is taken as being less than a thousandth of a degree.

or [taking into account § 142 eqns. (2) and (3)] with temperature and entropy by the equation

$$F = E - TS. \tag{1}$$

Those quantities which, in statistics, are designated statistical forces are known as *thermal forces* in thermodynamics (although their physical dimensions are by no means necessarily those of a force). Denoting them again, by U_1, U_2, *etc.*, and regarding *free energy as a function of the temperature and the parameters*, we have, by § 135 eqn. (1), the relations

$$U_1 = -\frac{\partial F}{\partial u_1}, \qquad U_2 = -\frac{\partial F}{\partial u_2}, \text{ etc.} \tag{2}$$

Again, in accordance with § 133 eqn. (11),

$$\frac{\partial F}{\partial \Theta} = \frac{F}{\Theta} - \frac{E}{\Theta} = -\log_e W,$$

so that

$$-\frac{\partial F}{\partial T} = S. \tag{3}$$

It is therefore a property of the free energy, considered as a function of the temperature and the parameters, that its negative partial derivatives with respect to the parameters and the temperature give the thermal forces and the entropy respectively.

It follows from § 133 eqn. (9) that free energy and *work* performed are connected by the relation

$$dF = \frac{\partial F}{\partial T} dT - dA, \tag{4}$$

or [by eqn. (3) and § 135 eqn. (2)],

$$dF = -S\,dT - U_1\,du_1 - U_2\,du_2 - \ldots. \tag{5}$$

Hence, if a process be carried out at constant temperature or (as we may say by analogy with the previously used term "isomodular") *isothermally*, we have simply

$$dF = -dA. \tag{6}$$

In isothermal processes, therefore, the *work performed* is *equal to the decrease in free energy.*

Again, in processes where *no external work is performed*, we have, by eqns. (4) and (3),

$$dF = \frac{\partial F}{\partial T} dT = -S\,dT. \tag{7}$$

Multiplying the first and third of these expressions by T, and bearing in mind eqn. (1), we obtain the relation

(8) $$E - F = -T\frac{dF}{dT}.$$

This equation was first established in 1882 by *Helmholtz* (to whom, moreover, the whole concept of free energy is due [1]), and is therefore generally known as Helmholtz's equation. It is of great importance on account of its applicability to all processes in which temperature changes occur without the performance of external work.[2]

Closely connected with the concept of free energy is that of *thermodynamic potential*. This is defined by the equation

(9) $$\Phi = F + U_1 u_1 + U_2 u_2 + \ldots.$$

Total differentiation leads to the expression

$$d\Phi = dF + U_1 du_1 + U_2 du_2 + \ldots + u_1 dU_1 + u_2 dU_2 + \ldots,$$

or, by eqn. (5),

(10) $$d\Phi = -S\,dT + u_1\,dU_1 + u_2\,dU_2 + \ldots.$$

If thermodynamic potential be regarded as a *function of the temperature and the thermal forces*, so that

(11) $$\Phi = \varphi(T, U_1, U_2, \ldots),$$

we obtain the relations

(12) $$-\frac{\partial\Phi}{\partial T} = S, \qquad \frac{\partial\Phi}{\partial U_1} = u_1, \textit{ etc.}$$

It is thus a common property of free energy and thermodynamic potential that, when they are given as functions of certain variables of state, the other quantities of state and the entropy are to be obtained by partial differentiation with respect to those variables.[3] In the case of a homogeneous system, we have simply

(13) $$\Phi = E - TS + pV.$$

§ 145. Changes of State in a Homogeneous System

In a *homogeneous system*, there is the possibility of further special cases of change of state in addition to *adiabatic* and *isothermal* ones, *viz.*, changes which take place at *constant*

[1] Of course, only in the thermodynamical sense.

[2] The most important example of the application of Helmholtz's equation, and one treated by Helmholtz himself, is afforded by the theory of *galvanic elements*.

[3] Free energy and thermodynamic potential are consequently termed *characteristic functions*—a definition introduced by *Massieu* (about 1870).

volume or *constant pressure.*[1] The quantity of heat required to raise the temperature of a body by one degree at constant volume or constant pressure is called its *heat capacity* (at constant volume or constant pressure). The quotient of the heat capacity and the mass is defined as the *specific heat*, and the quotient of the heat capacity and the number of gram-molecules as the *molar heat*.

The partial differential coefficient of the volume with respect to the temperature, at constant pressure and referred to unit-volume, is called the *expansion coefficient* of the body. Denoting it by α, we therefore have by definition

$$\alpha = \frac{1}{V}\left(\frac{\partial V}{\partial T}\right)_p. \tag{1}$$

(The subscript p, written after the bracketed derivative, indicates that the change of state is to be carried out at constant pressure.)

The partial differential coefficient of the pressure with respect to the temperature, at constant volume and referred to unit-pressure, is known as the *pressure coefficient*; it is thus defined by the equation

$$\beta = \frac{1}{p}\left(\frac{\partial p}{\partial T}\right)_V. \tag{2}$$

The *negative* partial derivative of the volume with respect to the pressure, at constant temperature and referred to unit-volume, is called the *compressibility coefficient*, and is hence defined by the relation

$$\kappa = -\frac{1}{V}\left(\frac{\partial V}{\partial p}\right)_T. \tag{3}$$

Between the three quantities α, β, and κ, there exists a simple relation which may easily be derived. Thus, if we take volume and temperature as the independent variables of state, differentiation of the equation of state gives

$$dp = \left(\frac{\partial p}{\partial V}\right)_T dV + \left(\frac{\partial p}{\partial T}\right)_V dT, \tag{4}$$

or, by eqns. (2) and (3)

$$dp = -\frac{1}{\kappa V} + \beta\, p\, dT.$$

[1] Equations describing changes of state at constant volume are also termed *isochores* or *isometrics*; those describing changes at constant pressure are called *isobars*.

In the case of a change of state at constant pressure, for which dp thus vanishes, we must therefore have

$$\left(\frac{\partial V}{\partial T}\right)_p = \beta\,\kappa\,p\,V.$$

Hence, by eqn. (1),

$$a = \beta\,\kappa\,p. \tag{5}$$

The expansion coefficient is equal to the product of the pressure and compressibility coefficients and the pressure. If two of the coefficients be known, the third therefore follows at once.[2]

In consequence of the existence of the equation of state, we can express the *internal energy* either as a function of the temperature and the volume, or as a function of the temperature and the pressure (the possible representation of the energy as a function of the volume and the pressure does not concern us). Inasmuch as we are dealing with homogeneous bodies, in the following considerations we shall refer the quantities which, in a statistical case, are proportional to the number of individuals, to the *unit of mass*, and denote these "specific" quantities by the corresponding small letters. Thus, e will denote specific energy, s specific entropy, v specific volume, and dq the heat introduced into the unit of mass. Specific heat is denoted by c, the subscripts v or p being added according as it is a question of specific heat at constant volume or at constant pressure.

In accordance with the previous paragraph, we may thus put either

$$e = \varphi(T, v), \tag{6}$$

so that

$$de = \frac{\partial \varphi}{\partial T}\,dT + \frac{\partial \varphi}{\partial v}\,dv, \tag{7}$$

or

$$e = \chi(T, p), \tag{8}$$

so that

$$de = \frac{\partial \chi}{\partial T}\,dT + \frac{\partial \chi}{\partial p}\,dp. \tag{9}$$

[2] For example, the expansion coefficient of mercury at 0° C. and atmospheric pressure amounts to 0·00018. The compressibility coefficient at atmospheric pressure is 0·0000039; *i.e.*, on an increase in the pressure by one atmosphere, the unit of volume decreases by that amount. It follows from eqn. (5) that the value of the pressure coefficient ($\beta = a/\kappa$) is 0·00018/0·0000039 = 46; *i.e.*, in order to maintain the volume of mercury constant while the temperature is raised 1°, a pressure of 46 atmospheres must be exerted. The accuracy of eqn. (5) is, in fact, confirmed by experiment.

Now, it follows from the *First Principal Law* that

$$(10) \qquad dq = de + p\,dv.$$

Hence, if temperature and volume be taken as the independent variables of state, the First Principal Law becomes

$$(11) \qquad dq = \frac{\partial \varphi}{\partial T}\,dT + \left(\frac{\partial \varphi}{\partial v} + p\right) dv\,.$$

If, on the other hand, pressure and temperature be chosen as independent variables, we have from the equation of state

$$(12) \qquad dv = \left(\frac{\partial v}{\partial T}\right)_p dT + \left(\frac{\partial v}{\partial p}\right)_T dp\,,$$

and therefore, by eqns. (9) and (10),

$$(13) \qquad dq = \left(\frac{\partial \chi}{\partial T} + p\,\frac{\partial v}{\partial T}\right) dT + \left(\frac{\partial \chi}{\partial p} + p\,\frac{\partial v}{\partial p}\right) dp\,.$$

Hence, we find for the specific heats

$$(14) \qquad c_v = \frac{\partial \varphi}{\partial T}\,,$$

and

$$(15) \qquad c_p = \frac{\partial \chi}{\partial T} + p\,\frac{\partial v}{\partial T}\,.$$

Putting, for brevity,

$$(16) \qquad \frac{\partial \varphi}{dv} + p = \zeta\,,$$

eqn. (7) becomes, when eqn. (14) is taken into account,

$$(17) \qquad de = c_v\,dT + (\zeta - p)\,dv\,.$$

Since e is completely determined by T and v, we must have

$$(18) \qquad \frac{\partial}{\partial v}\left(\frac{\partial e}{\partial T}\right)_v = \frac{\partial}{\partial T}\left(\frac{\partial e}{\partial v}\right)_T,$$

the partial differentiations being carried out with the help of eqn. (17). Thus we find

$$(19) \qquad \frac{\partial c_v}{\partial v} = \frac{\partial \zeta}{\partial T} - \frac{\partial p}{\partial T}\,,$$

where p is to be regarded as a function of v and T, so that it has to be differentiated with respect to these two quantities as independent variables.

Like the internal energy, the *entropy* can also be represented as a complete *function of volume and temperature.* According to the Second Principal Law, when the quantities are referred to unit-mass we have for the differential of the entropy

$$ds = \frac{dq}{T}. \tag{20}$$

Hence, in accordance with eqns. (11), (14), and (16),

$$ds = \frac{c_v}{T} dT + \frac{\zeta}{T} dv. \tag{21}$$

Here, again, the relation

$$\frac{\partial}{\partial v}\left(\frac{\partial s}{\partial T}\right)_v = \frac{\partial}{\partial T}\left(\frac{\partial s}{\partial v}\right)_T \tag{22}$$

must hold good, or, by eqn. (21),

$$\frac{1}{T}\frac{\partial c_v}{\partial v} = -\frac{\zeta}{T^2} + \frac{1}{T}\frac{\partial \zeta}{\partial T}. \tag{23}$$

On multiplying eqn. (23) by T and subtracting it from eqn. (19), we find

$$\zeta = T\frac{\partial p}{\partial T}, \tag{24}$$

or, bearing in mind eqn. (2),

$$\zeta = \beta p T. \tag{25}$$

Introducing this value for ζ into eqn. (11), we obtain finally, when eqn. (14) is taken into account, the following important relation which is a statement of the *First Principal Law* when *temperature* and *specific volume* are chosen as *independent variables of state* :—

$$dq = c_v\, dT + \beta\, p\, T\, dv. \tag{26}$$

In applying this equation to an *isothermal change of volume*, we have to put dT equal to zero. Eqn. (26) then shows that, on an isothermal increase in volume, a body absorbs or emits heat according as its *pressure coefficient* at the temperature in question is positive or negative. If we wish to apply eqn. (26) to an *adiabatic* change of state, dq must be equated to zero. An adiabatic increase in volume therefore always produces a cooling effect, provided the pressure coefficient be positive. Since this is generally the case, bodies are in general *heated by adiabatic compression.*

Again, let us investigate the form assumed by the First Principal Law when temperature and pressure are chosen as independent variables. We put, for brevity,

$$\frac{\partial \chi}{\partial p} + p\frac{\partial v}{\partial p} = \eta. \tag{27}$$

Eqn. (9) then becomes, when eqn. (15) is taken into account,

(28) $$de = \left(c_p - p\frac{\partial v}{\partial T}\right) dT + \left(\eta - p\frac{\partial v}{\partial p}\right) dp\,.$$

Since e can be represented as a function of the two independent variables, T and p, we must therefore have

$$\frac{\partial}{\partial p}\left(c_p - p\frac{\partial v}{\partial T}\right) = \frac{\partial}{\partial T}\left(\eta - p\frac{\partial v}{\partial p}\right),$$

or,

(29) $$\frac{\partial c_p}{\partial p} - \frac{\partial v}{\partial T} = \frac{\partial \eta}{\partial T}\,.$$

Moreover, we can make use of the fact that the entropy is also completely determined by T and p. Dividing eqn. (13) by T, and taking eqns. (15) and (27) into account, we find

(30) $$ds = \frac{c_p}{T}\,dT + \frac{\eta}{T}\,dp\,.$$

whence

$$\frac{\partial}{\partial p}\left(\frac{c_p}{T}\right) = \frac{\partial}{\partial T}\left(\frac{\eta}{T}\right),$$

or,

(31) $$\frac{1}{T}\frac{\partial c_p}{\partial T} = -\frac{\eta}{T^2} + \frac{1}{T}\frac{\partial \eta}{\partial T}\,.$$

On multiplying this equation by T and subtracting eqn. (29) from it, we obtain

(32) $$\eta = -T\frac{\partial v}{\partial T},$$

or, in view of eqn. (1),

(33) $$\eta = -\alpha\, v\, T.$$

If we insert this value in eqn. (13), taking eqns. (15) and (27) into account, we obtain finally the following relation as an expression of the *First Principal Law* when *temperature* and *pressure* are chosen as *independent variables of state* :—

(34) $$dq = c_p\, dT - \alpha\, v\, T\, dp.$$

In applying this equation to an isothermal process, dT is again to be equated to zero. Hence, in an *isothermal increase of pressure*, a body absorbs or emits heat according as its *expansion coefficient* at the temperature in question is negative or positive. In an adiabatic change of state ($dq = 0$), the temperature rises on an increase of pressure when the expansion coefficient is positive; when it is negative, an increase in pressure is accompanied by a fall in temperature. In the case of *water*, the expansion coefficient is positive above 4°, but negative between 0° and 4°. Hence, if at a temperature above 4° the pressure to which water is subjected be raised, the temperature of the water rises. If the pressure be raised at a temperature below 4°, the water is cooled.

In eqn. (34) the quantity p may also be negative. In that case we are dealing, not with a pressure, but with a *tension*. An example is afforded by the *extension of a rod by means of a suspended weight*, and an increase in the tension then means that dp is negative. In an adiabatic change of state, an increase in the tension thus produces a cooling effect when the expansion coefficient is positive, and a heating effect when that coefficient is negative. In the case of *rubber* α is negative, and a rubber cord stretched by a weight becomes, in fact, warmer when the tension is increased.

With the help of the formulæ previously derived, we can easily obtain a relation connecting the specific heat at constant pressure with that at constant volume. In doing this we commence with eqn. (12). Applying eqns. (1) and (3), eqn. (12) becomes

$$dv = \alpha\, v\, dT - \kappa\, v\, dp\,. \tag{35}$$

Introducing this value into eqn. (26), we find

$$dq = (c_v + \alpha\,\beta\, p\, v\, T)\, dT - \kappa\,\beta\, p\, v\, T\, dp\,. \tag{36}$$

In accordance with eqn. (5), we may also write this as

$$dq = \left(c_v + \frac{\alpha^2}{\kappa}\, v\, T\right) dT - \alpha\, v\, T\, dp\,. \tag{37}$$

Comparing this with eqn. (34), we obtain the relation

$$c_p - c_v = \frac{\alpha^2}{\kappa}\, v\, T\,. \tag{38}$$

κ is always positive, when the volume of a body decreases with increasing pressure. The specific heat at constant pressure is therefore always larger than that at constant volume, or

$$c_p > c_v\,. \tag{39}$$

In solid bodies, the quotient α^2/κ is very small, so that the difference between the two specific heats of solid bodies is generally negligible.

§ 146. The Joule-Thomson Effect.

An important application of the general laws of changes of state is to be found in a famous experiment performed by *William Thomson* and *Joule*, in 1853, and the results thereby obtained have been in many respects of great significance for theoretical physics. This experiment essentially consists in allowing a *gas* to flow over *adiabatically* from a vessel in which, at a pressure p', its specific volume is v' into another vessel in which, at a pressure p'', its specific volume is v''. If the most exact application possible of the principal laws of thermodynamics to this process is to be made, the passage of the gas must be so arranged that (according to § 134) *deviations from states of equilibrium are as small as possible.* This is effected in Thomson and Joule's experiment by *forcing* the gas very slowly *through a tube* in which is fitted a *plug of wool.* The difference in pressure can easily be measured, as can the specific volume, and any chance difference in the temperatures of the gas before and after the plug.[1]

If we assume the cross-section of the cylindrical tube to be f, then the length of tube occupied by unit-mass of the gas before it reaches the plug is v'/f, whilst, after passing the plug, it is v''/f. To a certain extent the plug divides the tube into two gas-holders I and II. When unit-mass of the gas passes out of container I, work is thereby performed over a distance v'/f *in the direction* of a force $p'f$. The work performed, or gained, is thus $p'v'$. On the other hand, since the unit-mass is compressed into container II, work is expended *against* a force $p''f$ over a distance v''/f. The work required to transfer unit-mass from the first to the second container is therefore given by the expression $(p''v'' - p'v')$, and, according to the Principle of the Conservation of Energy, this quantity must be equal to the decrease in internal energy (e) thereby experienced by the unit-mass, seeing that the transfer is an adiabatic one. We must therefore have

$$(1) \qquad e' - e'' = p''v'' - p'v' .$$

If we assume that the states of the gas in the two containers differ but little from each other, we may put

$$v'' - v' = \Delta v , \qquad p'' - p' = \Delta p , \qquad e'' - e' = \Delta e .$$

Neglecting small quantities of the second order, we can write eqn. (1) in the form

$$(2) \qquad \Delta e + v\,\Delta p + p\,\Delta v = 0 .$$

When pressure and temperature are chosen as independent variables of state, we have, by § 145 eqns. (10) and (34),

$$\Delta e + p\,\Delta v = c_p\,\Delta T - \alpha\, v\, T\,\Delta p .$$

[1] The fundamental idea of Thomson's experiment (the adiabatic transference of a gas from one vessel to another) had already been suggested by Gay-Lussac. The experimental method was perfected, first by Joule, and then by Thomson, who carried out his later researches conjointly with Joule.

Hence, it must follow from eqn. (2) that

$$c_p \Delta T + v (1 - \alpha T) \Delta p = 0,$$

or

$$(3) \qquad \Delta T = \frac{v (\alpha T - 1)}{c_p} \Delta p .$$

This equation determines the change in temperature experienced by the gas on passing through the plug of wool, where $(-\Delta p)$ represents the excess pressure under which the gas is forced through the tube. As we shall see later, the experimental results afford important clues to the nature of the gases used; but, in addition to that, they also permit of the *reduction* of the scale of a thermometer to a *thermodynamic temperature scale*, such as would be obtained by means of measurements on cyclic processes.

In order to effect this reduction, let us denote the temperature as measured on a thermometer by θ, and the thermodynamic temperature by T; then we know from the definition of the temperature that, at the two fundamental points (ice-point and boiling-point), T and θ coincide. Otherwise in general we have, approximately,

$$(4) \qquad \Delta T = \Delta \theta \frac{dT}{d\theta} .$$

In like manner, we distinguish by the symbols α and α' the expansion coefficients according as they refer to the thermodynamic or thermometric temperature, and, similarly, we distinguish between c_p' and c_p. Then, according to § 145 eqn. (1),

$$(5) \qquad \alpha' = \alpha \frac{dT}{d\theta}; \qquad c_p' = c_p \frac{dT}{d\theta} .$$

Hence, if c_p be brought across to the left-hand side, eqn. (3) may be written in the form

$$c_p' \frac{d\theta}{dT} \Delta\theta \frac{dT}{d\theta} = \left(v \alpha' T \frac{d\theta}{dT} - v\right) \Delta p .$$

On re-writing the equation with only the quotient dT/T on one side, and then integrating both sides between the ice-point (the same for both scales) and the unknown thermodynamic temperature T (corresponding to a given thermometric temperature θ), we find

$$(6) \qquad \log_e T - \log_e T_0 = \int_{\theta_0}^{\theta} \frac{v \alpha' \, d\theta}{v + c_p' \frac{\Delta\theta}{\Delta p}} .$$

Now all the quantities following the integration symbol can be measured in relation to the thermometric temperature. The value of the definite integral as a function of θ can hence be derived for arbitrary values of θ. If J be the value of the integral, we have therefore

(7) $$\log_e T - \log_e T_0 = J \, .$$

Again, we can evaluate the definite integral with the ice-point and boiling-points as integration limits ; let the value then obtained be J^*. Denoting the thermodynamic absolute temperature of the boiling-point by T_{100}, we thus have

(8) $$\log_e T_{100} - \log_e T_0 = J^* .$$

It follows therefore from eqns. (7) and (8) that

(9) $$T = T_0 \, e^J \, ,$$

and

(10) $$T_{100} = T_0 \, e^{J^*} .$$

In accordance with the definition of temperature, however, the difference between T_{100} and T_0 must be exactly 100°, so that

(11) $$100° = T_0 \, (e^{J^*} - 1) \, .$$

Hence

(12) $$T = \frac{100° \, . \, e^J}{e^{J^*} - 1} \, .$$

With the aid of this relation, the scale of the hydrogen thermometer can be reduced to the thermodynamic scale. As a matter of fact, the divergences are found to be very small (except at very low temperatures). Between the ice-point and the boiling-point, they do not exceed 1/20°. In that interval the hydrogen thermometer indicates too high a temperature ; above the boiling-point, the reading is too low. At 300° above the ice-point, the temperature as given by the hydrogen thermometer is nearly 1/2° too low.

§ 147. The Coexistence of Phases.

When a substance can exist in different modifications, between which transitions are necessarily accompanied by an absorption or a liberation of energy, a distinction is made in a thermodynamical sense between the different *phases* of the substance.[1] The most important example of such phases is afforded by the *states of aggregation* of a body.

[1] Phase literally means "form of appearance."

Allotropic modifications of a substance, however, also represent different phases. For instance, in order to transform rhombic crystalline sulphur into the monoclinic variety, heat must be introduced.

The quantity of heat required to transform unit-mass of a substance from a phase A into a phase B is called the *specific heat of reaction* of this transformation. When the heat of reaction is positive, B is called the *higher* and A the *lower phase*, and conversely when the heat is negative. Of the states of aggregation the highest is represented by the gaseous phase, and the lowest by the solid phase. The specific heats of reaction accompanying changes of the state of aggregation are known as the *heats of fusion, vaporization*, and *sublimation*.

We define as a *saturated complex of coexisting phases* a system which consists of different phases of a substance in thermodynamic equilibrium, its entropy thus representing a maximum for given values of the total mass, energy, and volume. The condition for the coexistence of phases in thermodynamic equilibrium is therefore

$$(1) \qquad \delta S = 0,$$

with the subordinate conditions

$$(2) \qquad \delta m = 0, \qquad \delta E = 0, \qquad \delta V = 0.$$

Let us consider, as a simple example, a system consisting of only *two* coexisting phases. We shall denote by s, v, and e respectively the values of the entropy, the volume, and the internal energy referred to unit-mass. If the total mass of the system be m, let mx be the mass of the higher phase and $m\,(1 - x)$ that of the lower. We shall distinguish the quantities determining the higher phase by one dash and those determining the lower phase by two dashes, so that we have the relations.

$$(3) \qquad s = x s' + (1 - x)\, s'',$$

$$(4) \qquad v = x v' + (1 - x)\, v'',$$

$$(5) \qquad e = x e' + (1 - x)\, e''.$$

Hence, since the total mass is constant, it follows from eqn. (1) that

$$(6) \qquad x\, \delta s' + (1 - x)\, \delta s'' + (s' - s'')\, \delta x = 0.$$

On the other hand, in accordance with the Second Principal Law,

$$(7) \qquad \delta s' = \frac{\delta e' + p'\, \delta v'}{T'},$$

and, likewise,

(8) $$\delta s'' = \frac{\delta e'' + p''\,\delta v''}{T''}.$$

We can, however, express $\delta e''$ and $\delta v''$ in terms of $\delta e'$ and $\delta v'$. For, on varying eqn. (4), we find, inasmuch as δV must vanish [according to eqns. (2)] and the total mass is constant, that

(9) $$x\,\delta v' + (1-x)\,\delta v'' + (v' - v'')\,\delta x = 0.$$

Similarly, since δE vanishes according to eqns. (2), variation of eqn. (5) gives

(10) $$x\,\delta e' + (1-x)\,\delta e'' + (e' - e'')\,\delta x = 0.$$

Solving eqns. (9) and (10) with respect to $(1-x)\,\delta v''$ and $(1-x)\,\delta e''$, and inserting the values thus obtained in eqn. (8), we find

$$(1-x)\,\delta s'' = -\frac{x\,\delta e'}{T''} - \frac{x\,p''\,\delta v'}{T''} - \left[\frac{e'-e''}{T''} + \frac{p''\,(v'-v'')}{T''}\right]\delta x.$$

Finally, introducing this relation into eqn. (6) and taking eqn. (7) into account, we obtain

(11) $$\left\{\begin{aligned} &x\,\delta e'\left[\frac{1}{T'} - \frac{1}{T''}\right] + x\,\delta v'\left[\frac{p'}{T'} - \frac{p''}{T''}\right] + \delta x\Big[s' - s'' \\ &\qquad\qquad - \frac{e'-e''}{T''} - \frac{p''\,(v'-v'')}{T''}\Big] = 0. \end{aligned}\right.$$

Since the variations are arbitrary and mutually independent, eqn. (11) can only be satisfied when each of the three expressions in square brackets vanishes in itself. We must therefore have

(12) $$T' = T'', \qquad p' = p'',$$

and, furthermore, if the common temperature be denoted by T and the common pressure by p,

(13) $$-s' + \frac{e'}{T} + \frac{v'p}{T} = -s'' + \frac{e''}{T} + \frac{v''p}{T}.$$

But, according to §144 eqn. (13), the thermodynamic potential referred to unit-mass is given by

(14) $$\varphi = e - Ts + v\,p.$$

Eqn. (13) may hence also be written in the form

(15) $$\varphi' = \varphi''.$$

Two coexisting phases can thus be in thermodynamic equilibrium only when both phases are at the *same pressure* and the *same temperature*, and when unit-masses of the phases have the *same thermodynamic potential.*

It follows from the definition of thermodynamic potential (§ 144) that φ' is a quite definite function of p' and T', and similarly that φ'' is a quite definite function of p'' and T''. Since, however, in accordance with eqn. (12) p' is equal to p'' and T' to T'', there must exist a perfectly definite connection between p and T in the form of an equation

$$\chi(p, T) = 0. \tag{16}$$

In the case of a saturated complex of two phases, every pressure is therefore associated with a quite definite *saturation temperature*, and, conversely, every temperature is associated with a quite definite *saturation pressure*. If we construct in a p-T-plane the curve corresponding to eqn. (16), we obtain the so-called *saturation curve.*

Since a transition between two phases is only possible as long as both phases are coexisting, a definite change of phase at a given pressure can only occur at a quite definite temperature (provided, indeed, that the transition can occur at all at that pressure). At a temperature which, at the given pressure, is lower than the corresponding saturation temperature, a substance can only exist in the lower phase. If the temperature be raised at constant pressure to the saturation temperature, the transformation of the substance from the lower to the higher phase commences. On a further addition of heat, this transformation proceeds *isothermally*. It is not until the substance has been completely transformed into the higher phase that (at constant pressure) a further rise in temperature becomes possible. It is at once evident why, at normal atmospheric pressure, every substance has a definite melting point, and why, during fusion, the temperature remains constant.[2]

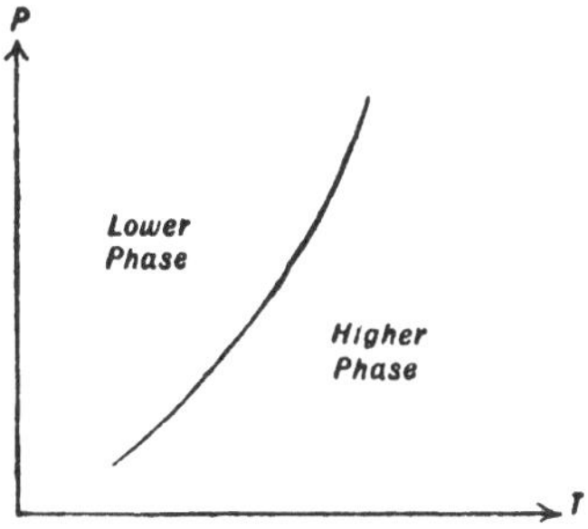

FIG. 119.

If we proceed parallel to the T-axis from a point on the saturation curve (Fig. 119) in the direction of increasing temperature, we enter the field of the higher phase. On the other hand, if we proceed from that point parallel to the p-axis in the direction of increasing pres-

[2] The constancy of temperature during fusion was first discovered, simultaneously, by *Deluc* and *Black* about 1750.

sure, we must accordingly enter the field of the lower phase. If, at a given temperature, the pressure be less than the corresponding saturation pressure, the substance can exist only in the higher phase. When the pressure is isothermally increased to the saturation pressure, a transformation of the substance into the lower phase at constant pressure commences. It is only when this transformation is completed that a further increase in pressure is possible at constant temperature. The saturation pressure thus represents the *maximum pressure* at which the higher phase can exist at the temperature in question.[3]

In accordance with the equation of state, the specific volume of a substance is determined by its pressure and temperature, so that, since the saturation pressure depends only on the saturation temperature, the *specific volume* of a phase in the saturated state can be a *function only of the saturation temperature.*[4] Since the mass contained in unit-volume is reciprocal to the specific volume, a cubic centimetre of air will always contain the same mass of water vapour when saturated at a given temperature: this quantity is called the *absolute humidity* at the temperature in question. In all vapours an increase in pressure is accompanied by an increase in density, and since the saturation pressure of water vapour is its maximum pressure, it follows that the saturation density of the vapour also represents its *maximum density* for the temperature concerned. The ratio between the density of unsaturated water vapour and the maximum density at that temperature is called the *relative humidity.*[5] The temperature which would correspond to the saturation temperature for the actual density of water vapour present is known as the *dew-point.*

Let us consider the transformation of unit-mass from the lower to the higher phase at constant pressure; then, for the (in this case positive) heat of reaction, the First Principal Law gives the relation

$$\lambda = e' - e'' + p\,(v' - v'')\,; \tag{17}$$

whence follows, according to eqns. (14) and (15), the simple formula

$$\lambda = T\,(s' - s'')\,. \tag{18}$$

If we remember that the phase transformation occurs

[3] The saturation pressure of water vapour is simply termed the *vapour tension*; unsaturated vapour is called *superheated*, and approximately obeys the gas laws (§ 148).

[4] Or, what amounts to the same thing, of the saturation pressure.

[5] It is generally expressed as a percentage.

at constant temperature, we obtain, with the help of §145 eqns. (26) and (2), the relation

$$\lambda = \frac{dp}{dT} T (v' - v''). \tag{19}$$

When the higher phase has the larger specific volume (or, the smaller density), dp/dT is positive, and the saturation temperature therefore increases with the pressure. If, on the other hand, the higher phase has the greater density, an increase in pressure produces a lowering of the transformation temperature. Since liquid water at 0° possesses a greater density than ice, the *freezing-point* is *lowered* by an increase in pressure,[6] whereas the *boiling-point* of water is *raised* thereby.

As long as the complex of two coexisting phases is saturated, it follows from what we have previously said that the state of each phase is completely defined if, besides the saturation temperature, the quantity x determining the mass ratio of the two phases be given ; for pressure and specific volume are alike fixed by the saturation temperature. For changes of state wherein the complex *remains saturated*, the quantities T and x may thus be regarded as *independent variables of state of the complex ;* the quantity x, in fact, also represents a parameter, a change in which involves an expenditure of work $\lambda\, dx$ in amount. If heat (of amount dq for unit-mass) be introduced into the complex, the state of saturation being maintained, the temperature of the complex will, on the one hand, be raised by the amount dT, whilst, on the other hand, the quantity x will be increased by the amount dx. Denoting by γ' and γ'' the *specific heats of the two phases in the saturated state*, we obtain from the First Principal Law the relation

$$dq = [x\gamma' + (1 - x)\gamma'']\, dT + \lambda\, dx. \tag{20}$$

Since, in accordance with the Second Principal Law, dq/T must be the complete differential of a function determined by the two independent variables of state, T and x, we must have

$$\frac{\partial}{\partial x}\left[\frac{x\gamma' + (1-x)\gamma''}{T}\right] = \frac{\partial}{\partial T}\left(\frac{\lambda}{T}\right),$$

or,

$$\gamma' - \gamma'' = \frac{\partial \lambda}{\partial T} - \frac{\lambda}{T}. \tag{21}$$

[6] An increase of one atmosphere in the pressure lowers the freezing-point by 0·0075° C. At a pressure of 1,000 atms., the freezing-point is no higher than − 7·5° C. On the other hand, for substances other than water, which, at the melting-point, have a greater density in the solid than in the liquid state, an increase in pressure raises the melting-point: *e.g.*, such is the case with paraffin wax.

This equation, which was first established by *Clausius* (1857), enables us, for example, to calculate the specific heat of saturated water vapour when the heat of vaporization (λ) is given as a function of the saturation temperature, and also the specific heat of water is known.

Hitherto we have considered the coexistence of only two phases. In thermodynamic equilibrium this is governed by the condition that, at equal pressure and temperature, φ' shall be equal to φ''. If *three phases* are to *co-exist*, then for equal temperature and pressure we must have

$$\varphi' = \varphi'' = \varphi''' . \tag{22}$$

Since thermodynamic potential is a function of pressure and temperature, we have two equations with two unknowns for the coexistence of three phases. The solution of these equations with respect to p and T yield a so-called *triple-point*, in which the three saturation curves intersect. More than three phases of a single chemical substance cannot thus coexist, because otherwise there would be more than the right number of equations to determine the unknowns p and T. Fig. 120 represents the three saturation curves of water (fusion, vaporization, and sublimation curves). They intersect at a triple-point (Δ), the pressure for which amounts to 0·006 atm., and the temperature of which is 0·0075°C.[7]

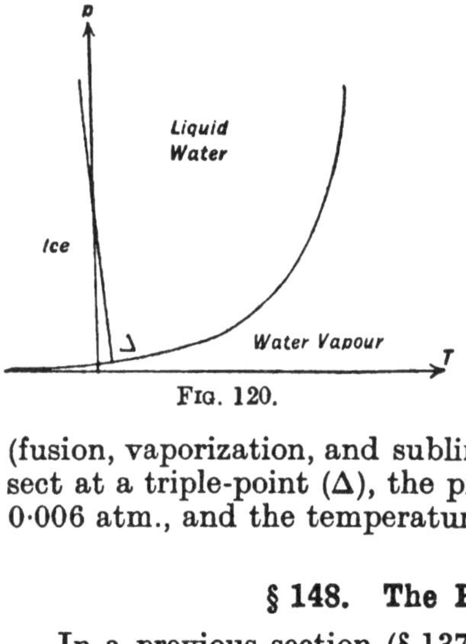

Fig. 120.

§ 148. The Perfect Gas.

In a previous section (§ 137) we considered, as a *special case of a canonical distribution of energy*, a distribution in which the individual energy is a pure quadratic function of some of the individual quantities of state, whilst the positional coordinates of the individuals (regarded as points) are

[7] Since there are various allotropic modifications of ice, this triple-point is by no means the only one for water.—For the coexistence of phases in *heterogeneous* systems, reference should be made to the special text-books of thermodynamics. Theory leads to the so-called *phase-rule of Gibbs*, which states that the largest number of phases which can coexist in one state is greater by two than the number of the chemical components.

among those of the individual quantities of state upon which the individual energy does *not* depend. If we consider a system of molecules to which these assumptions with respect to the distribution of heat energy apply, the following equations must then be satisfied :

(1) $$E = \frac{s}{2} N k T,$$

and

(2) $$p V = N k T.$$

In these equations N denotes the number of molecules, k Boltzmann's constant, and s the number of quantities of state between which and the molecular energy a quadratic relation exists.

We shall now replace the number of molecules by the product of the number (n) of gram-molecules and Loschmidt's number (L), and we define a new *universal constant* by the relation

(3) $$k L = R.$$

Eqns. (1) and (2) then become

(4) $$E = \frac{s}{2} n R T,$$

and

(5) $$p V = n R T.$$

Moreover, since the energy of a homogeneous body must, in general, be completely determined by volume and temperature, it follows from eqn. (4) that

(6) $$\left(\frac{\partial E}{\partial V}\right)_T = 0.$$

In isothermal processes, the internal energy must therefore be independent of the volume. Conversely, when changes in volume are not accompanied by any change in the internal energy E, and thus proceed *adiabatically and without the performance of external work*, the *temperature* of such a molecular system must remain *constant.*

The properties expressed by the last three equations (4, 5 and 6) have actually been confirmed in experimental physics in the case of *gases*, the agreement between theory and experiment being the more exact the more perfect the gas, *i.e.*, the further removed from liquefaction it is under the conditions of observation. We are therefore apparently justified in defining, in a theoretical-statistical sense, as an *ideal* or *perfect gas* a homogeneous body to whose *molecular distribution of energy* our original *hypotheses* apply. Any *picture* that we may draw of a gas satisfying these assumptions must then necessarily lead to eqns. (4), (5), and (6).

Now, the simplest picture of a gas whch we can in this way construct is the following :—We imagine the *molecules* to be executing *rapid translatory motions*, the distances between them, however, being so large that any eventual forces acting between the molecules can be neglected, and the molecules treated *as particles*. The molecular energy is then equal to the sum of the kinetic energy of translation and the *intramolecular energy*. If we suppose (as we have already done in explaining rotational and band spectra [1]) that the molecules *rotate rigidly*, then the intramolecular potential energy does not require to be considered at all. Moreover, the internal kinetic energy must obviously be representable as a quadratic function of the time-derivatives of those coordinates which describe the internal configuration of the molecules. The hypotheses previously made therefore hold good, for the molecular energy is independent of the positional coordinates of the molecules themselves.[2] The molecular energy is a quadratic function of a definite number (denoted by s) of independent individual quantities of state (in the sense of § 137) ; in point of fact we have, for *monatomic molecules*,

(7) $$s = 3 ,$$

whereas, for *polyatomic gases*,

(8) $$s > 3 .$$

The first law to be derived empirically for gases states that, at a given constant temperature, *pressure and volume* are *inversely proportional to each other*. This law was discovered in 1660 by *Boyle*, and is therefore known as *Boyle's Law*.[3] At the beginning of the 19th century (1802), *Gay-Lussac* observed that at constant volume the pressure, and at constant pressure the volume, increases linearly with the temperature. He also found that the expansion and pressure coefficients were independent of the special nature of the gas, and identical at equal temperatures. It does actually follow from eqn. (5), and from the equations defining

[1] *Vide* § 129.

[2] The coordinates meant are those determining the position of the centre of gravity of the molecules.

[3] This law is also wrongly named after Mariotte, who repeated Boyle's measurements with greater accuracy.

the expansion coefficient (α) and the pressure coefficient (β) [§ 145 eqns. (1) and (2)], that

$$(9) \qquad \alpha = \beta = \frac{1}{T}.$$

Consequently, if we observe by what fraction the volume increases at constant pressure, or the pressure increases at constant volume, when the temperature is raised from the ice-point to 1° C., this fraction is equal to the reciprocal value of the absolute temperature of the ice-point. Since a value of 1/273 was obtained for the fractional increase, the number resulting from the addition of the number 273 (more accurately, 273·1) to the measured temperature was defined as the *absolute temperature.* If the expansion coefficient be referred to the normal volume at the ice-point, instead of [as in eqn. (9)] to the volume at the temperature of observation, the expansion coefficient thus defined is equal to 1/273 at all temperatures. The same consideration applies, in corresponding fashion, to the pressure coefficient.

The universal constant R is called the *gas constant.* Its value is found by measurement to be

$$(10) \qquad R = 8{\cdot}315 \times 10^7 \text{ ergs/deg.},$$

or, using the calorie as unit of heat,

$$(11) \qquad R = 1{\cdot}98 \text{ cal./deg.}$$

With the help of the previously given value for the Loschmidt number [§ 88 eqn. (6)], we find for *Boltzmann's constant*

$$(12) \qquad k = \frac{R}{L} = 1{\cdot}37 \times 10^{-16} \text{ ergs./deg.}$$

It was first experimentally discovered by *Gay-Lussac,* and subsequently confirmed by *Joule,* that scarcely any change in temperature accompanies the expansion of a gas when no external work is performed. Since, by eqn. (9), αT does not differ from unity for an ideal gas, a temperature change could not be expected in a Joule-Thomson experiment on an *ideal* gas [*cf.* § 146 eqn. (3)]. It is just refinements in the experimental methods, however, that have demonstrated the divergences of actual gases from the ideal gas laws.

A very important conclusion, to be derived directly from eqn. (2), is to the effect that, *at equal pressure and equal temperature, equal volumes of different gases* must contain the *same number of molecules.* This important law was first established in 1811 by *Avogadro,* and is therefore known by his name. It may also be stated in the form that, at given temperature and given pressure, the *densities of different gases* are *in the ratios of their molecular weights.* By means of this law, the molecular weights of various substances can be determined, and, moreover, important conclusions as to the number of atoms in molecules can be drawn. Thus, Avogadro himself discovered that one litre of hydrogen and one litre of chlorine combine to form two litres of hydrogen chloride. In accordance with Avogadro's Law, there must therefore be twice as many hydrochloric acid molecules present after the reaction as there were hydrogen molecules before it. Since, however, each hydrochloric acid molecule must contain a hydrogen atom, the hydrogen molecules must have each consisted of two atoms before the reaction. Analogous considerations apply to the chlorine molecules. In like manner, the number of atoms in the molecules of other gases and vapours could be determined. Furthermore, eqn. (2) leads directly to the so-called *Dalton's Law* (1802), which states that the pressure exerted by a *gaseous mixture* is *equal to the sum of the partial pressures* which the gases would separately exert, were each of them severally to occupy the volume actually occupied by the mixture.

In accordance with eqn. (6), the *First Principal Law* becomes, in the case of an *ideal gas*,

$$(13) \qquad dQ = \frac{\partial E}{\partial T} dT + p\, dV .$$

According to the gas law expressed by eqn. (5),

$$p\, dV = d(pV) - V\, dp = n\, R\, dT = V\, dp.$$

Eqn. (13) may therefore also be written in the form

$$(14) \qquad dQ = \left(\frac{\partial E}{\partial T} + n\, R\right) dT = V\, dp .$$

Now, for changes of state at constant volume, dV is to be equated to zero in eqn. (13), whereas, for changes of state at constant pressure, dp vanishes in eqn. (14). If we denote the molecular heats (referred to the gram-molecule) at constant volume and constant pressure by C_v and C_p, we obtain from eqns. (13) and (14) the important relation

$$(15) \qquad C_p - C_v = R .$$

For the *ratio* (κ) *of the two specific heats*[4] (whether referred to

[4] The experimental determination of the specific heat at constant pressure is very much more easily effected than at constant volume; hence the latter is indirectly calculated when the ratio κ has been derived with the help of eqn. (21).

unit-mass or the gram-molecule is immaterial), we find from eqns. (13) and (14)

$$\kappa = 1 + \frac{nR}{\frac{\partial E}{\partial T}},$$

or, by eqn. (4),

$$\kappa = 1 + \frac{2}{s}. \tag{16}$$

In the case of *monatomic gases*, κ is hence [by eqn. (7)] equal to 5/3, while, quite generally, in accordance with eqn. (8) the relation

$$1 < \kappa \leq \frac{5}{3} \tag{17}$$

must hold good. In point of fact, measurements made on monatomic gases have yielded the value 5/3, and the accuracy of eqn. (17) has been confirmed for polyatomic gases.

Eqn. (13) may also be written in the form

$$dQ = n\,C_v\,dT + p\,dV,$$

or, utilizing the gas law [eqn. (5)],

$$dQ = n\,C_v\,dT + \frac{nRT}{V}\,dV. \tag{18}$$

It follows that, in the case of an *adiabatic change of state*,

$$C_v\frac{dT}{T} + R\frac{dV}{V} = 0. \tag{19}$$

On the other hand, by eqn. (15),

$$\frac{R}{C_v} = \kappa - 1, \tag{20}$$

and hence, for an adiabatic change of state of an ideal gas, we find the relation

$$T V^{\kappa-1} = \text{const.} \tag{21}$$

Using the gas law to replace temperature by volume and pressure, we can write eqn. (21) in the form

$$p V^{\kappa} = \text{const.} \tag{22}$$

Finally, it may be mentioned that the *mechanical equivalent of heat* may be easily derived from eqn. (15). For, according to the gas law, the gas constant is immediately given in absolute measure, if simply pressure and volume be observed at different temperatures. Specific heats are, however, actually measured in calories (not in ergs). If we denote the molecular heats measured in calories by C_v' and C_p', these quantities are equal to C_v/J and C_p/J, where J is the mechanical equivalent of heat. Hence

$$J = \frac{R}{C_p' - C_v'} = \frac{R}{C_p'\left(1 - \frac{1}{\kappa}\right)}. \tag{23}$$

By means of this relation, *Robert Mayer* was able, in 1842, to calculate the mechanical equivalent of heat from experimental data referring to the specific heats of gases.

As we have already mentioned, the concept which refers the heat content of gases to the rapid translatory motion of molecules in all directions was developed into an exact theory by *Clausius* in 1857.[5] This theory was based on the assumption that the molecules behave as solid and completely *elastic spheres*, and execute *rectilinear* motions with *constant velocity* until they collide with other molecules or a fixed wall. The collisions will thus cause no alteration in either the total momentum or the total kinetic energy. In virtue of this concept, *gas pressure* appears to be due to the continual impacts of the molecules with the walls of the containing vessel. The hypothesis led not only to a simple derivation of the ideal gas law, but also enabled Clausius to calculate the *molecular velocities* from the experimentally determined dependency of the pressure on the temperature.[6] The values of the velocities thus found for hydrogen are about 1,700, for nitrogen about 450, and for oxygen about 420 metres a second, referred to normal atmospheric pressure and the temperature of the ice-point.

Clausius' considerations were still based on the hypothesis that, not only may the collisions between the individual molecules be neglected, but also that all the molecules in a gas possess the same velocity. Neither of these two assumptions, however, is strictly true. In 1858 *Clausius* introduced the important new concept of the *mean free path.* By this he understood the average distance traversed by a

[5] Clausius' researches were preceded by those of Joule and Krönig by a few years. The first elements of the kinetic theory of gases, moreover, reach back to *Daniel Bernoulli*, who published in 1738 the fundamental concept of this theory in his " Hydrodynamics."

[6] If u be the molecular velocity and ϱ the gas density, we have simply

$$p = \frac{\varrho u^2}{3}. \tag{24}$$

molecule between two successive collisions.[7] With the aid of a simple (but not quite accurate) consideration, he succeeded in establishing a relation between the mean free path and, on the one hand, the number of molecules in unit-volume and, on the other hand, the diameter of the molecules which he assumed to be spherical.[8]

In 1860, starting from quite a different point of view, *Maxwell* actually succeeded in calculating quite definite values for the mean free path. Maxwell's fundamental relation concerned the *internal friction of gases*. If, within a gas, motions occur in which numerous molecules are involved, *e.g.*, currents, it is found by experience that a force, known as internal friction, is present, this force having the direction of the velocity gradient and being proportional to it. (By velocity, of course, we mean, not the molecular, but the stream velocity.) The proportionality factor between the force referred to unit-surface and the velocity gradient is called the coefficient of internal friction. Now, from the point of view of the kinetic theory of gases, Maxwell was able to interpret internal friction as a transfer of momentum from more to less rapidly moving layers of gas. With the help of this interpretation, he obtained an equation connecting the mean free path with the coefficient of internal friction (viscosity), the density of the gas, and the molecular velocity.[9] Since the molecular velocities had already been determined by Clausius, and the coefficient of viscosity for various gases could be very accurately measured, Maxwell was able to calculate definite values for the mean free path. For example, in the case of hydrogen he found $1{\cdot}85 \times 10^{-5}$ cm.; this would mean that, on an average, a molecule

[7] According to modern theory, there is, of course, no real sense in speaking of collisions. The molecules of a gas ought to be compared to miniature planetary systems, with electrical forces proceeding from their component parts. Electrical forces must therefore be acting between the molecules and constantly diverting them from their rectilinear paths, the deviations being larger the closer the mutual approach of two molecules. The molecular movements would thus be zig-zag in nature, but without any sharp corners.

[8] If λ be the mean free path, n the number of molecules in unit-volume, and σ the diameter of a molecule, we have

$$\lambda = \frac{3}{4\pi\, n\, \sigma^2}. \tag{25}$$

[9] If η be the coefficient of viscosity, we have

$$\eta = \frac{\varrho\, u\, \lambda}{3}, \tag{26}$$

the other symbols being as before. Since u depends only on the temperature, the viscosity is thus independent of the pressure. This result may at first appear remarkable, but it is completely confirmed by experiment.

experiences some ten thousand million collisions every second. This number could, of course, only have any real meaning if the distinctly arbitrary assumptions underlying the theory of gases in its usual form were actually valid.

Maxwell's paper, in which he first calculated the mean free path, has, however, been of the greatest importance for another reason. For, in this paper, he also replaced the earlier hypothesis which ascribed the same velocity to all the molecules, by his famous *law of distribution of molecular velocities.* He assumed the distribution of energy among the molecules to be canonical ; the concept of canonical distribution, so important for modern physics, was, indeed, suggested by Maxwell in connection with the problem of molecular velocities.[10] A uniform velocity was thus replaced by an average velocity.[11]

After Clausius and Maxwell had thus determined two of the three fundamental unknowns in the kinetic theory of gases, *viz.*, the molecular velocity and the mean free path, the most interesting problem left was the determination of the molecular diameter, *i.e.*, the evaluation of the *true size of the molecules.* This problem was first solved by *Loschmidt*, in 1865. Loschmidt started from the assumption that, when a gas is *liquefied*, the molecules occupy the whole of the volume of the liquid formed ; and, on the basis of this certainly very crude hypothesis, he was able to calculate an *upper limit for the diameter of a molecule.*[12] Thus, he found for the diameter of " air molecules " (*i.e.*, of nitrogen or

[10] The designation " canonical " was, indeed, not used by Maxwell himself, but was introduced later by Gibbs. Maxwell's law of velocity distribution is obtained from the general equation of the canonical distribution of energy. If $\dot{x}$, $\dot{y}$, $\dot{z}$ be the velocity components of the molecules, the number of molecules whose velocity components have values lying between $\dot{x}$ and $(\dot{x} + d\dot{x})$, $\dot{y}$ and $(\dot{y} + d\dot{y})$, and $\dot{z}$ and $(\dot{z} + d\dot{z})$ are given by the expression

$$(27) \qquad dN' = N \left(\frac{m}{2\pi k T}\right)^{\frac{3}{2}} e^{-\frac{m}{2kT}(\dot{x}^2 + \dot{y}^2 + \dot{z}^2)} \, d\dot{x}\, d\dot{y}\, d\dot{z} .$$

In this expression N denotes the total number of gas molecules, m the molecular mass, and k Boltzmann's constant, so that $3kT/2$ represents the energy of a molecule at the temperature T. For the number (dN'') of molecules with velocities lying between u and du, we find

$$(28) \qquad dN'' = 4\pi N u^2 \left(\frac{m}{2\pi k T}\right)^{\frac{3}{2}} e^{-\frac{m u^2}{2kT}} du .$$

[11] The ratio of the most probable to the mean velocity is $\sqrt{\pi/2}$ to 1. The mean value of the kinetic energy is 3/2 times as large as the kinetic energy corresponding to the most probable velocity.

[12] If γ be the ratio between the volume occupied by the gas before liquefaction to that occupied after liquefaction, Loschmidt found

$$(29) \qquad \sigma = 8\gamma\lambda .$$

oxygen molecules), a value of about 10^{-7} cm. Now, as was mentioned above, Clausius had already derived an equation connecting mean free path and molecular diameter with the number of molecules contained in unit-volume [Note 8, eqn. (25)]. Hence Loschmidt was also able to determine the order of magnitude of the number of molecules contained in 1 cm.^{3} of a gas at 0° C. and 1 atm., which number, in accordance with Avogadro's Law, must obviously be the same for all gases. He found as lower limit for this number about 10^{18}. From the known density and molecular weight of any gas, the number of molecules contained in a gram-molecule would thus follow as about 3×10^{22} as a lower limit, whereas the actual value of this constant, as we know now with great exactness (*vide* § 88) is about twenty times as large.

After Loschmidt had thus first calculated the true size of molecules, *Boltzmann*, a year later (1866), was able to give a statistical explanation of *irreversibility* and entropy on a gas-theoretical basis by means of his so-called "*H-theorem*."[13] Boltzmann's later investigations developed the kinetic theory of gases into a complete system. Two further interesting fields for the application of the kinetic theory were opened up, on the one hand by the theory of *dilute solutions*, in which the molecules of the dissolved substance behave as gas molecules, and, on the other hand, by the theory of *free electrons*, which likewise exhibit an analogous behaviour to that of gas molecules, and which are responsible for *electrical* and *thermal conductivity in metals*. The extension of the results of the theory of gases to free electrons led to the explanation of a law which had already been experimentally discovered by *Wiedemann* and *Franz*. This law is to the effect that the ratio between thermal conductivity and electrical conductivity is the same in all metals at the same temperature. A kinetic-theoretical investigation showed, further, that this ratio is proportional to the absolute temperature, and this conclusion was well confirmed by experiment.

Finally, the kinetic theory of matter has had to face in recent years a whole series of new problems concerned with the *phenomena of molecular fluctuations*. The origin of these lies in the general fact that an average value, which is obtained from only a small number of individual values, may experience relatively considerable variations, not through any external cause, but simply because of *individual irregularities*. Spontaneous fluctuations must therefore be

[13] The quantity denoted by H by Boltzmann is identical with the logarithm of the statistical probability, taken with the opposite sign.

perceptible in all physical quantities, the value of which depends on only a small number of individual values. Conversely, the observation of such fluctuational phenomena may be regarded as a certain proof of the individual composition of matter and physical processes.

The most thoroughly investigated phenomenon of molecular fluctuations is the *Brownian movement.* This originates in the irregular molecular impacts experienced by a small body suspended in a gas or a liquid. Since even a small body experiences millions of impacts in a minute fraction of a second, the effects of these will cancel out in consequence of the statistical regularity due to their large number, inasmuch as they take place in all directions. The smaller the suspended body, however, the greater is the part played by individual accidental irregularities. The mutual compensation of the impacts occurring in all directions becomes no longer perfect, and, if the body be sufficiently small, it will be forced into a zig-zag motion which will be the more lively the smaller the body. Now in 1827, *Brown*, the botanist, actually discovered remarkable irregular movements executed by the pollen of plants in liquids, the linear dimensions of the pollen amounting to about a two-hundredth of a millimetre. It was not until half a century later, however, that *Christian Wiener* recognized the true cause of this phenomenon in the state of motion which, in accordance with the kinetic theory of matter, must be supposed to exist within the liquid. A Brownian movement in *gases* was not demonstrated until much later, by *Ehrenhaft.* The exact theory of the Brownian movement was developed in 1905 simultaneously by *Einstein* and *Smoluchowski. Smoluchowski* has also thoroughly investigated other phenomena of molecular fluctuations, notably *molecular fluctuations of density.*

The theory of all these phenomena has already advanced so far that it has been possible, in several instances, to calculate the *Loschmidt number on the basis of phenomena of molecular fluctuations.* Thus, accurate measurements on the Brownian movement [14] have led to a value of $5{\cdot}94 \times 10^{23}$, and measurements on the concentrations of *emulsions* [15] at

[14] Erich Schmid, *Sitz. Ber. der Wien. Akad.*, 1921.

[15] Emulsions are distinguished from solutions by the fact that, in them, matter is not resolved into molecules but is suspended as particles in the liquid. These particles are composed of many molecules and can be observed in a microscope, and their size can also be measured. Such emulsions are also known as *colloidal solutions.* Under the influence of gravity, a definite distribution is formed in emulsions, and to this is due the variation of the concentration with the height.

different heights [16] have given a value of $6{\cdot}04 \times 10^{23}$, for the Loschmidt number. The perfection of experimental methods will probably give rise to countless new problems for the kinetic theory of matter. The great *historical* importance of the *kinetic theory of gases*, however, is due on the one hand to its exact demonstration of the fruitfulness and accuracy of the *atomistic* concept, and on the other hand to its responsibility for the introduction of *statistical* methods into physics.

§ 149. The Equation of State of van der Waals.

The equation of state of the ideal gas is only approximately true for actual gases. At ordinary temperature and pressure the deviations are smallest for those gases which were formerly regarded as non-liquefiable, or "permanent." Chief among these were hydrogen, oxygen, nitrogen, and (later) the rare gases; whilst large deviations from the ideal gas law were observed, for instance, in the case of carbon dioxide. The *deviations from the ideal gas law* were first described by *van der Waals*, in 1873, in terms of an equation of state which he obtained with the help of comparatively simple theoretical considerations, and which agree fairly well with experiment. Van der Waals took into account two circumstances neglected in the derivation of the ideal gas law: first, the fact that *forces of attraction* obviously *act between the molecules* in consequence of *cohesion*,[1] and secondly, that, because of its molecular structure, the *volume of a gas cannot be diminished below a definite limit.*

The effect of the mutual attraction of the molecules on a gas is evidently the same as that of an external pressure. We can therefore allow for this internal attraction by supposing the *external pressure* to be *increased by an additive term.* The denser the gas, the larger must be this term; it must, indeed, evidently be proportional to the *square* of the density if, as seems quite plausible, we suppose the undetermined law of internal attraction to be similar in form to Newton's Law of Gravitation or Coulomb's Law, for in them the force varies as the *product* of two masses, or of two quantities of electricity, or of two magnetic pole strengths. Since, however, the density is inversely proportional to the specific

[16] Westgren, *Zeitschr. f. anorg. Chem.* **95** (1915), p. 231.

[1] Debye has made an interesting attempt to calculate these cohesion forces, and thus the constant a in van der Waals's equation, on the basis of the electron theory. *Vide Physikal. Zeitschr.* **21** (1920), pp. 178–187.

volume (v), the additive term may be given the form a/v^2, where a denotes a constant characteristic of the gas concerned.

The lowest limit to which the volume of a gas can ever be diminished would represent a second constant characteristic of the gas, and this is generally denoted by b. It seems then reasonable to replace the specific volume in the equation of state by the difference between it and this constant. The improved equation of state therefore runs

$$\left(p + \frac{a}{v^2}\right)(v - b) = R\,T\,, \tag{1}$$

for which we may also write

$$p\,v^3 - (b\,p + R\,T)\,v^2 + a\,v = a\,b. \tag{2}$$

Since this equation is of the *third degree* with respect to v, we recognize that, at constant temperature, each value of p is associated with three values of the specific volume, of which either all three, or else only one, are real. Now, if the state of a gas be represented graphically in a p-v-plane, the so-called *isotherms* for an ideal gas must represent *rectangular hyperbolas*, because at constant temperature the product $p\,v$ is constant. According to van der Waals's equation of state, however, these rectangular hyperbolas are replaced by *curves of the third degree* which, when the equation has three real roots, are *intersected at three points* by a straight line parallel to the v-axis. To a definite pressure there correspond three values v_1, v_2, v_3, where $v_1 < v_2 < v_3$ (Fig. 121).

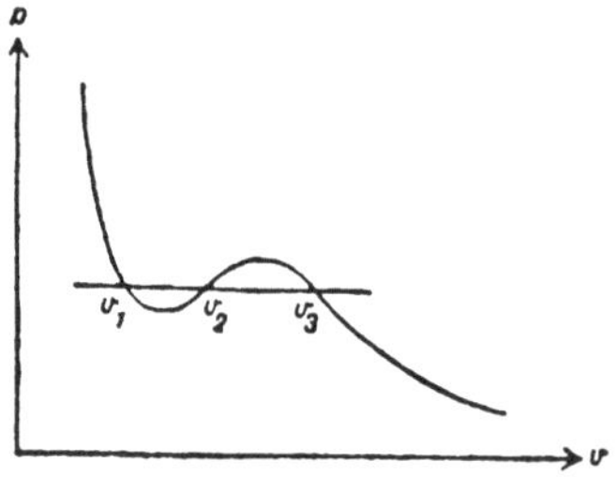

FIG. 121.

The significance of these three values can be explained, in agreement with experimental facts, by stating that, at any specific volume smaller than v_1, the body is liquid, but that, at any specific volume larger than v_3, it is gaseous, whereas, for values between v_1 and v_3, both the liquid and the gaseous states of aggregation are possible. Now the part of the isotherm lying between v_1 and v_3 shows a minimum and a maximum. It is, indeed, not possible to realize the segment of the curve lying between the minimum and the maximum, for the volume would then have to increase with increasing pressure. Under certain circumstances, however, the parts of the curve extending from v_1 to the minimum and from the maximum to v_3 can be realized. The first part represents the state of *superheated liquid ;* the second that of *super-cooled* or, as we may say, of *supersaturated vapour.* When, as is generally the case, liquid and gaseous phases coexist in thermodynamic equilibrium, the curve is replaced by a straight line between v_1 and v_3, owing to the constancy of the pressure.[2]

If we draw a set of isotherms in accordance with van der Waals's equation (Fig. 122), we see that, with increasing temperature, the minimum and maximum of the curves draw closer and closer together. We may thus construct a further curve, represented by a broken line in Fig. 122, which passes through the minima of all the isotherms on one branch, and through the maxima on the other branch. This curve possesses a maximum at a certain point, and the state of the gas represented by that point is termed its *critical state.* This state is defined by a *critical temperature*, a *critical pressure*, and a *critical specific volume.* At temperatures above the critical one no liquefaction at all is possible, for all higher isotherms can give only a single real root for v.

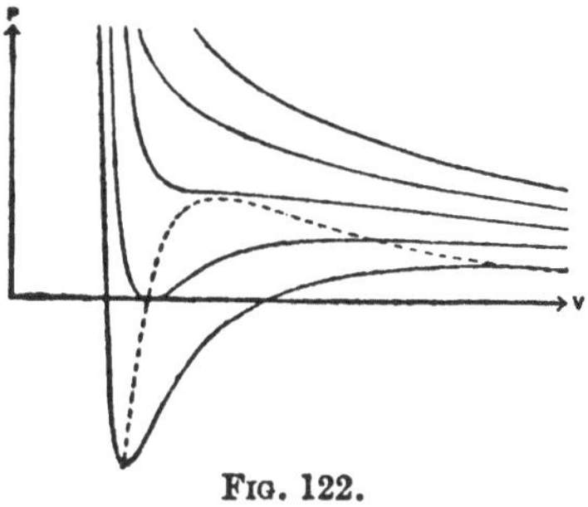

Fig. 122.

At the minimum and maximum on each isotherm the tangent to the isotherm is parallel to the v-axis, so that dp/dv vanishes. At every point along the broken curve in Fig. 122 the relation

$$\frac{dp}{dv} = 0 \tag{3}$$

must therefore be satisfied. Now, by eqn. (1),

$$p = \frac{RT}{v-b} - \frac{a}{v^2}. \tag{4}$$

If we differentiate this equation with respect to v (T being kept constant, since we are considering only a single isotherm), and equate the differential coefficient to zero, we find in accordance with eqn. (3)

$$\frac{RT}{(v-b)^2} = \frac{2a}{v^3}. \tag{5}$$

[2] *Vide* § 147 eqn. (12).

From this, together with eqn. (4), we obtain as equation of the broken curve in Fig. 122

$$p = \frac{2a(v-b)}{v^3} - \frac{a}{v^2},$$

or,

(6) $$p = \frac{a}{v^2} - \frac{2ab}{v^3}.$$

The maximum of the curve, which represents the critical state, is further characterized by the fact that the tangent to the curve is there horizontal. At the maximum, therefore, in addition to eqns. (4) and (6) which, of course, hold good, the differential coefficient of the right-hand side of eqn. (6) with respect to v must vanish. In the critical state, we must therefore have

$$\frac{d}{dv}\left(\frac{a}{v^2} - \frac{2ab}{v^3}\right) = 0.$$

Hence, denoting the critical quantities of state by asterisks, it follows that

(7) $$v^* = 3b.$$

Introducing this value into eqn. (6), we obtain

(8) $$p^* = \frac{a}{27b^2},$$

and, finally, we find from eqn. (5) for the critical temperature,[3]

(9) $$T^* = \frac{8a}{27bR}.$$

The constants a and b, which determine the deviations from the ideal gaseous state, are smallest in the case of *helium*, and are also very small in the case of *hydrogen*, which explains why these two gases are so peculiarly suitable for thermometric purposes. The lowest critical temperature is that of helium, *viz.*, − 268° C. For hydrogen it amounts to − 240° C., for nitrogen to − 147° C., and for oxygen to

[3] On solving eqns. (7) and (8) with respect to a and b, we find

$$b = \frac{v^*}{3}, \quad a = 3p^* v^{*2}, \quad R = \frac{8p^* v^*}{3T^*}.$$

Introducing, for the sake of brevity, the so-called *reduced quantities of state,*

$$\pi = \frac{p}{p^*}, \quad \omega = \frac{v}{v^*}, \quad \tau = \frac{T}{T^*},$$

we can write van der Waals's equation of state, as may readily be seen, in a form which is quite independent of all special properties of the gas concerned. The reduced equation of state thus becomes

(10) $$\left(\pi + \frac{3}{\omega^2}\right)(3\omega - 1) = 8\tau.$$

$-119°$ C. For carbon dioxide the critical temperature is $+31°$ C., so that carbon dioxide can be liquefied at room temperatures. It was, indeed, with this gas that the deviation from Boyle's Law was first thoroughly investigated by *Andrews*, who thereby first arrived at the important concept of critical temperature.[4]

§ 150. The Specific Heat of Solid Bodies.

From the statistical point of view we have seen that the heat content of gases is to be regarded as a form of translational energy, and, similarly, statistical theory refers the *heat content of solid bodies to molecular oscillations.* In doing this, use is made of the conclusion, which follows from the quantum principle, that oscillatory energy is composed of *energy-elements* which are proportional to the frequency.

The simplest possible concept of this kind is the assumption that all the molecules in a solid body oscillate with the *same frequency*, which may be denoted by ν^*. The quantity $h\nu^*/k$, which must obviously have the dimensions of a temperature, may then be defined as the *characteristic temperature* (T^*) of the body concerned. Since the modulus of distribution of the oscillatory energy among the molecules is to be equated to the heat modulus, *i.e.*, to the product of k and the temperature T, we find as mean value for the energy of an oscillation [according to § 139 eqn. (5)]

$$\bar{\epsilon} = \frac{k\,T^*}{e^{\frac{T^*}{T}} - 1}. \tag{1}$$

If we consider an *ideal monatomic solid body*, the oscillations of each molecule can be resolved into *three* mutually perpendicular and independent partial oscillations. On the above assumptions, therefore, the oscillatory energy of a gram-molecule is given by

$$E = 3L\,\bar{\epsilon}, \tag{2}$$

where L denotes the Loschmidt number. Introducing the value for $\bar{\epsilon}$ following from eqn. (1), and differentiating with respect to the temperature, we obtain the specific heat referred to a gram-atom, or the so-called *atomic heat* (denoted by C). As a result of the differentiation we find

$$C = \frac{3L\,k}{\left(e^{\frac{T^*}{T}} - 1\right)^2} \cdot \left(\frac{T^*}{T}\right)^2 \cdot e^{\frac{T^*}{T}}. \tag{3}$$

[4] Half a century before Andrews' time, Cagniard de la Tour had already experimentally realized the critical state.

This equation assumes an especially simple form when T is so much larger than T^* that, by way of approximation, we may put

$$e^{\frac{T^*}{T}} - 1 = \frac{T^*}{T}, \qquad e^{\frac{T^*}{T}} = 1$$

[making use of the formula according to which, for small values of y, e^y is approximately equal to $(1 + y)$]. In that case, because the product $k\,L$ represents the gas constant,

(4) $$C = 3R\,.$$

Since the value of the gas constant in thermal measure is 1·98 gram-calories [§ 148 eqn. (11)], it follows that, when the special assumptions for the validity of eqn. (4) are satisfied, the *atomic heats of all solid substances must have the same value of* 5·94 *calories.*

This law was, in fact, experimentally discovered by *Dulong* and *Petit* as long ago as 1819. The determinations then carried out showed that, for the majority of solid elements, the atomic heats at constant pressure differed by less than 10 per cent. from an average value of 6·4 cal., and a reduction of this value to constant volume [1] yields just 5·9 calories. On the other hand, no agreement was found to exist for some elements of low atomic weight, such as beryllium, boron, and especially *diamond*, whose atomic heat at room temperature proved to be only 1·7 cal., whilst at − 50° C., as was discovered in 1875 by *H. F. Weber*, it falls to as low as 0·7 cal.

No explanation for these *deviations* from Dulong and Petit's Law could be given on the basis of "classical" statistics whose deductive processes, indeed, necessarily led to that law as being generally valid (we cannot enter into further details here).[2] From the point of view of the quantum theory, however, the decrease in the specific heat with the temperature immediately became intelligible, as did the fact that these deviations are particularly noticeable for elements of low atomic weight. For, if two oscillating particles are bound to their position of rest by equally large forces, it follows from the formulæ of mechanics that the frequency of the lighter particle will be greater than that of

[1] Cf. § 145 eqn. (38).

[2] Classical statistics ascribed an amount of energy $k\,T/2$ to each degree of freedom of an oscillation, and further assumed the potential energy to be equal to the kinetic energy.

the heavier,[3] and the higher frequency is associated with a higher characteristic temperature. Thus the characteristic temperature of one element may be far below room temperature, and that of another element far above; and examples of the latter case, involving the largest deviations from Dulong and Petit's Law, may well be expected with the lightest elements.

In the light of modern atomic theory, Dulong and Petit's Law appears simply as a *special case* of a more general expression for the atomic heat, which is approximately represented by eqn. (3). This equation was derived in 1907 by *Einstein*, and formed the first application of the quantum principle to mechanical oscillations. Exact determinations carried out by *Nernst* and other investigators subsequently to the formulation of Einstein's law have shown, in fact, that the atomic heat of all substances converge at sufficiently high temperatures towards $3R$, and at sufficiently low temperatures towards zero.[4] The atomic heat of diamond (C_v) at about 1,000° C. amounts to 5·5 cal. On the other hand, the atomic heat of copper, for example, falls to 3·38 at 88° K.,[5] to 0·54 at 33° K., and to 0·22 cal. at 23° K. The decrease is certainly most marked with diamond. At 86° K. the atomic heat is but 0·03 cal., and at 23° K. it was found by measurement to be 0·00 cal.; *i.e.*, at these low temperatures the idea of temperature is practically meaningless in the case of diamond.

In order to be able to test quantitatively the accuracy of Einstein's formula, it is above all necessary to determine the characteristic temperature, and thus the characteristic frequency of various substances, and, in this connection, several available paths for research have been found. In the first place, as *Lindemann* has shown, a simple relation exists between the characteristic frequency and the *temperature of the melting-point*, if we make the plausible assumption that, at the melting-point, the amplitude of the atomic oscillations becomes equal to the distances between the atoms, so that

[3] If the force attracting a particle of mass m to its position of rest from a distance x be $(-\gamma^2 x)$, it follows from § 6 that the frequency of the oscillation is given by

$$\nu = \frac{\gamma}{2\pi\sqrt{m}}.$$

[4] In the neighbourhood of the melting-point, the value $3R$ is generally actually exceeded, owing to the commencement of further oscillations.

[5] Absolute temperatures are generally denoted by the letter K ("Kelvin").

atomic collisions occur.[6] Furthermore, *Madelung* and then *Einstein* have demonstrated that the *elasticity constants*, particularly the *modulus of compressibility*, are determined by the characteristic frequency. Finally, *Nernst* has drawn attention to the connection existing between the characteristic frequencies and the wave-lengths (measured by *Rubens*) of the so-called *residual rays*[7] from *diatomic crystals* such as potassium chloride.

The values of the characteristic frequencies derived in these different ways agree quite well with each other; the values obtained are of the order of magnitude of 10^{12} sec.$^{-1}$, in diamond as high even as 10^{13} sec.$^{-1}$, so that the frequencies all lie well within the *infra-red*. They are associated with characteristic temperatures which, for example, for lead, silver, copper, and aluminium, amount to 100°–400° K., and for diamond to about 2,000° K. It was found, however, that, using these values, Einstein's formula does not reproduce the temperature course of the atomic heats accurately; at low temperatures it invariably gives too *small* values. *Nernst* and *Lindemann* therefore formulated an *empirical* expression which represents the actual temperature course very exactly. This formula was obtained by adding to the expression on the right-hand side of Einstein's formula [eqn. (3)] another term of precisely similar form, but with the fraction T^*/T everywhere replaced by the fraction $T^*/2T$. *Einstein* himself realized that the imperfection of his formula was to be ascribed to the arbitrary assumption of a uniform frequency. A better agreement was only to be expected with a theory which took into account a *multiplicity of frequencies*, as, in a first rough approximation, was done in the Nernst-Lindemann formula, a frequency $\nu/2$ being introduced besides the principal frequency ν.

In the development of this idea, the kinetic theory of solid bodies has been extended in two directions. *Born* and *Kármán*, and *Thirring*, have (like Bragg[8]) regarded solid bodies as *space-lattices*, and have calculated the elastic and thermal properties for them, finding these to be in good agreement with experiment. On the other hand, in 1912 *Debye* developed a simple theory by *disregarding any atomistic structure of solid bodies*. He viewed them in the light of a

[6] The formula proposed by Lindemann contains, in addition to the frequency and the temperature of the melting-point, only the atomic weight and the atomic volume; a proportionality factor was empirically determined.

[7] *Vide* § 86.

[8] *Vide* § 106.

continuum filled with stationary elastic waves, and applied to these waves the principle which, it is true, was first deduced by *Rayleigh* and *Jeans* (*vide* § 138) for stationary electromagnetic waves.

§ 151. Nernst's Heat Theorem.

Let us now consider a body filled with *stationary elastic waves* in the sense of Debye's concept; then we must obviously distinguish between *longitudinal* waves with one degree of freedom, and *transverse* waves with two degrees of freedom. We shall denote the velocity of the longitudinal waves by v_1, and that of the transverse waves by v_2, both these quantities (according to § 49) being determined by the density and the elasticity constants. The number of longitudinal vibrations whose frequencies lie between ν and $(\nu + d\nu)$ is given by Rayleigh and Jeans's principle [§ 138 eqn. (12)], and we have

$$(1) \qquad dZ_1 = \frac{4\pi V \nu^2}{v_1^3} d\nu ,$$

where V is the volume of the body. Similarly, the number of transverse vibrations whose frequencies lie between ν and $(\nu + d\nu)$ is given by

$$(2) \qquad dZ_2 = \frac{4\pi V \nu^2}{v_2^3} d\nu .$$

Debye now assumes that the number of vibrations is not unlimited, but that there is a *maximum frequency* $(\nu_{\max})$ determined by the condition

$$(3) \qquad \int_0^{\nu_{\max}} dZ_1 + 2\int_0^{\nu_{\max}} dZ_2 = 3N ,$$

where N is the total number of atoms in the body; for the number of degrees of freedom of each of the longitudinal vibrations is one, of each of the transverse vibrations two, and of each of the atomic vibrations three. If we put, for the sake of brevity,

$$(4) \qquad \frac{1}{v_1^3} + \frac{2}{v_2^3} = \frac{3}{v^3},$$

where v may be called the mean wave velocity, we have

$$3N = \frac{2\pi V}{v^3} \int_0^{\nu_{\max}} \nu^2 \, d\nu ,$$

or,

$$(5) \qquad \nu_{\max} = v \sqrt[3]{\frac{3N}{4\pi V}}$$

As was mentioned above, the mean wave velocity can be calculated from the density and the elasticity constants. The quotient N/V represents the number of atoms in unit-volume, and we know this number from the density and the molecular weight of the substance. The maximum frequency can therefore be found from eqn. (5), and the values thus obtained agree in order of magnitude with those which had been calculated for the characteristic frequency on the basis of the more primitive assumption of a uniform frequency.

For the total *energy density* we find [according to § 139 eqn. (8)] the value

$$(6) \qquad \eta = \frac{12\pi h}{v^3} \int_0^{\nu_{\max}} \frac{\nu^3 \, d\nu}{e^{\frac{h\nu}{kT}} - 1} .$$

Now, if the temperature be sufficiently low, $h\nu/kT$ becomes relatively large, and hence the integrand in eqn. (6) decreases very rapidly with increasing ν. In this special instance of sufficiently low temperature, we may therefore replace the integration limit ν_{max} in eqn. (6) by ∞. If, however, we do this, the laws derived in § 139 become valid. Even if we admit the possibility of the internal energy of a body possessing a component which is independent of the temperature, the differential coefficients of the energy, the pressure, and the free energy with respect to the temperature must at any rate still be proportional to the cube of the temperature (according to § 139).

The differential coefficient of the energy with respect to the temperature simply represents the *specific heat;* at *low temperatures,* this must therefore be *proportional to the cube of the temperature.* This law was derived by *Debye* from his theory, and measurements carried out by Nernst and other investigators on diamond, copper, and other substances have actually confirmed it.

Entropy is connected with energy and free energy by the relation (frequently quoted)

$$(7) \qquad S = \frac{E - F}{T}$$

[§ 144 eqn. (1)].

The following two equations must therefore be satisfied :—

$$(8)\qquad \lim_{T\to 0} \frac{dF}{dT} = 0\,; \qquad \lim_{T\to 0} S = 0\,.$$

According to these equations, therefore, both the *temperature coefficient of the free energy and the entropy vanish at the absolute zero.* On these two relations is based an important theorem which—independently of the quantum theory—was derived by *Nernst* in 1906, and is hence generally known as *Nernst's Heat Theorem.* It is assumed to hold good not only for solid bodies, but quite generally for "*condensed*" (*i.e.*, both solid and liquid) systems.[1]

We cannot here further discuss Nernst's Heat Theorem, but it leads to a series of important thermodynamical and, in particular, thermochemical results. One of these may be briefly mentioned. As we have seen, entropy depends only on the temperature, and hence, at the lowest temperatures, we must have

$$(9)\qquad \left(\frac{\partial S}{\partial p}\right)_T = 0\,.$$

Again, multiplying § 145 eqns. (30) and (33) by the mass of the body (m), we find

$$(10)\qquad dS = \frac{m\,c_p}{T}\,dT - \frac{\partial V}{\partial T}\,dp\,.$$

Eqn. (9) can thus only be satisfied when

$$(11)\qquad \frac{\partial V}{\partial T} = 0\,.$$

At very low temperatures, the volume must therefore be independent of the temperature, so that the *expansion coefficient vanishes.* The accuracy of this theorem has actually been proved by experiments of *Grüneisen.*

§ 152. Cavity Radiation.

Stationary vibrations may be of an electromagnetic as well as of a mechanico-elastic nature. In that case, however, the number of degrees of freedom will only be two, instead of three, because electromagnetic waves are always purely transversal, longitudinal waves being impossible.

[1] It is true that, for ideal gases, it follows from § 148 eqn. (1) that the specific heat is independent of the temperature. On the other hand, real gases exhibit "degradation phenomena" at extremely low temperatures, and these point to a dependency of the specific heat on the temperature (in the sense of Nernst's Heat Theorem).

A system of stationary electromagnetic waves filling an enclosed space is defined as *cavity radiation*. The theorems derived in § 139 will, of course, be applicable to a system of this kind, especially as there is no question of any limiting maximum frequency, which only had to be taken into account in considering atomic oscillations.

Now *electromagnetic energy* can be *absorbed* and *emitted by atoms*. An *interchange of energy* between the cavity radiation and a body composed of molecules (which may be either a portion of the boundary or exist within the enclosure) is therefore possible. When, despite the possibility of an energy interchange, the energy values of neither the cavity radiation nor of the body alter, *thermodynamic equilibrium* is said to exist. In accordance with the general principles of statistics (§ 134), this is only possible when the modulus of distribution of the electromagnetic energy is equal to the heat modulus of the body; the modulus of the cavity radiation must thus be equal to the product of Boltzmann's constant k and the temperature of the body with which the radiation is in thermodynamic equilibrium. This temperature may then be sufficiently defined as the *temperature of the cavity radiation*. Since the suppositions of § 139 hold good here, the energy density can depend only on the modulus of distribution, being proportional to its fourth power. In *thermodynamic equilibrium*, therefore, the *radiation density* can be a *function only of the temperature*, and must be *proportional to the fourth power of the temperature*.

Again, we must regard the equilibrium of a body which is *both radiating and receiving radiation* in accordance with a concept developed in 1809 by *Prévost*, *viz.*, as consisting in the *emission and absorption of equal amounts of energy* in any given time.[1] The amount of energy radiated by unit-surface of the body in unit-time is known as the *emissive power* of the body. Denoting it by K', we find for the quantity of energy emitted by a surface-element df in an interval of time dt

$$(1) \qquad dE = K' \, df \, dt \, .$$

Again, the total quantity of energy *incident* on the surface-element df in the time dt is given by

$$(2) \qquad dE' = f(T) \, . \, df \, dt \, ,$$

for the radiation density depends only on the temperature, so that the quantity by which $df \, . \, dt$ is multiplied on the

[1] Prévost, it is true, regarded heat as a material substance.

right-hand side of the equation can only be a function of the temperature. Of the incident energy a fraction $\varrho \,.\, dE'$ will, in general, be reflected. The quantity $(\gamma = 1 - \varrho)$ is called the *absorption coefficient*, for the quantity of energy $\gamma\, dE'$ is absorbed by the irradiated body. In thermodynamic equilibrium we must have

$$dE = \gamma\, dE', \tag{3}$$

and hence, by eqns. (1) and (2),

$$\frac{K'}{\gamma} = f(T). \tag{4}$$

On the supposition of thermodynamic equilibrium, therefore, the *ratio of the emissive power to the absorption coefficient*

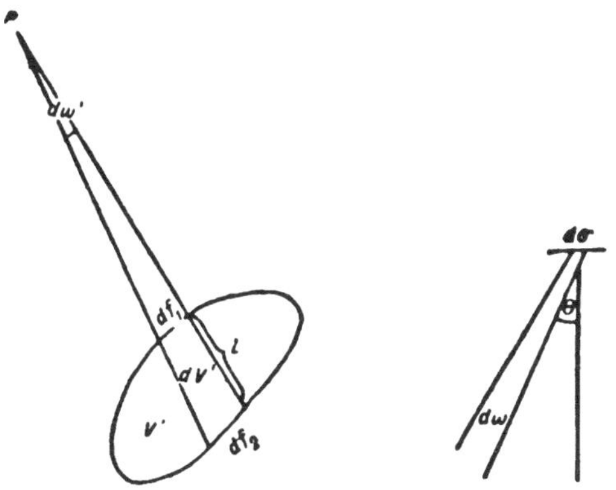

FIG. 123. FIG. 124.

at a given temperature must be the *same for all bodies.* This important theorem was first propounded in 1859 by *Kirchhoff.* If, in particular, γ is equal to unity, so that the body has no reflective power at all, the body is said to be *perfectly black.*[2] The *emissive power of a perfectly black body* therefore *depends only on the temperature.*

Again, it follows from eqns. (2) and (4) that the quantity of energy incident in unit-time on unit-surface of an arbitrary body is, in thermodynamic equilibrium, equal to the emissive power of a black body at the same temperature, which may be denoted simply by K.

The exact relation between emissive power and radiation density may be derived with the help of a purely geometrical

[2] The nearest approach to a perfectly black body among actual substances is soot. When γ is equal to zero, the body is said to be perfectly reflecting.

consideration. We suppose a small volume (V') of arbitrary shape to be delimited within the enclosure, and around this as centre a sphere of radius r to be constructed (Fig. 123). We consider now an arbitrary surface-element ($d\sigma$) of the spherical surface at the point P, the linear dimensions of this element being small in comparison with those of the delimited volume V'. Then we imagine a quite arbitrarily directed *elementary cone* of apical angle $d\omega$ (*cf.* Fig. 124) to be constructed, the surface-element $d\sigma$ lying at its apex. The direction of this cone is determined, on the one hand, by the angle θ included between its axis and the surface-normal to $d\sigma$, and, on the other hand, by the angle φ included between the axis of the cone and any arbitrarily chosen plane.

Let dE_0 be the quantity of energy penetrating into the elementary cone through the surface-element $d\sigma$ in the time dt. The quantity dE_0 must at any rate be proportional to the quantities $d\sigma$, dt, and $d\omega$, and also, as a simple consideration will at once show us, to the *cosine* of the angle θ. For, as is immediately evident from Fig. 125, the surface-element $d\sigma$, the normal to which makes an angle θ with the direction of a pencil of rays (to be regarded as parallel because of the smallness of $d\sigma$), admits in a given interval of time the same quantity of energy as would a surface-element of area $d\sigma \cos\theta$ *normal* to the direction of the rays. Conversely, if we incline the surface-element $d\sigma$ at an arbitrary angle to the direction of the rays, the quantities of energy admitted in a given element of time will be proportional to the cosine of the angle θ.

FIG. 125.

Since, therefore, the quantity dE_0 above introduced into our considerations is proportional to the quantities $d\omega$, $d\sigma$, dt, and $\cos\theta$, we can put

$$dE_0 = C \cos\theta \, d\omega \, d\sigma \, dt, \tag{5}$$

where C is the proportionality factor. The apical angle of the elementary cone, however, *i.e.*, the area of the section cut by it out of a sphere of unit-radius,[3] is given by

$$d\omega = \sin\theta \, d\theta \, d\varphi, \tag{6}$$

and hence

$$dE_0 = C \cos\theta \sin\theta \, d\theta \, d\varphi \, d\sigma \, dt. \tag{7}$$

Again, if $d\sigma$ were not the element of a purely geometrical surface but rather the element of the spherical surface of a

[3] The section is a rectangle with sides $R\ d\theta$ and $R \sin\theta\ d\phi$, where R (the radius of the unit-sphere) is equal to unity.

physical body, the quantity of energy incident on this element from outside in unit-time would be $K\,d\sigma$, where K is the emissive power of a black body at the temperature concerned. Conversely, the total energy penetrating into the interior of the sphere in the time-element dt through the purely geometrical surface-element $d\sigma$ must therefore be equal to $K\,d\sigma\,dt$. We must thus obtain the expression $K\,d\sigma\,dt$ if we integrate the right-hand side of eqn. (7) over all values of θ from zero to a right-angle, and over all values of φ from 0 to 2π. Hence we find the relation

$$(8) \qquad K = C\int_0^{\frac{\pi}{2}} \sin\theta\cos\theta\,d\theta\int_0^{2\pi} d\varphi\,,$$

or,

$$(9) \quad K = 2\pi C\int_0^{\frac{\pi}{2}} \sin\theta\,d\,(\sin\theta) = 2\pi C\left|\frac{\sin^2\theta}{2}\right|_0^{\frac{\pi}{2}} = \pi C\,.$$

The value of the quantity dE_0 then follows from eqn. (5):

$$(10) \qquad dE_0 = \frac{K}{\pi}\cos\theta\,d\omega\,d\sigma\,dt\,.$$

This is the amount of energy which penetrates into an elementary cone of apical solid angle $d\omega$ in the element of time dt through the surface-element $d\sigma$, the normal to the surface making an angle θ with the axis of the cone.[4]

Let us now return to Fig. 123, and suppose an elementary cone of apical solid angle $d\omega'$ to be constructed at right angles to $d\sigma$ at its apex, this cone cutting out a volume-element dV' of length l from the volume V'. The time required by the rays to traverse this length is given by

$$(11) \qquad \tau = \frac{l}{c}\,,$$

where c denotes the velocity of light.

If, now, rays were to penetrate into the volume-element dV' *only* through the surface-element $d\sigma$, the quantity of energy contained in the volume-element dV' would evidently always be equal to that penetrating through the surface-element $d\sigma$ into the elementary cone in the time τ. Denoting this quantity of energy by dE', we have by eqn. (10)

$$(12) \qquad dE' = \frac{K}{\pi}\tau\,d\omega'\,d\sigma$$

(for $\cos\theta$ is equal to unity, seeing that the axis of the elemen-

[4] In this derivation it is taken for granted that the value of the proportionality factor C is independent of the direction.

tary cone which passes through the centre of the sphere is at right angles to $d\sigma$).

Let df_1 and df_2 be the surface-elements cut by the elementary cone out of the bounding surface of the volume V', and let df' be their equally large projections at right-angles to the axis of the cone. Then, since the distance of P from the delimited small volume V' may be put equal to r,

$$(13) \qquad d\omega' = \frac{df'}{r^2}.$$

Again

$$(14) \qquad dV' = l\, df',$$

and hence, if the values following from eqns. (13), (14), and (11) be inserted in eqn. (12), the latter becomes

$$(15) \qquad dE' = \frac{K}{\pi c r^2}\, d\sigma\, dV'.$$

We find the *total* energy contained in the volume V' by integrating dV' in eqn. (15) over the whole volume V', and $d\sigma$ over the whole surface of the sphere of radius r.[5] Since the first integral gives V' and the second $4\pi r^2$, the total energy (E') contained in the volume V' is given by

$$(16) \qquad E' = \frac{K}{\pi c r^2} \cdot 4\pi r^2 V'.$$

Now V' was taken as being small, and thus the quotient of E' and V' simply represents the energy density (η) at the point where V' was chosen. Hence

$$(17) \qquad \eta = \frac{4K}{c},$$

or conversely,

$$(18) \qquad K = \frac{c\eta}{4}.$$

Since the modulus of distribution of the radiation is equal to kT, we have by § 139 eqn. (14)

$$(19) \qquad \eta = aT^4,$$

and therefore

$$(20) \qquad K = \frac{ac}{4} T^4,$$

[5] For only such rays as have penetrated the surface of the sphere of radius r can enter the volume V'.

where (because there are two degrees of freedom) the constant proportionality factor

$$(21) \qquad a = \frac{48\pi\alpha k^4}{h^3 c^3}.$$

The *emissive power of a black body* is therefore *proportional to the fourth power of the absolute temperature.* This important law was first derived in 1879 by *Stefan* from the experimental observations of other investigators, and was later proved by *Boltzmann* on the basis of theoretical thermodynamical considerations (quite different, however, from those applied in the present derivation). Measurements by *Kurlbaum* have given for the constant a the value

$$(22) \qquad a = 7{\cdot}06 \times 10^{-15} \frac{\text{erg.}}{\text{cm.}^3\,\text{deg.}^4}.$$

§ 153. Planck's Law of Radiation.

Like energy density, *emissive power* can also be *spectrally resolved* with respect to frequency intervals. The share taken in the total emissive power by the spectral region between ν and $(\nu + d\nu)$ is found from a combination of § 139 eqn. (6) with § 152 eqn. (18). Denoting this share by $K_\nu \,.\, d\nu$, we have, inasmuch as the number of degrees of freedom is two,

$$(1) \qquad K_\nu = \frac{2\pi h \nu^3}{c^2} \cdot \frac{1}{e^{\frac{h\nu}{kT}} - 1}.$$

Now the spectrum as a whole can also be divided up according to wave-length, instead of to frequency. Let us consider the interval between a frequency ν' and a frequency ν'', the corresponding wave-lengths being λ' and λ''. Let $(\nu'' - \nu')$ be equal to $\Delta\nu$, and $(\lambda'' - \lambda')$ to $\Delta\lambda$. The share taken in the total emissive power by the small spectral region considered may then be equated to either $K_\nu\,\Delta\nu$ or $(-K_\lambda\,\Delta\lambda)$. (For the smaller wave-length corresponds to the larger frequency.) Now ν and λ are connected by the elementary relation

$$(2) \qquad \nu = \frac{c}{\lambda}$$

[§ 42 eqn. (9)]. Differentiating this, we find

$$(3) \qquad d\nu = -\frac{c}{\lambda^2}\,d\lambda.$$

Hence

(4) $$K_\lambda = \frac{c}{\lambda^2} K_\nu ,$$

and therefore, by eqns. (1) and (2),

(5) $$K_\lambda = \frac{2\pi h c^2}{\lambda^5} \cdot \frac{1}{e^{\frac{ch}{k\lambda T}} - 1} .$$

This equation is an expression of *Planck's Law of Radiation*. Planck derived it in 1900 by introducing into theoretical physics, as a radical innovation, the hypothesis of the *elementary quantum of action*. The confirmation of Planck's Radiation Law by accurate measurements constituted a brilliant justification for the new hypothesis, whereas all the earlier laws derived from the " classical theory " proved to be of only limited validity.[1]

Planck's Law determines the specific, or " monochromatic," emissive power of a small region of the spectrum as a function of two variables, the *temperature* and the *wave-length*. On differentiating partially with respect to the wave-length, we obtain the relation

(6) $$\frac{\partial K_\lambda}{\partial \lambda} = - \frac{10\pi h c^2}{\lambda^6\left(e^{\frac{ch}{k\lambda T}} - 1\right)} + \frac{2\pi h c^2}{\lambda^5\left(e^{\frac{ch}{k\lambda T}} - 1\right)^2} \cdot \frac{ch}{k\lambda^2 T} \cdot e^{\frac{ch}{k\lambda T}} .$$

From this equation we can calculate the *wave-length* for which the *specific emissive power* represents a *maximum*.

[1] The two laws of radiation which had been previously derived form special cases of Planck's Law according as the product λT is taken to be small or large. In the first case, the exponential expression in eqn. (5) is large compared with unity. (It follows from the values of h and k that the expression ch/k is approximately equal to $1\frac{1}{2}$.) Owing to the magnitude of the exponential expression, the number 1 may therefore be omitted in the special case considered, and eqn. (5) then assumes the special form

(5b) $$K_\lambda = \frac{2\pi h c^2}{\lambda^5} \cdot e^{-\frac{ch}{k\lambda T}} .$$

This radiation law, which has thus only a *limited validity, viz.* in the *ultra-violet* or at *low temperatures*, was propounded by *Wien* in 1896.—If, on the other hand, the product λT is large, so that we are dealing with large wave-lengths or high temperatures, the value of the exponential in Planck's Radiation Law is small, and for these small values we can use the approximation formula in virtue of which, for small values of x, e^x may be put equal to $(1 + x)$. Eqn. (5) then becomes

(5c) $$K_\lambda = \frac{2\pi c k T}{\lambda^4} .$$

This equation is an expression of *Rayleigh's radiation law*, which is likewise of limited validity, holding *only* in the *infra-red* or at *high temperatures*. In passing it may be mentioned that it follows directly from Rayleigh and Jeans's principle, on the assumption that the mean value of the energy of a single characteristic vibration is equal to kT.

(That such a wave-length must exist follows from eqn. (5), for at every temperature K_λ must vanish both when $\lambda = 0$ and when $\lambda = \infty$.) For this particular wave-length, which may be denoted by λ_{max}, $\partial K_\lambda/\partial\lambda$ must be equal to zero. If we put, for the sake of brevity,

(7) $$\frac{c\,h}{k\,\lambda_{max}\,T} = \beta\,,$$

the relation

$$-5 + \frac{\beta\,e^\beta}{e^\beta - 1} = 0\,,$$

or

(8) $$\frac{\beta}{5} + e^{-\beta} = 1$$

must be satisfied. The quantity β defined by eqn. (7) must thus be a constant, which is given as the root of the transcendental eqn. (8). Its value is

(9) $$\beta = 4{\cdot}9651\,.$$

In accordance with eqn. (7), the product of the absolute temperature and the wave-length corresponding to the maximum of the specific emissive power must likewise be a *constant*. Denoting this constant by b, we may put

(10) $$\lambda_{max}\,.\,T = b\,.$$

With increasing temperature the wave-length corresponding to the maximum specific emissive power is therefore displaced in the direction of smaller wave-lengths, and thus in the visible spectrum from the red to the violet end.[a] The law expressed by eqn. (10) was derived in 1893 on purely thermodynamic grounds by *Wien*, before the birth of the quantum theory, and it is known as *Wien's displacement law*. Exact measurements carried out by *Lummer* and *Pringsheim* have confirmed the law, and lead to a value for the constant b of

(11) $$b = 0{\cdot}294 \text{ cm. deg.}$$

[a] In this lies the explanation of the well-known experimental fact that, on heating, bodies first glow red; then at higher temperatures, yellow; and at very high temperatures, white. Moreover, the reason why all bodies begin to glow at the same temperature—525° C.—as *Draper* first observed, is explained by eqn. (10), for this contains no quantities depending on the chemical nature of the radiating body. It should, however, be expressly observed that Planck's and Wien's laws only hold for radiation which takes place at the cost of the heat content of the radiating body. (This is not the case with so-called luminescence phenomena.)

According to eqn. (7), this constant is connected with the universal constants h and k by the simple relation

$$b = \frac{ch}{\beta k}. \tag{12}$$

Now, by combining this equation with § 152 eqn. (21), Planck was able to evaluate the constants h and k by solving the two equations with respect to these two quantities as unknowns. The constants a and b of Stefan's and Wien's Laws are actually known from measurements, as is, of course, the velocity of light; apart from these, the two equations only contain pure numbers. In this way Planck found

$$h = 6{\cdot}548 \times 10^{-27} \text{ erg. sec.}; \qquad k = 1{\cdot}346 \times 10^{-16} \text{ erg./deg.},$$

and hence for the Loschmidt number, as quotient of the gas-constant and k,

$$L = 6{\cdot}175 \times 10^{23}.$$

Finally, dividing the known Faraday constant by L, he obtained

$$e = 4{\cdot}69 \times 10^{-10} \text{ E.S.U.}$$

These values, which were derived in 1900 by Planck from his radiation law, and which were the *first exact values* obtained for these universal constants, agree extraordinarily well with those resulting from later and much more accurate investigations of a different kind.

Modern physics thus affords at least seventeen different and independent equations for the four fundamental constants of atomistics, *viz.*, h, e, L, and m (mass of an electron). These equations contain only these constants as unknowns in various combinations, as may be seen from the summary in Table XVIII. (in which k is replaced by R/L).

In whatever way the four unknowns are derived from these seventeen equations, the same values (within the limits of error) are always obtained. If the fact that the sensations of sight, hearing, and taste all indicate the same objects is to be taken as an argument for the existence of the world around us, then theoretical physics may regard as a *certain proof of the actual existence of electrons, atoms, and the elementary quantum of action* the fact that the *same values* are always obtained for the characteristic constants of atomistics even by fundamentally different methods.

TABLE XVIII

Methods of Derivation of the Fundamental Constants of Atomistics

Method.	Reference.	Relevant combination between e, h, m, and L.
1. Individual observation of small electric charges .	§ 87 eqn. (7)	e
2. Electrolysis	§ 88 eqn. (5)	Le
3. Deflection of cathode rays .	§ 89 eqn. (4)	e/m
4. Deflection of α-rays . .	§ 90 eqn. (1)	Le
5. Scintillation of α-particles .	§ 90	e
6. Photo-electric effect . .	§ 91 eqn. (2)	e/h
7. Rydberg's constant . .	§ 96 eqn. (8)	e^4m/h^3
8. Displacement towards the violet of helium lines .	§ 98 eqn. (25)	Lm
9. Fine structure of spectral lines	§ 100 eqn. (3)	e^2/h
10. Stark effect	§ 101 eqn. (2)	h/em
11. Zeeman effect	§ 102 eqn. (8)	e/m
12. Excitation of Röntgen rays .	§ 107 note (4)	e^4l^3/h
13. Formation of helium from radium	§ 111	L
14. Stefan's Law	§ 152 eqn. (21)	h^3L^4
15. Wien's Displacement Law .	§ 153 eqn. (12)	Lh
16. Brownian movement . .	§ 148	L
17. Emulsions	§ 148	L

PART V

THEORY OF RELATIVITY

Chapter XVIII

SPECIAL THEORY OF RELATIVITY

§ 154. Michelson's Experiment.

In § 13 it was shown that the theory of relative motion leads to the *mechanical principle of relativity*, according to which all mechanical equations have the same form for two coordinate systems *moving uniformly* with respect to each other. Moreover, it must appear inherently improbable that the same principle should not also apply to the equations of *electrodynamics* and *optics*, but a brief consideration will show that such a supposition meets with great difficulties.

Let us consider two parallel coordinate systems, of which the second is moving with a velocity **u** with respect to the first (in a direction coinciding with the common x-direction). Distinguishing the coordinates in the first system by single dashes, and those in the second system by double dashes, then in accordance with the usual laws of kinematics the coordinates in the two systems will be connected by the relations

(1) $$x'' = x' - u\,t,\; y'' = y',\; z'' = z' .$$

Hence, if the form of a mechanical equation in the first system be

(2a) $$\varphi\,(x', y', z', t) = 0\,,$$

it follows from the mechanical principle of relativity that the *same* function must also be zero when x' is replaced by $(x' - u\,t)$, so that

(2b) $$\varphi\,(x'', y'', z'', t) = 0\,.$$

We see at once, however, that the equations whose form remains unaltered on the transition from the first to the second coordinate system *cannot include* the equation describing the *uniform propagation of light in spherical waves.* If this equation, which states that

(3) $$x^2 + y^2 + z^2 - c^2\,t^2 = 0\,,$$

holds good in the first coordinate system, it cannot do so in the second, for, when u differs from zero, it follows inevitably from eqn. (1) that

$$(4) \qquad x''^2 + y''^2 + z''^2 = c^2 t^2 \neq x'^2 + y'^2 + z'^2 - c^2 t^2 .$$

In contradistinction to the mechanical principle of relativity, which puts the idea of an absolute motion beyond the bounds of possibility, the above inequality appears to demand the existence of *only one single system* in which light is propagated uniformly in all directions with the same velocity. If this were the case, however, there could evidently be no question of a uniform propagation of light in all directions on the moving earth. A confirmation of inequalities in this propagation thus presented to physicists the enticing possibility of demonstrating, or even of actually measuring, the motion of the earth relatively to that privileged single system, *i.e.*, the *motion of the earth relatively to the so-called æther*.

Among the manifold experiments undertaken with this object [1] in view, by far the greatest importance attaches to the famous *experiment of Michelson* (1881). In this experiment, a ray of light from a source L (Fig. 126) strikes a half-silvered mirror S at an angle of 45°. In this way it is split into two mutually perpendicular rays which are reflected by two other mirrors, S_1 and S_2, the distances between these and S being approximately the same (l). At P the interference phenomena due to the superposition of the rays coming from the mirrors S_1 and S_2 are observed through a telescope. For, since the intervals SS_1 and SS_2 differ, in general, by a very small amount, the same equidistant

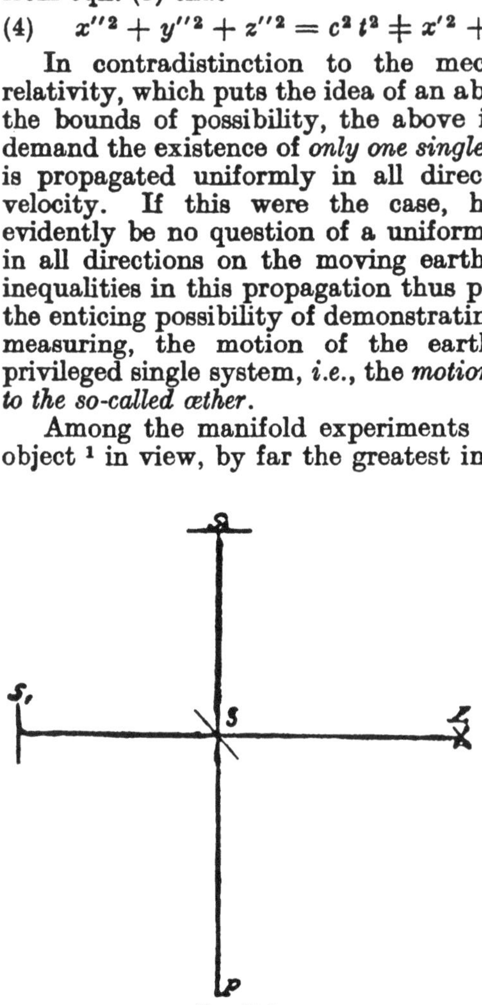

FIG. 126.

[1] An extensive summary of all these experiments is to be found in the first chapter of "*Relativitätstheorie*," by Laue (3rd edition; Brunswick, 1921).

interference fringes will be observed in the telescope as are produced, with suitable apparatus, by reflection at plane-parallel plates (we cannot enter into further details here).

On the assumption that an æther really exists, in which light is propagated in all directions with the same velocity c, we shall suppose the translatory velocity of the earth relatively to this æther to be $\mathbf{u}$. Let $\mathbf{v}_a$ be the velocity of a light ray of given direction with respect to the æther, and $\mathbf{v}_e$ its velocity with respect to an observer on the earth. Then, according to the formula for relative velocity (and neglecting the rotation of the earth),

$$\mathbf{v}_a = \mathbf{v}_e + \mathbf{u}, \tag{5}$$

where the magnitude of the vector $\mathbf{v}_a$ must always be the same (c).

We shall now suppose the apparatus to be so arranged that the straight line SS_1 lies in the direction of the vector $\mathbf{u}$. We shall denote the magnitude of the velocity of light (as observed on the earth) along the interval SS_1 by v_1', and along the interval S_1S by v_1''. Since, in these two special cases, the vectors $\mathbf{v}_e$ and $\mathbf{u}$ are similarly or oppositely directed respectively, we obtain from eqn. (5) the relations

$$\begin{cases} c = v_1' + u, \\ c = v_1'' - u. \end{cases} \tag{6}$$

Let v_2' be the velocity of light with respect to the earth along the interval SS_2, and v_2'' the velocity along the interval S_2S. Since the vectors $\mathbf{v}_e$ and $\mathbf{u}$ are mutually perpendicular in these two special cases, it follows from the Theorem of Pythagoras that the square of the magnitude of their geometrical sum is simply equal to the sum of the squares of the individual magnitudes. Hence we find from eqn. (5) the relations

$$\begin{cases} c^2 = v_2'^2 + u^2, \\ c^2 = v_2''^2 + u^2. \end{cases} \tag{7}$$

If we denote by t_1 the time required by the light to travel along the path SS_1 and back, and by t_2 the time required to travel along the path SS_2 and back, we have, by eqns. (6) and (7),

$$t_1 = \frac{l}{v_1'} + \frac{l}{v_1''} = \frac{l}{c-u} + \frac{l}{c+u} = \frac{2l}{c}\left(1 - \frac{u^2}{c^2}\right)^{-1}, \tag{8}$$

and

$$t_2 = \frac{l}{v_2'} + \frac{l}{v_2''} = \frac{2l}{\sqrt{c^2 - u^2}} = \frac{2l}{c}\left(1 - \frac{u^2}{c^2}\right)^{-\frac{1}{2}}. \tag{9}$$

With the aid of the binomial theorem, we can develop eqns. (8) and (9) into series. For that purpose we use the formula

$$(10)\qquad (1+x)^n = 1 + \frac{n}{1}x + \frac{n(n-1)}{1\cdot 2}x^2 + \ldots ,$$

which is generally valid provided that

$$-1 < x < +1 .$$

According to this formula,

$$(11)\qquad \left(1-\frac{u^2}{c^2}\right)^{-1} = 1 + \frac{u^2}{c^2} + \frac{u^4}{c^4} + \ldots ,$$

and

$$(12)\qquad \left(1-\frac{u^2}{c^2}\right)^{-\frac{1}{2}} = 1 + \frac{1}{2}\frac{u^2}{c^2} + \frac{3}{8}\frac{u^4}{c^4} + \ldots .$$

Hence, as a first approximation,

$$(13)\qquad t_1 - t_2 = \frac{l u^2}{c^3} .$$

Now if, in consequence of the earth's motion, the quantity $(t_1 - t_2)$ differed from zero, this would obviously result in a *displacement of the interference fringes* relatively to the position which they would occupy, were the earth at rest with respect to the æther. Let r' be the magnitude of this displacement, r the distance between two adjacent interference fringes, and τ the period of the monochromatic light used; then the relation

$$(14)\qquad r' : r = (t_1 - t_2) : \tau$$

must evidently exist. Hence, denoting the wave-length of the light by λ, and remembering that λ is equal to $c\,\tau$, we have

$$(15)\qquad \frac{r'}{r} = \frac{l}{\lambda}\cdot\frac{u^2}{c^2} .$$

If now we *rotate* the whole apparatus *through a right angle* in its own plane, the time t_1 required by the light to travel along the path SS_1 and back again then becomes equal to the previous value t_2. On the other hand, the time t_2 required for traversing the interval SS_2 and back again is now equal to the original t_1. Hence we obtain for the difference $(t_1 - t_2)$ the amount

$$(16)\qquad t_1 - t_2 = -\frac{l u^2}{c^3} .$$

The displacement of the interference fringes with respect to the positions which they would occupy were the earth at rest must now be equal, but opposite in sense, to the original displacement. In consequence of the rotation of the apparatus, the interference fringes must therefore appear displaced by an interval $2r'$ relatively to their position with the original orientation of the apparatus. If we express this total displacement due to the rotation of the apparatus as a fraction (α) of the interval between two adjacent fringes, we have therefore

$$\alpha = 2\frac{l}{\lambda}\frac{u^2}{c^2}. \tag{17}$$

Admittedly, the translatory velocity of the earth is only the ten-thousandth part of the velocity of light, so that u^2/c^2 is equal to 10^{-8}. On the other hand, however, the length l in Michelson's experiment measured a few metres, and, in a later and improved experiment carried out by Michelson and *Morley* (1887), was increased to 11 m.[2] Since λ is approximately 6×10^{-5} cm., the resulting value of α was 0·4. The rotation of the apparatus should thus have been accompanied by a displacement of the interference fringes through about half the distance between two neighbouring fringes. Of the expected and quite considerable displacement, however, nothing was to be observed.[3]

§ 155. Einstein's Principle of Relativity.

Michelson and Morley's experiment necessarily led to the conclusion that, despite the motion of the earth, light is nevertheless propagated with the same velocity in all directions relatively to a terrestrial observer. Thus the *principle of relativity* does actually appear to be as valid for *electrodynamical and optical phenomena* as it is for mechanical phenomena. *All equations* which describe any kind of physical process with reference to a definite system must, therefore, *retain the same form* when the original system is replaced by another *moving uniformly* with respect to it. We are led to this conclusion not only by Michelson's experiment; for it complies with the concept of physical

[2] Each ray of light was reflected several times to and fro.

[3] The great importance of Michelson's experiment lies in the fact that the expected effect depends on the *square* of the ratio u/c. Lorentz, indeed, had previously shown (we cannot enter into further details here) that the absence of any effect could always be explained on the basis of simple assumptions, if it depended only on the first power of the ratio u/c.

uniformity, with which any limitation of the validity of the relativity principle to a single branch of physics would be incompatible. On the other hand, however, an extension of that principle to the fields of electrodynamics and optics must necessarily stand in contradiction to the inequality expressed by § 154 (4); and thus theoretical physics was faced with the difficult problem of solving this incompatibility.

The first attempt at a solution was made by *Lorentz* and *Fitzgerald*, who assumed that a body moving through the æther experiences a *contraction* in the direction of its motion and depending on the magnitude of its velocity, and this *ad hoc* hypothesis certainly afforded an explanation of the negative result of Michelson's experiment. It was, however, due to the critical ingenuity of *Einstein* that the contradiction was first solved from a much more general point of view. Einstein accomplished this by liberating physics from a deeply rooted prejudice which, owing principally to *Newton's* influence, had become a fundamental concept of theoretical physics; this was the concept of *absolute time.*[1]

Einstein pointed out that it was entirely arbitrary and unjustifiable to ascribe the same scales of time to two systems moving relatively to each other. On the contrary, it must be postulated that observers in the two systems shall use *different time-scales* for their measurements of time, exactly as they would use different spatial coordinates in stating the position of any given body.

Thus, whereas the two sides of the inequality (4) in § 154 must always differ so long as the concept of an absolute time is retained, a transformation of the inequality into an equation becomes possible when on the left-hand side t', and on the right-hand side t'', is substituted for t; t' and t'' are the times obtained in measurements by the observers in the first and second systems respectively. The simple relation between x'' and x' expressed by § 154 eqn. (1) must then, of course, be replaced by a more complicated relation, in which absolute time will not appear, as it did in the original equation.

By uprooting, in this way, the prejudice of an absolute time and demolishing the great obstacle to an extension of the concept of relativity to the whole territory of physics,

[1] The first definition in Newton's "*Philosophiæ naturalis principia mathematica*" runs: "Tempus absolutum verum et mathematicum in se et natura sua sine relatione ad externum quodvis æqualiter fluit." Newton's conception of absolute motion has been resolutely opposed, notably by *Mach.*

Einstein was able, in 1905, to propound his famous so-called *special Principle of Relativity*. According to this principle, every physical event which is described by a definite equation in a definite reference system must be described, in any other system *moving uniformly and rectilinearly* with respect to the first, by the *same equation*, provided that the spatial and temporal coordinates of the first system be replaced by those of the second.

Conversely, the spatial and temporal coordinates in two systems moving uniformly and rectilinearly with respect to each other must be connected, quite generally, by a *universal relation depending only on the mutual relative velocity*. The form of this relation is determined by the postulate that, *in both systems*, the *velocity of light* shall appear to have the *same value in all directions*, this value being a universal constant.

§ 156. The Minkowski-World.

Einstein's conception of the relativity of time renders possible a transformation into an equation of the inequality expressing the apparent incompatibility of Michelson's experiment with classical theoretical physics [§ 154 formula (4)]. Hence it permits also of an extension of the validity of the relativity principle from mechanics to the fields of electrodynamics and optics. Absolute time (t) is replaced by the relative time (t') for the first system, and by the relative time (t'') for the second system. The sign of inequality becomes the symbol for an equation; and the inequality itself is replaced by the *fundamental relation of the theory of relativity* :

$$(1) \quad x''^2 + y''^2 + z''^2 - c^2 t''^2 = x'^2 + y'^2 + z'^2 - c^2 t'^2 .$$

In 1908 this equation was given a *geometrical interpretation* by *Minkowski*,[1] and this allows of a considerable simplification of all deductions made from Einstein's Principle of Relativity. Let us consider for a moment the analytical geometry of a plane surface; it is then a property of the *sum of the squares of the coordinates* x and y that its *value* is the *same* for all arbitrary rectangular coordinate systems based on the same origin. This sum is found to be independent of the orientation of the coordinate system. Analogous considerations apply to the sum of the squares

[1] A paper published by Poincaré in 1906 touched, indeed, on Minkowski's concepts.

of the *three* coordinates x, y, z in a three-dimensional space. Hence, as Minkowski recognized, Einstein's Principle of Relativity as expressed by eqn. (1) may also be interpreted geometrically as follows: The square of the product of the *time* and the velocity of light, taken with a negative sign, and the squares of the three *spatial* coordinates stand in the same relation to each other as do the squares of the *four coordinates in a four-dimensional geometry.*

The four-dimensional manifold, which appears as a *linking together of space and time* by these four coordinates, is known as the *Minkowski-world.* The "world-coordinates" are therefore the four quantities x, y, z, and $i\,c\,t$ (where i, as usual, represents $\sqrt{-1}$). The location of a moving particle at a given time t^* at a given point (x^*, y^*, z^*) may thus be expressed by the fixation of a so-called *world-point*, whose coordinates are equal to x^*, y^*, z^*, and $i\,c\,t^*$. Quantities which can be represented by directed intervals in the Minkowski-world are called *world-vectors.*

The *motion* of a particle may thus be represented by a so-called *world-line*, which expresses the dependency of the path on the time, and whose projection in three-dimensional space gives the course of the path.[2] The form of a world-line and the length of its elements are therefore independent of the reference system. If we denote the length of a world-line-element by ds, we have (by an extension of Pythagoras' Theorem) for an arbitrary reference system the equation

$$ds^2 = dx^2 + dy^2 + dz^2 - c^2\,dt^2\,. \tag{2}$$

Again, the *velocity* (v) of the particle, whose motion is represented by the world-line, with respect to the system used is given by the equation

$$v^2 = \left(\frac{dx}{dt}\right)^2 + \left(\frac{dy}{dt}\right)^2 + \left(\frac{dz}{dt}\right)^2. \tag{3}$$

Hence

$$ds^2 = (v^2 - c^2)\,dt^2\,, \tag{4}$$

and therefore, since $1/i = -i$,

$$\frac{dt}{ds} = -\frac{i}{c\sqrt{1-\frac{v^2}{c^2}}}\,. \tag{5}$$

[2] This is in no way to be taken as implying that, conversely, every world-line may be regarded as the representation of a motion. *Cf.* § 159 note 1.

Finally, let us see what is the meaning of the four *directional cosines* associated with a *world-line-element;* we shall denote them by $\alpha_1, \alpha_2, \alpha_3, \alpha_4$. We have

$$\alpha_1 = \frac{dx}{ds} = \frac{dx}{dt} \cdot \frac{dt}{ds},$$

where dx/dt is equal to v. On the other hand, inasmuch as the fourth coordinate is $i\,c\,t$,

$$\alpha_4 = i\,c\frac{dt}{ds}.$$

On substituting for dt/ds its value as given by eqn. (5), we obtain

$$(6) \qquad \alpha_1 = -\frac{i}{c}\frac{v_x}{\sqrt{1-\frac{v^2}{c^2}}}, \qquad \alpha_4 = \frac{1}{\sqrt{1-\frac{v^2}{c^2}}}.$$

The formulæ for α_2 and α_3 are similar to that for α_1, except that v_x is replaced by v_y or v_z respectively. Conversely, we can derive from eqns. (6) the velocity of the particle with respect to the system used, when the cosines of the angles included between the world-line-element of the particle and the coordinate axes are given as equal to $\alpha_1, \alpha_2, \alpha_3, \alpha_4$. The *magnitude* of the velocity follows directly from the value of α_4 alone.

If, in particular, the four-dimensional coordinate system be so arranged that its *time-axis coincides* with the direction of the *world-line,* we have in respect to this system

$$(7) \qquad \alpha_1 = \alpha_2 = \alpha_3 = 0, \quad \alpha_4 = 1.$$

In accordance with eqns. (6) v is then equal to 0. Relatively to this system, the particle is therefore *at rest.*

§ 157. The Lorentz-Transformation.

When we pass from one four-dimensional coordinate system $(x, y, z, i\,c\,t)$ to a second $(x', y', z', i\,c\,t')$, four equations govern the accompanying *transformation* [in accordance with § 9 eqns. (1) and (3)]. We shall, however, write down only two of these, *viz.,*

$$(1) \quad \begin{cases} x' = x\cos(x', x) + y\cos(x', y) + z\cos(x', z) \\ \qquad\qquad + i\,c\,t\cos(x', t), \\ i\,c\,t' = x\cos(t', x) + y\cos(t', y) + z\cos(t', z) \\ \qquad\qquad + i\,c\,t\cos(t', t). \end{cases}$$

We shall only consider the simple special case in which

the y- and z-axes are common to both four-dimensional coordinate systems; the transformation from one system to the other is then produced by a *rotation in the x-t-plane*. If φ be the angle through which the x-axis must be turned in order to make it coincide with the x'-axis, the angle between t and t' is also equal to φ. The angle between x' and t will be equal to $(90° - \varphi)$, so that its cosine is equal to $\sin\varphi$, and the angle between x and t' will be $(90° + \varphi)$, so that its cosine is equal to $(-\sin\varphi)$. Of the other twelve cosines occurring in the four eqns. (1) $\cos(y', y)$ and $\cos(z', z)$ are equal to unity, whilst all the others are zero. In the special case considered, therefore, eqns. (1) (on division of the last equation by $i\,c$) assume the simpler form

$$(2) \qquad \begin{cases} x' = x\cos\varphi + i\,c\,t\sin\varphi\,, \\ y' = y, \quad z' = z\,, \\ t' = \dfrac{i}{c}\,x\sin\varphi + t\cos\varphi\,. \end{cases}$$

If, now, we consider a particle which is at rest with respect to the coordinate system (x', y', z'), it follows from the concluding remarks of § 156 that its world-line must coincide with the t'-axis. With respect to the coordinate system (x, y, z), however, it will be moving with a velocity $\mathbf{v}$, which can be determined from § 156 eqns. (6) when we put

$$(3) \qquad \begin{cases} \alpha_1 = \cos(t', x), \quad \alpha_2 = \cos(t', y)\,, \\ \alpha_3 = \cos(t', z), \quad \alpha_4 = \cos(t', t)\,. \end{cases}$$

Since the angles between t' on the one hand and y and z on the other are right angles, v_y and v_z vanish, so that v_x is equal to v. Inasmuch as α_1 is equal to $(-\sin\varphi)$, and α_4 to $\cos\varphi$, we have α_1/α_4 equal to $(-\tan\varphi)$. Hence, introducing the values given in § 156 eqns. (6) for α_1 and α_4, we find

$$(4) \qquad \tan\varphi = i\,\frac{v}{c}\,.$$

Again, it follows directly from § 156 eqns. (6) [seeing that α_1 is equal to $(-\sin\varphi)$, and α_4 to $\cos\varphi$], or from eqn. (4) with the help of elementary trigonometrical formulæ, that

$$(5) \qquad \begin{cases} \sin\varphi = \dfrac{i}{c}\,\dfrac{v}{\sqrt{1 - \dfrac{v^2}{c^2}}}\,, \\ \cos\varphi = \dfrac{1}{\sqrt{1 - \dfrac{v^2}{c^2}}}\,. \end{cases}$$

Now if, as we have assumed, a particle is at rest with respect to the second coordinate system, whilst with respect to the first it is moving with the velocity v in the common x-direction, this means that the *spatial coordinate system* (x', y', z') is *moving with a velocity* v in the direction of the common x-axis relatively to the *spatial coordinate system* (x, y, z). A *rotation of four-dimensional coordinate axes* through an angle φ in the x-t-plane is therefore *equivalent to a transformation from one spatial coordinate system to another* which is *moving uniformly in the common x-direction with a velocity v relatively to the first system.* In this rotation, the tangent of the angle φ is equal [according to eqn. (4)] to the quotient of the velocity v and the velocity of light, multiplied by the imaginary quantity i.

Eqns. (2) and (5) enable us at once to answer the question as to how the *spatial coordinates and the time are transformed in accordance with the Principle of Relativity,* when we pass from a given spatial coordinate system to another system parallel to the first and *moving uniformly* relatively to it, with a velocity v in a direction which we suppose to coincide with the common x-axis. For, substituting in eqns. (2) the values given by eqns. (5), we find

$$(6) \qquad \begin{cases} x' = \dfrac{x - v\,t}{\sqrt{1 - \dfrac{v^2}{c^2}}}, \\ y' = y, \quad z' = z, \\ t' = \dfrac{-\dfrac{v}{c^2}\,x + t}{\sqrt{1 - \dfrac{v^2}{c^2}}}. \end{cases}$$

The totality of the substitutions expressed by eqns. (6) are known as a *Lorentz-transformation* [because the first attempts at deriving eqns. (6) were made by Lorentz in a paper published in 1904]. If, in particular, the *velocity v is small as compared with the velocity of light,* the Lorentz-transformation assumes the special form

$$(7) \qquad x' = x - v\,t, \quad y' = y, \quad z' = z, \quad t' = t.$$

In three of these relations we recognize the substitutions which, according to the previously quoted laws of so-called " classical " kinematics, connect the coordinates of two systems moving uniformly with respect to each other. These laws were stated in § 154 eqns. (1), and the fourth relation was taken for granted in classical mechanics, the same time-scale being assigned to both systems.[1]

Since of two spatial coordinate systems moving uniformly with respect to each other neither, of course, is in any way to be preferred to the other, it must be possible, quite generally, to derive the "undashed" variables (x, y, z, t) from the "dashed" by means of a Lorentz-transformation, thus by means of eqns. (6), provided only that the quantity ($+ v$) be replaced by the quantity ($- v$). For, if the second system be moving with a velocity v relatively to the first, the relative translational velocity of the first with respect to the second system must be ($- v$) in the common x-direction.

On forming the sum of the squares of the four "dashed" world-coordinates, we find by eqns. (6)

$$x'^2 + y'^2 + z'^2 - c^2 t'^2 = y^2 + z^2 + \frac{x^2 - 2 x v t + v^2 t^2 - \frac{v^2 x^2}{c^2} + 2 x v t - c^2 t^2}{1 - \frac{v^2}{c^2}}$$

$$= y^2 + z^2 + \frac{x^2\left(1 - \frac{v^2}{c^2}\right) - t^2 (c^2 - v^2)}{1 - \frac{v^2}{c^2}} = x^2 + y^2 + z^2 - c^2 t^2 .$$

Bearing in mind § 156 eqn. (1), we see that the Lorentz-transformation does actually satisfy Einstein's Principle of Relativity.

In point of fact, the equations expressing the Lorentz-transformation can be directly derived from § 156 eqn. (1), without having to introduce the Minkowski-world. The calculations, however, are then much more complicated and tedious than those just performed.

§ 158. Relativity of Scales of Time and Length.

Let us consider, with reference to the undashed system, two events, one of which occurs at the point (x_1, y_1, z_1) at the time t_1, and the other at the point (x_2, y_2, z_2) at the time t_2. Then, in accordance with the equations expressing the Lorentz-transformation, the times at which the two

[1] The special form of the Lorentz-transformation expressed by eqn. (7) is known as a Galileo-transformation, because it leads to the formulæ of classical mechanics established by Galileo.

events occur relatively to the dashed system are given by the formulæ

$$(1)\qquad t_1' = \beta\left(t_1 - x_1\frac{v}{c^2}\right),\quad t_2' = \beta\left(t_2 - x_2\frac{v}{c^2}\right),$$

where, for brevity,

$$(2)\qquad \beta = \frac{1}{\sqrt{1 - \frac{v^2}{c^2}}}.$$

If, in particular, we assume that

$$t_1 = t_2,$$

it does not in any way follow that the quantities t_1' and t_2' are identical, provided that x_1 and x_2 differ. For an observer in the undashed system the two events, which occur at different places, are *simultaneous;* but to an observer in the dashed system they appear to happen *at different times.* In general, therefore, two events which are simultaneous in reference to one system are not simultaneous when referred to another system moving relatively to the first. Thus, in the theory of relativity, the concept of *simultaneity*, like that of time, has *not an absolute, but only a relative meaning.*

When, in particular, we put

$$x_1 = x_2$$

in eqns. (1), this means that both events occur at the same place for an observer in the undashed system. In this special case, it follows from eqns. (1) and (2) that

$$(3)\qquad t_2' - t_1' = \frac{t_2 - t_1.}{\sqrt{1 - \frac{v^2}{c^2}}}.$$

Since the denominator of the fraction is always smaller than unity, the intervals of time between the two events appears larger to the observer in the dashed system than to the observer in the undashed system. Conversely, among all systems in a state of uniform relative motion, that one in which the two events occur at the same place is characterized by the fact that the interval between the two events appears shortest to an observer in that system. A process which, with reference to a given system, occurs at one place alone (which, as we may say, is " fixed in space " with respect to this system) appears to run its course *most rapidly* to an observer in this system, but *more slowly* to an observer in any other. The process, indeed, will appear the slower the more quickly that any system is moving relatively to the first system.[1] This conclusion from the Principle of Relativity is commonly stated in the form of a concise, but not very clearly expressed, rule, *viz.*, that a moving clock goes more slowly than one at rest.

The difference in time-scales for two systems in relative motion suggests that we should *compare scales of length* in the two systems. Now the *length of a rod* which is moving in an arbitrary system can, in general, evidently be determined only by measuring the coordinates (x_1, y_1, z_1) of one end of the rod at an arbitrary time t_1, and *simultaneously* (or, more accurately, at a time t_2 which, with respect to the system, is identical with t_1) measuring the coordinates (x_2, y_2, z_2) of the other end. The length of the rod in the system is then determined by the relation

$$(4) \qquad l = \sqrt{(x_2 - x_1)^2 + (y_2 - y_1)^2 + (z_2 - z_1)^2}\,.$$

If, as we shall assume, the rod always remains parallel to the x-axis, the equation becomes simply

$$(5) \qquad l = x_2 - x_1\,.$$

We shall now so arrange the undashed coordinate system that, with respect to it, the rod is *at rest* and lying in the direction of the x-axis. Then, since x_1 and x_2 are invariable, it is obviously quite unnecessary that the times t_1 and t_2, at which the coordinates x_1 and x_2 are measured, should be the same. These times may be quite arbitrary without the value obtained for l being in any way affected.

If, next, we pass from the undashed to the dashed system, which is connected with the former by a Lorentz-transformation, then with respect to the latter system, relatively to which the rod is moving with a velocity of magnitude v,[2] the length of the rod is evidently given by the relation

$$(6) \qquad l' = x_2' - x_1'\,.$$

In this equation x_1' and x_2' are the dashed coordinates corresponding to the undashed coordinates x_1 and x_2 in accordance with the Lorentz-transformation, whilst t_1' and t_2' must be identical. Now

$$(7) \qquad x_1' = \beta\,(x_1 - v\,t_1), \qquad x_2' = \beta\,(x_2 - v\,t_2),$$

[1] It must be remembered that, in accordance with eqn. (3), the degree of retardation is independent of the sign of v.

[2] The vectorial value of this relative velocity is $(-\mathbf{v})$.

so that

(8) $$x_2' - x_1' = \beta (x_2 - x_1) - \beta v (t_2 - t_1) .$$

Again, since t_2' and t_1' must be identical, it follows from the last of § 157 eqns. (6) that

$$-\frac{v}{c^2} x_1 + t_1 = -\frac{v}{c^2} x_2 + t_2 ,$$

or,

(9) $$t_2 - t_1 = \frac{v}{c^2} (x_2 - x_1) .$$

Introducing this value into eqn. (8), we obtain

$$x_2' - x_1' = \beta (x_2 - x_1) \left(1 - \frac{v^2}{c^2}\right) ,$$

or, by eqn. (2),

(10) $$l' = l \sqrt{1 - \frac{v^2}{c^2}} .$$

To an observer who sees the rod *in motion*, it therefore appears *shortened* ; [3] and, among all systems in a state of uniform relative motion, that one in which the rod is at rest is distinguished from all others by the fact that, in it, the rod appears longer than in any other system.[4] For example, a metre rod lying on the earth in the direction of its motion would appear to an observer on the sun to be shortened by 5×10^{-7} cm.

If the direction of the rod be that of the y- or the z-axis, its length will be the same with respect to both the dashed and the undashed system, for $y' = y$ and $z' = z$, and the time-coordinates are independent of y and z, or of y' and z', as the case may be. A body which is at rest and appears spherical to an observer in the undashed system must hence appear as an *oblate spheroid* to an observer in the dashed system. The theory of relativity recognizes *no absolute*, but only a *relative shape*.

§ 159. Composition of Velocities.

Minkowski's representation enables us to supply a ready answer to the question as to the *composition of velocities* from the point of view of the relativity theory ; in the following

[3] The *contraction hypothesis* of Lorentz and Fitzgerald (*cf.* § 155) thus actually appears as a necessary consequence of Einstein's Principle of Relativity.

[4] As eqn. (10) shows, the contraction is independent of the sign of v. A rod which is at rest in the dashed system will appear shortened to an observer in the undashed system, in exactly the same way as a rod at rest in the undashed system does to an observer in the dashed system.

discussion, however, we shall only consider the especially simple case in which the two component velocities have the same direction. Let v' be a velocity with respect to a system which is itself moving in the same direction with a velocity v relatively to a standard system ; we wish to find the resultant velocity (u) with respect to the latter. We shall suppose the common direction of the velocities v, v', and u to coincide with the common x-axis of the coordinate systems.

We shall denote the original coordinate system (in which the velocity is u) by I ; that in which the velocity is v' by II ; and, finally, one in which the velocity vanishes by III. Now if the velocity vanishes in III, whilst in I it is equal to u, it follows from § 157 eqn. (4) that the relation

$$(1) \qquad i\frac{u}{c} = \tan\psi$$

must exist, where ψ is the angle through which system I must be rotated in the x-t-plane in order to obtain system III.

On the other hand, however, we can also pass from system I to system III by first transforming to system II, which is moving with a velocity v with respect to system I, and thence to system III, which is moving with a velocity v' relatively to system II. In a four-dimensional representation the first transformation appears as a rotation through an angle φ, and the second as a rotation through an angle φ', where

$$(2) \qquad i\frac{v}{c} = \tan\varphi, \qquad i\frac{v'}{c} = \tan\varphi'.$$

The transformation from I to III is therefore represented as a sequence of two rotations through the angles φ and φ', and hence

$$(3) \qquad \psi = \varphi + \varphi'.$$

But, according to an elementary trigonometrical formula,

$$\tan(\varphi + \varphi') = \frac{\tan\varphi + \tan\varphi'}{1 - \tan\varphi\tan\varphi'}.$$

Hence, by eqns. (1), (2), and (3),

$$i\frac{u}{c} = \frac{i\dfrac{v}{c} + i\dfrac{v'}{c}}{1 - i^2\dfrac{vv'}{c^2}},$$

or,

$$u = \frac{v + v'}{1 + \frac{v v'}{c^2}}. \tag{4}$$

For the special case in which v' is equal to c, we have

$$u = \frac{v + c}{1 + \frac{v}{c}} = c. \tag{5}$$

Thus, according to the theory of relativity, the addition of an arbitrary velocity to the velocity of light always gives the quantity c again. Hence the *velocity of light* plays a similar part in the theory of relativity to that played by *infinity* in mathematics. It appears from the relativity theory that *velocities larger than that of light* are *impossible.*[1]

[1] From the impossibility of velocities exceeding that of light it also follows that a *world-vector* drawn between two world-points can *only represent a motion* when the *square of the distance* between the two world-points is *negative*. For, if the world-vector is to represent a motion, the relation

$$\Delta r < c\,\Delta t \tag{6}$$

must hold, where Δr denotes the spatial interval between the two world-points as defined by the equation

$$(\Delta r)^2 = (x_2 - x_1)^2 + (y_2 - y_1)^2 + (z_2 - z_1)^2. \tag{7}$$

Furthermore,

$$(\Delta s)^2 = (\Delta r)^2 - c^2 (\Delta t)^2. \tag{8}$$

Using a coordinate system in which the particle is at rest, we can take the world-vector as our *time-axis*, so that Δr then vanishes, and the two world-points joined by the world-vector *coincide in their positions in space*. If we pass from the reference system in which the time-axis coincides with the world-vector to any other system moving with a velocity v' relatively to the former, we have in this latter system

$$(\Delta r'^2) = v'^2 (\Delta t')^2. \tag{9}$$

Hence, since v' cannot be larger than c, $(\Delta s)^2$ must be *negative in every reference system* if the inequality (6) holds for any system.

The other possible case is that

$$\Delta r > c\,\Delta t. \tag{10}$$

There can then be no motion representable by the world-vector connecting the two world-points, for the velocity in the reference system considered would have to exceed that of light, and that is impossible. Hence it is also impossible to take the world-vector as a time-axis by a transformation into any other reference system. If the inequality (10) holds for only a single reference system, $(\Delta s)^2$ is *positive* in *every* reference system. The world-vector can then be used only as a *spatial axis*. If we suppose its direction to be that of the x-axis, the y- and z- as well as the time-coordinates are the same for both world-points, and therefore, when the inequality (10) holds good, the two world-points can be made *simultaneous*.

In the two opposite cases of the inequalities (6) and (10), the world-vector connecting the two world-points has different properties. In the first case it can be taken only as a time-axis, in the second case only as a space-axis. In the first instance we speak therefore of a *temporal*, and in the second instance of a *spatial, world-vector*. In the first case the world-points connected by the

§ 160. Fizeau's Experiment.

In the formula which, according to the theory of relativity, holds for the addition of two similarly directed

world-vector can be made to coincide in space, and in the second case they can be made simultaneous; the reverse is possible in neither case.

The difference between temporal and spatial world-vectors can more readily be appreciated if, instead of the imaginary time-coordinate (ict), we introduce a *real coordinate* equal to ct and denoted by u. For simplicity's sake we shall at first consider only a single spatial coordinate, and we construct a rectangular coordinate system in the x-u-plane. The Lorentz-transformation then appears graphically (Fig. 127) as a transition to the new axes Ox' and Ou', where the angle included between the new and the old axes is ψ and the angle between the new axes themselves is $(90° - 2\psi)$; the tangent of the angle ψ is equal to the quotient of the relative velocity of the two reference systems and the velocity of light. The magnitude of the units of length and time in the new system can be very simply derived from this graphical representation, as we shall show (without, however, giving the proof here). To do this we construct the two rectangular *hyperbolas*

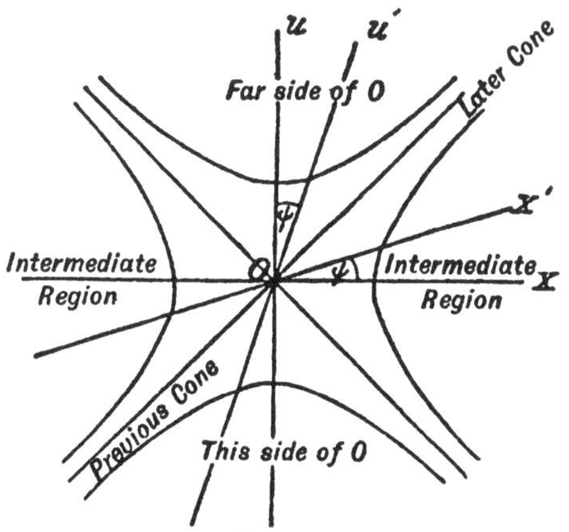

FIG. 127.

$$x^2 - u^2 = \pm 1.$$

The intercepts of the hyperbolas on the x'- and u'-axes then represent the new units of length and time (the latter multiplied by c), referred to the old system. The *pair of asymptotes*

$$x^2 - u^2 = 0$$

represents the propagation of a light wave. If we imagine a y-axis to be introduced, the pair of asymptotes is replaced by a double cone, and the further addition of a z-axis leads to a double conical super-surface which is called a *light-cone*. The light-cone consists of a *previous cone* and a *later cone*. The former ($u < 0$) represents the propagation of a light-wave which, at the time $t = 0$, arrives at the point ($x = y = z = 0$); the latter represents the propagation of a wave which, at the time $t = 0$, departs from the point ($x = y = z = 0$).—The light-cone divides the four-dimensional x-y-z-u-manifold (which, however, does not possess the invariant properties of the Minkowski-world) into *three zones*. The first zone lies on the *far side* of O, and, in the plane of the figure, is bounded by the two asymptotic branches for which u is positive. To every world-point in this zone a temporal vector can be drawn leading to a world-point which is *later* than the origin. The second zone lies on the *near side* of O, and, in the plane of the figure, is bounded by the two asymptotic branches for which u is negative. To every world-point in this zone a temporal vector can be drawn leading to a world-point which is *earlier* than O. The third zone, consisting of the rest of the manifold, is the double-branched *intermediate zone*. Spatial vectors can be drawn from the origin to the world-points in it, so that the points in the intermediate zone can be made simultaneous with the origin.

velocities [§ 159 eqn. (4)], we shall take v' as being the *velocity of light in a medium* whose *refractive index* is n; thus, let v' be equal to c/n. We shall suppose the quantity v in § 159 eqn. (4) to denote the velocity with which the medium itself is moving, as a *flowing liquid* or gas. Then, in accordance with § 159 eqn. (4),

$$(1) \qquad u = \frac{v + \frac{c}{n}}{1 + \frac{v}{c\,n}}.$$

We multiply numerator and denominator by the expression $(1 - v/c\,n)$, neglecting the very small quantity $(v/c\,n)^2$ in comparison with unity. Thus we find

$$(2) \quad u = \left(v + \frac{c}{n}\right)\left(1 - \frac{v}{c\,n}\right) = v + \frac{c}{n} - \frac{v^2}{c\,n} - \frac{v}{n^2} = v - \frac{v}{n^2} + \frac{c}{n}\left(1 - \frac{v^2}{c^2}\right).$$

If again we neglect the very small fraction v^2/c^2 as compared with unity, we obtain the approximation formula

$$(3) \qquad u = v\left(1 - \frac{1}{n^2}\right) + \frac{c}{n}.$$

Now, if the propagation of light be unaffected by the flow of the liquid or gas, u must be equal to c/n. If, on the other hand, the light-waves be simply carried along by the fluid, u must be equal to the sum of v and c/n. Eqn. (3) shows, however, that neither is the case. The *light-waves* are, indeed, affected by the *motion of the fluid*, but not to the full extent of its velocity. The extent to which they partake in that motion is given by a *fraction* $\alpha\,v$, whose magnitude depends on the refractive index of the fluid; for, according to eqn. (3), the value of the quantity α, which is known as the *dragging coefficient*, is given by

$$(4) \qquad \alpha = 1 - \frac{1}{n^2}.$$

As long ago as 1851, a famous experiment performed by *Fizeau* demonstrated that light-waves partake in the flow of a liquid with only a certain fraction of the velocity of flow, and he derived empirically the value given by eqn. (4) for this fraction.[1] Whereas the interpretation of Fizeau's remarkable experimental result had previously demanded the most complicated hypotheses, his experiment finds a simple explanation in Einstein's Principle of Relativity.

§ 161. The Doppler Effect and the Aberration of Light.

An application of the formulæ expressing a Lorentz-transformation to the equations of a light-wave leads to results which are of importance in theoretical optics. Let us consider a system in which a *source of light* (L) is advancing *with constant velocity* (v) in a direction which we take to be that of the x-axis (Fig. 128). We wish to know by what equations in this system the waves proceeding from the source of light can be described.

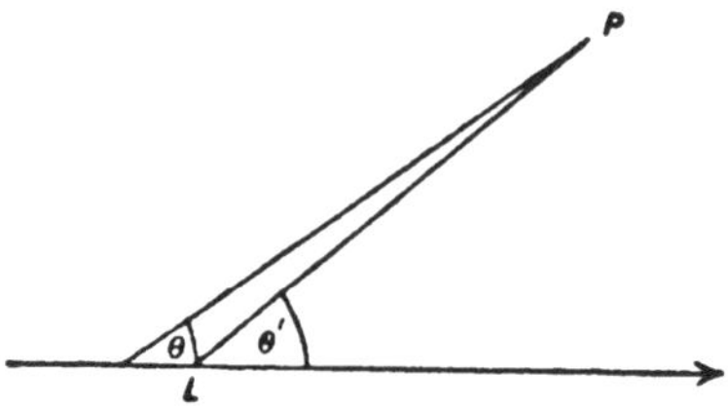

FIG. 128.

In order to answer this question, we shall introduce a second (dashed) reference system, relatively to which the source of light is *at rest*, and which must therefore be connected with the undashed system by a Lorentz-transformation. In the dashed reference system (which is thus moving with the velocity v in the x-direction with respect to the undashed system), the propagation of the light, emanating from a source at rest, is undoubtedly described [according to § 80 eqn. (4)] by equations of the form

$$(1) \qquad S = A' \cos \left\{ \frac{2\pi}{\tau'} \left[t' - \frac{r'}{c} \right] + \delta' \right\}.$$

In this expression S denotes any arbitrary one of the components of the vibrating electric or magnetic field-strengths, A' the amplitude, τ' the period, δ' the phase-constant, and finally r' denotes a distance measured in the direction of a wave-normal.

[1] Since the refractive index differs very little from unity in the case of gases, an appreciable dragging effect is not to be observed with them. The quantity a is also known as *Fresnel's* dragging coefficient, for Fresnel, on the basis of a theoretical investigation of the aberration of light, had come to the conclusion that only a portion of the light æther partakes in the earth's motion. This portion was supposed to be that which existed in matter, and it was assumed to partake only to the extent of a fraction a of the velocity, the value found for a by Fresnel being the same as that given in eqn. (4).

Let us now consider an arbitrary point (P) rigidly connected with the undashed coordinate system, and let us suppose an observer to be situated there, so that for him the velocity of the source of light is v. If we choose the plane determined by the direction of motion of the light source and the point P as the x-y-plane, and hence also as the x'-y'-plane, we have

$$(2) \qquad r' = x' \cos \theta' + y' \sin \theta',$$

where θ' is the angle included between the normal to a wave travelling from L to P and the x'-direction (or the x-direction, which is the same). Since the direction of this wave-normal is undoubtedly LP in the system in which the source of light is at rest, θ' is thus also the angle included between the x-axis and the line joining source and observer. Moreover, since the frequency (ν') is the reciprocal of the period τ', we can write eqn. (1) in the form

$$(3) \qquad S = A' \cos \left\{ 2\pi\nu' \left[t' - \frac{x' \cos \theta'}{c} - \frac{y' \sin \theta'}{c} \right] + \delta' \right\}.$$

On replacing x', y', z', t' by x, y, z, t with the help of the Lorentz-transformation, and using the abbreviated notation introduced in § 158 eqn. (2), we find

$$(4) \qquad \begin{cases} S = A' \cos \left\{ 2\pi\nu' \left[\beta \left(t - \frac{vx}{c^2} \right) - \frac{\beta}{c}(x - vt) \cos \theta' \right. \right. \\ \qquad\qquad \left. \left. - \frac{y \sin \theta'}{c} \right] + \delta' \right\}. \end{cases}$$

It is easy to see that this equation can also be written in a form analogous to that of eqn. (3):

$$(5) \qquad S = A \cos \left\{ 2\pi\nu \left[t - \frac{x \cos \theta}{c} - \frac{y \sin \theta}{c} \right] + \delta \right\},$$

provided we equate A to A', and δ to δ', and further put

$$(6) \qquad \nu = \nu' \beta \left(1 + \frac{v}{c} \cos \theta' \right)$$

and

$$(7) \qquad \cos \theta = \frac{\cos \theta' + \frac{v}{c}}{1 + \frac{v}{c} \cos \theta'}.$$

For, by a simple calculation, we find from eqn. (7) that

$$(8) \qquad \sin \theta = \sqrt{1 - \cos^2 \theta} = \frac{\sin \theta'}{1 + \frac{v}{c} \cos \theta'} \sqrt{1 - \frac{v^2}{c^2}};$$

and, on inserting the values given by eqns. (6), (7), and (8) in eqn. (5), we obtain eqn. (4).

Light from a moving source will therefore appear of a *different colour* to an observer than light from a source which is at rest with respect to him. The change in frequency is determined by eqn. (6). When, in particular, the source of light is approaching directly towards the observer, or receding directly away from him, cos θ' becomes equal to $\pm$ 1. In these two special cases we have, therefore,

$$(9) \qquad \nu = \nu' \frac{1 \pm \frac{v}{c}}{\sqrt{1 - \frac{v^2}{c^2}}}.$$

Neglecting the term of the second order in this formula, we may write, *by way of approximation*,

$$(10) \qquad \nu = \nu'\left(1 \pm \frac{v}{c}\right).$$

Spectral lines arising from *moving sources of light* must thus experience a *displacement*, towards the *violet end* in the visible region of the spectrum when the source is *approaching the observer*, and towards the *red* end when it is *receding* from him. This effect, which was known to physicists long before the development of the theory of relativity, though only in the form of the approximately correct eqn. (10),[1] is known as the *Doppler effect* after its discoverer (who, indeed, knew nothing of spectral lines, and hence, in 1842, spoke only generally of colour).[2]

Whilst eqn. (6) shows us that the frequency changes, we learn from eqn. (7) that the *normal* to a wave approaching an observer from a source of light is *differently directed* according as the source is at rest or in motion with respect to the observer. Whereas, in the first case, the direction of the

[1] In contradistinction to the older theory, the theory of relativity demands a displacement of spectral lines also in the case in which the observer is looking *at right angles* to the direction of motion of the source of light. For, since cos θ' then vanishes, ν must be equal to $\beta\,\nu'$ [by eqn. (6)]. On account of its smallness, however, this effect has not yet been observed.

[2] Displacements of the spectral lines emitted by *stars* were first observed by *Huggins* (1868). Exact measurements of these displacements enabled the *radial components* of the velocities of many stars to be determined. In 1905, *Stark* succeeded in demonstrating the Doppler effect also in *canal rays*. Moreover, the *finite width of spectral lines* can be explained on the basis of the Doppler effect, for, according to the kinetic theory of gases, the atoms of a gas which emit the light are moving in all directions. At low temperatures the breadth of the lines must decrease with the molecular velocities, and this has been proved by experiment to be the case.

wave-normal is that of the line joining source and observer, in the second case the light appears to come in a different direction ; hence the observer sees a source of light which is moving relatively to him in a different place from that where it is actually.

If, in particular, $\theta' = 0$, and thus $\cos \theta' = 1$, so that the observer is looking in the direction of the relative motion, it follows from eqn. (7) that also $\theta = 0$. On the other hand, when $\cos \theta' = 0$, so that the observer is looking *at right angles* to the direction of the relative motion, we find by eqn. (7) that

$$\cos \theta = \frac{v}{c}. \tag{8}$$

The apparent change in direction of the rays of light coming from a moving source is known as the *aberration of light*. This phenomenon was discovered in 1727 by *Bradley*, who showed that it was responsible for an apparent periodic annual motion of all the fixed stars, owing to the motion of the earth around the sun.

§ 162. Relativistic Dynamics.

According to the theory of relativity, an absolute significance attaches to neither space nor time, but only to the Minkowski-world as an intermingling of the two. In relativistic physics, therefore, the relations between spatial vectors (as known to classical physics) must be replaced by such relations between *world-vectors* that the classical relations can be derived, as special cases, from them.

The theory of relativity is faced with a fundamental problem in the derivation of a more general law to replace *Newton's Second Law of Motion*, which must itself result as a *special case* of that more general law, holding with great approximation for motions in which the velocity is small as compared with the velocity of light.

If, now, there be a privileged system for which the more general relation becomes *identical* with the relation of classical mechanics, this system can evidently only be the one in which the particle considered is *at rest*. We shall define this system briefly as the *proper system* of the particle,[1] and the time corresponding to this system as the *proper time*.

[1] When the motion is not uniform and rectilinear, we can, of course, only speak in general of a proper system for a definite small length of path.

Denoting the latter by τ, we have, by § 156 eqn. (4) (since the particle is at rest in the proper system),

$$ds = i\,c\,d\tau\,. \tag{1}$$

Our previous considerations have shown that the four directional cosines of a world-line-element, which we denoted by α_1, α_2, α_3, α_4, form the components of a four-dimensional vector.[2] We shall denote by q_x, q_y, q_z, $i\,q_t$ the components of the world-vector obtained by multiplying the directional cosines by $i\,c$. Furthermore, as in § 158, we shall put, for brevity,

$$\frac{1}{\sqrt{1-\frac{v^2}{c^2}}} = \beta\,. \tag{2}$$

Then, according to § 156 eqn. (6),

$$q_x = \beta\,v_x\,, \tag{3}$$

and

$$q_t = \beta\,c\,. \tag{4}$$

For the *magnitude* of the world-vector in question, which we define as being the *Minkowski-velocity*, we find from eqns. (3) and (4) the relation

$$q^2 = q_x{}^2 + q_y{}^2 + q_z{}^2 - q_t{}^2 = \beta^2\,(v^2 - c^2)\,. \tag{5}$$

Hence, replacing β by its value as given in eqn. (2), we obtain

$$q^2 = -\,c^2\,. \tag{6}$$

If, in accordance with our earlier assumption, we suppose that Newton's Second Law of Motion in its classical form holds exactly for the *proper system*, we obtain for this system three equations of the form

$$K_x = \frac{d}{d\tau}\left(\mu\,v_x\right), \tag{7}$$

etc., where μ denotes the mass, and K_x the x-component of the acting force.

This relation suggests that we might *define* a *world-vector*, which we shall call the *Minkowski-force*, by the equations

$$\begin{cases} P_x = \dfrac{d}{d\tau}\left(\mu\,q_x\right), \\ P_t = \dfrac{d}{d\tau}\left(\mu\,q_t\right). \end{cases} \tag{8}$$

[2] For, in three-dimensional coordinates also, the directional cosines form the components of a vector.

Now, if A be an arbitrary quantity, we have by eqn. (1) and § 156 eqn. (5)

$$\text{(9)} \qquad \frac{dA}{d\tau} = i\,c\,.\,\frac{dA}{dt}\,.\,\frac{dt}{ds} = \beta\,\frac{dA}{dt}\,.$$

Hence, on substituting for q_x and q_t the values given in eqns. (3) and (4), we find

$$\text{(10)} \qquad \begin{cases} P_x = \beta\,\dfrac{d}{dt}\left(\beta\,\mu\,v_x\right), \\ P_t = \beta\,c\,\dfrac{d}{dt}\left(\beta\,\mu\right). \end{cases}$$

For the proper system (for which β is equal to unity), the first of eqns. (10) actually becomes identical with eqn. (7), and thus we see that the spatial components of the Minkowski force coincide in the proper system with the components of the " Newtonian " force. Consequently, the first of eqns. (10) and the two analogous expressions for P_y and P_z represent a *generalization of Newton's Second Law of Motion*, in virtue of which the *force* is *equal to the time-derivative of the momentum*, as measured by the product of mass and velocity. In this generalization we have, on the one hand, to put

$$\text{(11)} \qquad P_x = \beta\,K_x\,,$$

and, on the other hand, we must relinquish the classical conception of the constancy of mass, assuming, instead, a *dependency of the so-called momental mass on the velocity* according to the relation

$$\text{(12)} \qquad m = \beta\,\mu = \frac{\mu}{\sqrt{1 - \dfrac{v^2}{c^2}}}\,.$$

Under these circumstances, the first of eqns. (10) is transformed, quite generally, into the relation which expresses Newton's Second Law of Motion :

$$\text{(13)} \qquad \mathbf{K} = \frac{d}{dt}\left(m\,\mathbf{v}\right).$$

When $v = 0$, m becomes equal to μ, and μ is therefore known as the *stationary mass*. The momental mass is always *larger* than the stationary mass. For motions whose velocity is small as compared with that of light, however, the difference between momental and stationary mass is quite inappreciable.

If, in accordance with eqn. (12), we take into account the

dependence of the mass on the velocity, and thus on the time, we can write eqn. (13) in the form

$$\mathbf{K} = m\frac{d\mathbf{v}}{dt} + \mathbf{v}\,\mu\frac{d\beta}{dt}, \tag{14}$$

where

$$\frac{d\beta}{dt} = \frac{\frac{v}{c^2}}{\left(1-\frac{v^2}{c^2}\right)^{\frac{3}{2}}}\frac{dv}{dt} = \beta^3\frac{v}{c^2}\frac{dv}{dt}. \tag{15}$$

We shall now suppose the force to be resolved into two mutually perpendicular components in the plane determined by the directions of the force and the velocity. One of the two components shall have the direction of the motion, and therefore the components may be distinguished as *longitudinal* and *transversal*. We shall denote the former by the subscript 1, and the latter by the subscript 2. If we resolve the velocity in a similar manner, v_1 is equal to v and v_2 equal to zero. It follows from eqn. (14) that the acceleration (determined by the differential coefficient $d\mathbf{v}/dt$) must lie in the plane determined by the vectors $\mathbf{K}$ and $\mathbf{v}$; let its two components be b_1 and b_2. According to the generally valid and purely vectorial relation expressed in § 3 eqn. (30), the longitudinal component of the time-derivative of the velocity vector is then equal to the time-derivative of the magnitude of the velocity; hence

$$b_1 = \frac{dv}{dt}. \tag{16}$$

On resolving eqn. (14), therefore, into a longitudinal and a transversal partial equation, we obtain [bearing in mind eqn. (12)]

$$K_1 = \beta\mu\, b_1 + \beta^3\frac{v^2}{c^2}\mu\, b_1 = \beta\mu\, b_1\left(1+\frac{v^2}{c^2}\beta^2\right).$$

But

$$1+\frac{v^2}{c^2}\beta^2 = 1+\frac{v^2}{c^2\left(1-\frac{v^2}{c^2}\right)} = 1+\frac{v^2}{c^2-v^2} = \frac{c^2}{c^2-v^2} = \beta^2.$$

Hence

$$\frac{K_1}{b_1} = \beta^3\mu = \beta^2 m; \tag{17}$$

whilst it follows from eqn. (14), since v_2 is equal to 0, that

$$(18) \qquad \frac{K_2}{b_2} = \beta \mu = m .$$

Whereas in classical mechanics, therefore, both K_1/b_1 and K_2/b_2 are equal to the mass, in relativistic mechanics this is only the case for the transversal component.[3] Since β must invariably be larger than unity, we have always

$$\frac{K_1}{b_1} > \frac{K_2}{b_2},$$

or

$$\frac{b_2}{b_1} > \frac{K_2}{K_1}.$$

Thus, in general, *force* and *acceleration* are *not similarly directed*, but the direction of the *acceleration* always includes a *larger angle* with the direction of motion than the direction of the force.

In previous sections reference has been made on several occasions to the *experimental confirmations in atomic physics* of the *mass formula* of the relativity theory. As was mentioned in § 89, measurements of the *magnetic deflection of β-rays* have demonstrated the actual validity of the assumption (demanded by the formula) of the decrease in the specific charge with increasing velocity. A further striking confirmation (mentioned in § 100) is afforded by the complete agreement between the actually observed *frequency difference of the hydrogen doublet* and that calculated on the basis of the relativity formula and the Bohr-Sommerfeld theory.[4]

§ 163. The Inertial Mass of Energy.

In order to form the *scalar product* of the *world-vectors* representing the *Minkowski-force* and the *Minkowski-velocity*, we shall first write § 162 eqn. (8) in the form

$$(1) \qquad P_x = \mu \frac{dq_x}{d\tau} + q_x \frac{d\mu}{d\tau}.$$

[3] The quotient K_1/b_1 was formerly known as the " longitudinal mass," and the quotient K_2/b_2 as the " transversal mass." Nowadays " mass " is generally understood to mean " momental mass."

[4] Furthermore, the relation between the magnetic β-spectrum of a radio-element and the Röntgen spectrum of the isotopic (ordinary) element (discussed in § 118) presupposes the validity of the relativistic mass formula, and this relation has actually been found to be accurately obeyed.

Then

(2) $$P_x q_x = \mu \frac{d}{d\tau}\left(\frac{q_x^2}{2}\right) + q_x^2 \frac{d\mu}{d\tau}.$$

On forming the analogous equations for the products $P_y q_y$, $P_z q_z$, and $i^2 P_t q_t$, and then adding them together, we find [bearing in mind the fact that, according to § 162 eqn. (6), q^2 is equal to $(-c^2)$, and thus represents a universal constant]

(3) $$P_x q_x + P_y q_y + P_z q_z - P_t q_t = -c^2 \frac{d\mu}{d\tau}.$$

Finally, taking § 162 eqns. (3), (9), and (11) into account, we may write this in the form

(4) $$\beta^2 \mathbf{K}\,\mathbf{v} - P_t q_t = -\beta c^2 \frac{d\mu}{d\tau}.$$

Now the scalar product of the three-dimensional vectors **K** and **v** simply represents the *work* performed in unit-time. Moreover, in classical mechanics the proper mass μ is, of course, regarded as being constant, so that the right-hand side of eqn. (4) vanishes. Hence, provided we neglect the infinitesimal difference of β from unity for an arbitrarily slow motion, the product $P_t q_t$ simply represents the *time-derivative of the energy.*

If, therefore, we consider a process in which the *energy* of a body *changes without the performance of external work*, *e.g.*, owing to the absorption or emission of radiation, or to the introduction or abstraction of heat, and since we may assume the motion to be arbitrarily slow, we find, denoting the energy by E and putting β equal to unity, that eqn. (4) becomes

(5) $$\frac{dE}{dt} = c^2 \frac{d\mu}{dt}.$$

Thus the theory of relativity leads to the extremely important conclusion that every *change in the energy of a body* involves an *alteration in its mass*, by an amount which is equal to the energy change *divided by the square of the velocity of light.*[1] The *Law of the Inertial Mass of Energy*, expressed in eqn. (5), was derived in 1905 by *Einstein* for a special case, and was subsequently placed on a more general basis by *Planck.*

The theory of relativity, therefore, ascribes mass to all

[1] For a more exact, yet simple, proof of the Law of the Inertia of Energy, *cf.* Frank, "*Ein einfacher Beweis für die Trägheit der Energie*," in the scientific journal "*Lotos*" (Prag), 70 (1922), p. 301.

energy, and, conversely, energy to every mass.[2] Mass and energy become identical concepts, differing only by a universal proportionality factor ; Lavoisier's Law of the Conservation of Mass and the Mayer-Joule Law of the Conservation of Energy are united by the theory of relativity into one single principle.

The changes of mass associated with observable alterations in energy are admittedly far too small to be detected. For example, even in the comparatively enormous development of heat by radium, the loss in mass due to the evolution of heat by one gram of radium in a radium preparation in a year would only amount to the 20,000th part of a milligram —an amount which could scarcely be observed.[3]

An important application of the Law of the Inertial Mass of Energy is to problems of *atomic nuclear structure.* The *mass defect in the atoms* due to the so-called *packing effect* has already been discussed in an earlier section (§ 116).

Another very interesting application has been made by *Nernst*[4] to an important astrophysical problem, *viz.*, the question of the *age of the sun.* In every second the sun radiates approximately 1×10^{26} cal., or an energy quantity of about 4×10^{33} erg.[5] According to the Law of the Inertia of Energy, the *mass of the sun* must therefore *decrease by some four million tons* (4×10^{12} gm.) *every second.* Since the mass of the sun amounts to 2×10^{33} gm., this means that its mass would diminish by about 1 part in 10^{12} in ten years : 10^{13} years would at any rate represent an upper limit for the sun's age.[6]

[2] The energy associated with every mass, obtained by multiplying the proper mass by c^2, is known as the *proper energy* of the mass. As compared with the proper energy, the so-called kinetic energy (which naturally has only a relative significance) is in general vanishingly small. Even the kinetic energy of a discharged rifle bullet would only represent about 1 part in 10^{12} of its proper energy.

[3] The loss in mass due to the evolution of heat must, of course, not be confused with the disappearance of the radium consequent on the formation of emanation.

[4] W. Nernst, *Das Weltgebäude im Lichte der neueren Forschung* Berlin (Springer), 1921.

[5] The distance of the earth from the sun is about $1{\cdot}5 \times 10^{13}$ cm. ; a sphere described about the sun with this radius has a superficial area of some 3×10^{27} cm^2. A square centimetre of the earth's surface receives 2·0 cal. in a minute through direct solar radiation, and thus a sixtieth of that amount in a second.

[6] As a matter of fact, the upper limit is probably much lower, perhaps a hundred thousand million years, if only a small fraction of the sun's mass represents the equivalent of the energy which can be radiated. A *lower limit* for the age of our solar system, however, follows from the fact that, as was explained in § 113, the age of uranium minerals has been found from their lead content to be anything up to fifteen hundred million years.

§ 164. World-Tensors.

In complete analogy to the relations holding for a three-dimensional system (*vide* § 26), we define 16 quantities as the components of a *world-tensor* when, for a change of the four-dimensional coordinate system, they are *transformed* in the same way as the *products formed by multiplying together the components of two world-vectors in pairs.* If, in particular, the components remain unaltered for an interchange of indices, so that

$$(1) \qquad t_{hk} = t_{kh}$$

(where both h and k may be any one of the four numbers 1, 2, 3, 4), the world-tensor is said to be *symmetrical.* If, on the other hand, the sign is reversed for an interchange of the indices, so that

$$(2) \qquad t_{hk} = - t_{kh} ,$$

the world-tensor is defined as being *skew-symmetrical.* In that case the tensor components of the first kind (*i.e.*, those in which the indices are identical) must be equal and opposite to themselves, which is only possible if they vanish. Hence, for a skew-symmetrical tensor,

$$(3) \qquad t_{hh} = 0 .$$

Thus, whereas *ten* quantities are necessary for the determination of a symmetrical world-tensor, only *six* are required to define a skew-symmetrical tensor, because four of the components vanish, and in value the other twelve are alike by pairs.

The conception of a skew-symmetrical tensor is, of course, in no way peculiar to four-dimensional geometry alone. In three-dimensional geometry, skew-symmetrical tensors are simply the "supplementary auxiliary vectors" which represent directed plane-magnitudes (*vide* § 3), and which, as "axial" vectors, are contrasted with the so-called "polar" vectors. Quite generally, in a geometry of an arbitrary number of dimensions, the number of the components of a vector is equal to the number of coordinate *axes*, whereas the number of the components of a skew-symmetrical tensor is equal to the number of coordinate *planes.* In three-dimensional geometry both numbers are the same, and to that fact is due the adoption of the careless form of expression which erroneously describes the skew-symmetrical tensors of three-dimensional geometry as axial vectors.

Now we may regard as the components of a spatial axial vector the three out of the six components of a skew-symmetrical world-tensor which do not contain the time index. On the other hand, the three components containing the time index may be regarded as the components, multiplied by i, of a spatial polar vector. Hence we can define a spatial axial vector (**C**) and a spatial polar vector (**D**) by means of the equations

$$(4) \quad C_x = t_{23} = -t_{32}, \quad C_y = t_{31} = -t_{13}, \quad C_z = t_{12} = -t_{21};$$

and

$$(5) \quad \begin{cases} i D_x = t_{41} = -t_{14}, & i D_y = t_{42} = -t_{24}, \\ i D_z = t_{43} = -t_{34}. \end{cases}$$

The *vector divergence* of any world-tensor can be found quite generally according to the scheme

$$(6) \qquad q_1 = \frac{\partial t_{11}}{\partial x_1} + \frac{\partial t_{12}}{\partial x_2} + \frac{\partial t_{13}}{\partial x_3} + \frac{\partial t_{14}}{\partial x_4}$$

[*cf.* the analogous § 36 eqns. (4)]. We shall put

$$q_1 = q_x, \quad q_2 = q_y, \quad q_3 = q_z, \quad q_4 = i\, q_t,$$

and regard the three components q_x, q_y, q_z as those of a spatial vector **q**. Bearing in mind that x_4 is equal to $i\,c\,t$, we find for the vector divergence of the skew-symmetrical world-tensor considered

$$q_x = \frac{\partial C_z}{\partial y} - \frac{\partial C_y}{\partial z} - \frac{1}{c}\frac{\partial D_x}{\partial t},$$

and

$$q_t = \frac{\partial D_x}{\partial x} + \frac{\partial D_y}{\partial y} + \frac{\partial D_z}{\partial z}.$$

Hence

$$(7) \qquad \mathbf{q} = \operatorname{rot} \mathbf{C} - \frac{1}{c}\frac{\partial \mathbf{D}}{\partial t},$$

and

$$(8) \qquad q_t = \operatorname{div} \mathbf{D}.$$

The generalized operation of *forming a rotation* can be readily found from the formula

$$(9) \qquad B_{hk} = \frac{\partial A_k}{\partial x_h} - \frac{\partial A_h}{\partial x_k}.$$

According to § 32, the partial derivatives of a vector with respect to the coordinates represent the components of a tensor, and therefore the quantities B_{hk} determined by eqn. (9) are also the components of a world-tensor. This, as we can recognize at once, must be a *skew-symmetrical world-*

tensor which is thus derived as the four-dimensional *rotation* of any world-vector.

On partial differentiation of eqn. (9) with respect to an arbitrary coordinate, we obtain

$$(10) \qquad \frac{\partial B_{hk}}{\partial x_r} = \frac{\partial^2 A_k}{\partial x_h\, \partial x_r} - \frac{\partial^2 A_h}{\partial x_k\, \partial x_r}\,.$$

Cyclic interchanges of the three indices h, k, r give us the two analogous equations

$$(11) \qquad \begin{cases} \dfrac{\partial B_{kr}}{\partial x_h} = \dfrac{\partial^2 A_r}{\partial x_k\, \partial x_h} - \dfrac{\partial^2 A_k}{\partial x_r\, \partial x_h}\,, \\ \dfrac{\partial B_{rh}}{\partial x_k} = \dfrac{\partial^2 A_h}{\partial x_r\, \partial x_k} - \dfrac{\partial^2 A_r}{\partial x_h\, \partial x_k}\,. \end{cases}$$

Adding eqns. (10) and (11) together, we find

$$(12) \qquad \frac{\partial B_{hk}}{\partial x_r} + \frac{\partial B_{kr}}{\partial x_h} + \frac{\partial B_{rh}}{\partial x_k} = 0\,.$$

This relation is characteristic of every tensor which, in its turn, forms the rotation of a vector.[1]

If we wish to write down individually the equations comprised in eqn. (12),[2] we have to replace h, k, r by 1, 2, 3, and then effect a cyclic interchange of the four indices 1, 2, 3, 4. We find

$$(13) \qquad \begin{cases} \dfrac{\partial B_{12}}{\partial x_3} + \dfrac{\partial B_{23}}{\partial x_1} + \dfrac{\partial B_{31}}{\partial x_2} = 0\,, \\ \dfrac{\partial B_{23}}{\partial x_4} + \dfrac{\partial B_{34}}{\partial x_2} + \dfrac{\partial B_{42}}{\partial x_3} = 0\,, \end{cases}$$

etc.

As before, let us now denote the three spatial components of the skew-symmetrical tensor, represented by the rotation, by C_x, C_y, C_z, and the other three by $i\,D_x$, $i\,D_y$, $i\,D_z$, in accordance with eqns. (4) and (5). Eqn. (13) then becomes

$$(14) \qquad \begin{cases} \dfrac{\partial C_z}{\partial z} + \dfrac{\partial C_x}{\partial x} + \dfrac{\partial C_y}{\partial y} = 0\,, \\ -\dfrac{i}{c}\dfrac{\partial C_x}{\partial t} - i\,\dfrac{\partial D_z}{\partial y} + i\,\dfrac{\partial D_y}{\partial z} = 0\,, \end{cases}$$

etc. Thus we may also write

[1] When r is equal either to h or to k, one of the three terms in eqn. (12) vanishes.

[2] Eqn. (12) is the four-dimensional generalization of the three-dimensional equation

$$\text{div rot } \mathbf{A} = 0$$

[§ 32 eqn. (24)].

(15) $$\operatorname{rot} \mathbf{D} = -\frac{1}{c}\frac{\partial \mathbf{C}}{\partial t},$$

and

(16) $$\operatorname{div} \mathbf{C} = 0.$$

These two equations apply to every world-tensor which itself forms the rotation of a world-vector, and whose components can be grouped together in three dimensions to form an axial vector **C** and a polar vector **D**.

§ 165. The Electromagnetic Field Tensor.

Of the four *Maxwell's equations*, two contain only the electric field-strength (**E**) and the magnetic field-strength (**H**), but not the charge or current density. According to § 67 eqns. (1), these two have the form

(1) $$\operatorname{rot} \mathbf{E} = -\frac{1}{c}\frac{\partial \mathbf{H}}{\partial t}, \quad \operatorname{div} \mathbf{H} = 0.$$

A comparison of these equations with § 164 eqns. (15) and (16) shows us at once that the components of the magnetic and electric field-strengths represent, in a four-dimensional system, the components of a *skew-symmetrical world-tensor* which, in its turn, must be the *rotation of a world-vector*. This world-vector is called the *four-potential*,[1] and the skew-symmetrical world-tensor derived from it by the operation of rotation is known as the *electromagnetic field tensor*. The components of this tensor are given by the relations

(2) $$F_{23} = -F_{32} = H_x; \quad F_{41} = -F_{14} = i\,E_x, \text{ etc.}$$

The pair of Maxwell's equations which we have not yet considered, containing the charge and current densities

[1] A simple calculation from Maxwell's equations shows us that the x-, y-, and z-components of the four-potential are identical with the components of Maxwell's vector-potential. The time-component is equal to the so-called *field-potential*, which was introduced into classical electrodynamics by *Lorentz*. Lorentz was led to the concept of the field-potential by the following simple consideration. Denoting Maxwell's vector-potential by **V**, we have

$$\mathbf{H} = \operatorname{rot} \mathbf{V},$$

and, since the order of the operations of forming a rotation and partial differentiation can be reversed,

$$\operatorname{rot}\left(\mathbf{E} + \frac{1}{c}\frac{\partial \mathbf{V}}{\partial t}\right) = 0.$$

Again, according to § 32 eqn. (23), every vector whose rotation vanishes can be regarded as the gradient of a scalar function of the coordinates. Hence we may put

$$\mathbf{E} + \frac{1}{c}\frac{\partial \mathbf{V}}{\partial t} = -\operatorname{grad} \phi,$$

d ϕ represents Lorentz's field-potential.

besides the electric and magnetic strengths of field, can be formulated in the following way (according to § 67), if we write the density of the convection current in the form $\varrho\,\mathbf{v}$:—

$$(3)\qquad \frac{4\pi}{c}\varrho\,\mathbf{v} = \operatorname{rot}\mathbf{H} - \frac{1}{c}\frac{\partial \mathbf{E}}{\partial t}, \quad 4\pi\,\varrho = \operatorname{div}\mathbf{E}.$$

In order to arrive at the four-dimensional significance of these two equations, we shall commence with the concept of a quantity of electricity. Since electrical charge is obviously a quantity which has nothing whatsoever to do with a state of motion, we may reasonably assume that a quantity of electricity will not be affected by a Lorentz-transformation. *Density of charge*, however, will be so affected, because the *volume* is altered by a Lorentz-transformation. For, since a length x is reduced to x/β by the contraction in the direction of motion,[a] whilst there is no contraction in the y- and the z-direction, the volume (V) itself is not a constant quantity, but

$$(4)\qquad \beta\,V = \text{invariant},$$

i.e., independent of the reference system. Now the density of charge (ϱ) is the quotient of the invariant quantity of electricity and the volume, and therefore

$$(5)\qquad \frac{\varrho}{\beta} = \text{invariant}.$$

Let us now multiply the components of the Minkowski-velocity [§ 162 eqns. (3) and (4)] by this scalar (which is independent of the reference system), and also by the universal constant $4\pi/c$. In this way we obtain the components of a world-vector, which is known as the *four-current*. Its components have the values

$$(6)\qquad s_x = \frac{4\pi}{c}\varrho\,v_x, \qquad s_t = 4\pi\,\varrho.$$

Resolution of the four-current into its spatial components and its temporal component therefore separates it into convection current and charge (referred to unit-volume). This resolution, however, is of course only relative, and, like the resolution of the electromagnetic field tensor into its electric and magnetic components, depends only on the point of view of the observer describing it. An absolute

[a] The quantity β, used for brevity's sake, has here again the same meaning as defined by § 158 eqn. (2).

significance, independent of the observer's point of view, attaches only to the four-current and the electromagnetic field tensor.

On substituting the values given by eqns. (6) in eqns. (3), we recognize complete agreement with § 164 eqns. (7) and (8), except that **q**, **C**, **D** are now replaced by **s**, **H**, **E** respectively.

In a four-dimensional representation, therefore, Maxwell's equations may be summarized in the extremely simple statement that the *four-current* is the *vector divergence of the electromagnetic field tensor*, which, in its turn, represents the *rotation* of a world-vector. The transition from four to three dimensions is effected by splitting the four-current into convection current and charge, and the field tensor into electric and magnetic strengths of field.

§ 166. The Matter Tensor.

The *proper mass* (μ), like a quantity of electricity, is evidently independent of the reference system, as, according to § 165 eqn. (4), must also be the case for the quotient $\mu/\beta V$, where V denotes the volume. According to § 162 eqn. (12), however, the product $\beta \mu$ represents the momental mass (m), and this divided by the volume gives the so-called *mass density*. For simplicity's sake, we shall denote the mass density, like density of charge, by ϱ. We have, therefore,

$$(1) \qquad \frac{m}{\beta^2 V} = \frac{\varrho}{\beta^2} = \text{invariant}\,.$$

We derive a world-tensor, which is of great importance in the theory of relativity, by multiplying the products (taken *in pairs*) of the directional cosines of the *world-line-element* representing the motion by the scalar described by eqn. (1). We find, according to § 156 eqns. (6),

$$(2) \qquad \begin{cases} T_{11} = -\dfrac{\varrho\, v_x^2}{c^2}\,, \quad T_{12} = T_{21} = -\dfrac{\varrho\, v_x\, v_y}{c^2}\,, \\ T_{14} = T_{41} = -\,i\dfrac{\varrho\, v_x}{c}\,, \quad T_{44} = \varrho\,; \end{cases}$$

with analogous expressions for the other components.

The purely temporal component of this tensor thus represents nothing else than the *density of matter* ; in subsequent discussions this tensor will therefore be briefly termed the *matter tensor*. The space-time-components (T_{14}, T_{24}, T_{34}) have also a simple meaning, for, apart from a universal factor, they represent the *density of momentum.* When the mechanical velocity is small as compared with the velocity of light, the purely spatial components become small quantities of the second order, and hence, in this case, we find the sum of the four tensor components of the first kind (important for later considerations)

$$T_{11} + T_{22} + T_{23} + T_{44} = \varrho \,. \tag{3}$$

CHAPTER XIX

THEORY OF GRAVITATION

§ 167. The Law of Gravitation as a Generalization of the Law of Inertia.

As a result of the Principle of Relativity established by Einstein in 1905, the concept of absolute rest has lost all meaning in physics. In spite of this, however, the concept of absolute motion was, at that time, by no means completely banished by this principle, for the equivalence of two systems moving with respect to each other was, in the first instance, confined to the *special case of a uniform translatory motion.* Although no single privileged reference system can exist, according to the Principle of Relativity, yet there remains a manifold of privileged coordinate systems which are all moving uniformly with respect to each other. Those are the fundamental reference systems of mechanics, previously mentioned in § 13, which are also called *inertial systems.* It is only with regard to such systems that the motion of a body appears uniform, when the body is subject only to the influence of its inertia. On the other hand, an observer who refers the process to another system will detect the presence of so-called *inertial forces*, examples of which are *centrifugal force* and the *Coriolis force* (§ 13). The effect of such inertial forces is manifested in the jerk experienced in a railway train on a sudden application of the brakes, or on travelling round a sharp curve, and also in the phenomenon of *Foucault's pendulum experiment*, or in the *oblateness of the earth*, which is referred to the earth's rotation.

Thus we feel tempted to assume that, whereas absolute velocities do not exist, yet there are *absolute accelerations* and *absolute rotations*. In the long run, however, such an assumption would signify a return to the prejudices vanquished by the Principle of Relativity, to the Newtonian hypothesis of an absolute space and an absolute time. We can understand that such a partial and self-contradictory relativity could not permanently satisfy theoretical physicists, least of all the creator of the theory of relativity himself. An ingeniously simple consideration brought

Einstein on to the right path to the solution of the problem, although he had still enormous difficulties to surmount before reaching his destination.

In order to understand Einstein's fundamental ideas, we shall imagine a spatial *inertial system* (x, y, z), with reference to which an observer (A) is at rest, and we shall suppose this observer to describe in terms of this system the physical processes that he observes. A particle which is moving, say, in the z-direction solely under the influence of its *inertia* will appear to observer A to be moving *uniformly and rectilinearly*; and this will be the case for any body moving with any velocity in any direction, provided no external forces act upon it.

Let us now consider a second spatial coordinate system (x', y', z') arranged in such a way that the two z-axes coincide, whilst the two x- and the two y-axes constantly remain parallel; and we shall suppose the origin of the second system to be moving with a *constant acceleration* γ in the direction of the common positive z-axis away from the origin of the first system. We shall imagine a second observer (B) to refer to this second system the processes observed by him, including the motion of the previou ly mentioned particle. To observer B this motion will, of course, appear to be *uniformly accelerated*, the acceleration being of magnitude γ and in the direction *opposite* to that of the positive z-axis. Let us now consider various bodies of different mass and different chemical nature, bodies composed, say, one of lead, one of stone, one of cork, and so on; they may be moving in any direction and with arbitrary velocities, but only under the influence of their own inertia. As regards observer B, all these bodies will obviously exhibit the *same acceleration* $(-\gamma)$, *independently of their mass and chemical nature.* Provided he knows nothing of observer A, and of his own acceleration with respect to A, he will obviously interpret this result as *indicating* that he is situated in a *homogeneous gravitational field.* For it has been known to physicists since the time of Galileo that it is characteristic of such a field for all bodies situated in it to experience the same acceleration, independently of all their properties.

In describing the physical processes which take place in the neighbourhood of the two observers, it is thus quite immaterial whether we take the point of view of observer A, who refers the processes to his own coordinate system, or of observer B, who makes use of the other coordinate system and, in addition, assumes the presence of a homogeneous gravitational field. Physical processes offer no possibility of deciding between the two points of view, which must therefore be regarded as being equally justified.[1] Recognition of this fact, which is generally known as *Einstein's principle of equivalence*, suggests that the problem of gravitation must obviously be closely linked with the problem of general relativity.

Guided in the right direction by the equivalence principle, we can now form some idea of the fundamental notions of the general theory of relativity. Like gravitational forces, *inertial forces* of every kind, being characteristic of the behaviour of a moving body at a definite point in space and at a definite time, are not a consequence of absolute accelerations, but are *determined by the space-time distribution of matter in the universe.* Conversely, however, this also opens up the possibility of a conception, according to which the *law of gravitation* appears as nothing else than the *generalization of Galileo's Law of Inertia demanded by a truly general principle of relativity.*

How, then, is a generalization of this law possible? The Law of Inertia teaches us that every body, upon which no external forces are acting, moves with constant velocity in a straight path, so that not only is the path thus rectilinear, but also the three spatial coordinates are linear functions of the time. Translated from three into four dimensions, the Law of Inertia therefore states that the *world-line* describing a *motion unattended by the application of force* must be a *straight line.* Now, is this statement capable of further generalization? As a matter of fact it is, on the basis of a *generalization of the geometrical foundations of physics.*

In order to understand this, we shall, for simplicity's sake, first consider two-dimensional geometry. There can hardly be any doubt as to when a line drawn on a plane sheet of paper is to be regarded as a straight line. A *two-dimensional geometry*, however, is just as possible *on a spherical surface* or, indeed, *on any curved surface whatsoever*, as on a plane; but the question as to when a line drawn on

[1] The example given by Einstein himself, of a physicist enclosed in a large chest somewhere in space, is particularly illuminating. The chest is suspended by a rope which passes over a pulley, and the other end of the rope is pulled downwards with constant acceleration. If he knows nothing of the world outside the chest, and nothing about the pulley and the rope, the physicist inside the chest will naturally observe the presence of a gravitational field (which, if we imagine the chest to be on the earth instead of in space, will be superimposed on the gravitational field of the earth).

such a surface is to be regarded as a straight line is not so easily answered. Let us consider two points on a globe, which lie on the same parallel of latitude, but differ widely in longitude. Since we are dealing with two-dimensional geometry, we are, of course, unable to leave the surface of the globe. Now what do we mean by the straight line connecting the two points on the surface of the globe ? Is it the portion of the parallel of latitude between them ? If this question is to have any meaning at all, we must obviously first replace the vague conception of a " straight line " by a more general and well-defined one. Such a conception is already long known in geometry, *viz.*, that of a *geodetic line*, which is *defined* as being the *shortest line joining two points on the surface*. On the basis of this definition (to illustrate by an example) we find, by means of a geometrical investigation, that the segment of the parallel of latitude between the two points in the above consideration in no way represents a geodetic line. We do, however, obtain such a line by constructing a great circle on the sphere to pass through both points.[2] In this lies the explanation of the apparently paradoxical fact that an aviator, who wishes to fly by the shortest route from London to Australia, should go *viâ* Petrograd.

Now we can clearly conceive a curved surface as a two-dimensional manifold " embedded " in one of three dimensions. On the other hand, we cannot form any conception of a curved space, for this in itself is of three dimensions. The limitations of the human imagination, however, offer no barrier to an abstract mathematical treatment. After *Gauss*, in 1827, had developed a *general theory of surfaces*, in 1854 the great mathematician, *Riemann*, showed how our usual *three-dimensional solid geometry* can also be regarded as a *special case* of more general three-dimensional geometry, in which space is " curved " just as a surface may be in two-dimensional geometry. Moreover, as Riemann himself recognized, what holds for three-dimensional geometry must also apply to a geometry of *any number of dimensions :* it must therefore also hold for the four-dimensional manifold which, linking time and space together, is known as the *Minkowski-world*. When we base our mathematical treatment of it upon a general instead of a special geometry, we must regard the Minkowski-world also as being *curved*,

[2] We can temporarily transfer the pole to one of the points, and choose from the manifold of meridians passing through it the meridian which passes through the second point.

whilst the degree of curvature may vary, of course, continuously from place to place.

We owe to Einstein the ingenious concept of the curved Minkowski-world and of connecting the curvature with the space-time distribution of matter. Einstein's *generalized Law of Inertia*, which represents at the same time the new *Law of Gravitation*, simply states that every particle always moves in such a way that its *world-line is a geodesic in the curved Minkowski-world*, the *curvature* being *caused by matter*.[3]

This conception of gravitation leads directly to a very important conclusion. Let us suppose a particle with any velocity to be situated at any position in space at any time. We shall consider its mass, however, to be only a "test mass," *i.e.*, to be so small that the field of matter and thus the curvature of the Minkowski-world are not affected by it (just as in electrostatics we consider proof charges which have no influence on the electric field). According to our previous remarks, the motion of the test mass will be represented by the geodesic passing through the world-point considered in the direction corresponding to that of the given velocity. Now let us suppose a test body to be situated at the same place again, at the same time, and to possess the same velocity, but with a *mass half as large* as was originally the case. The motion will be represented by *precisely the same geodesic* as before. On the other hand, the acceleration is completely determined by the world-line for the given reference system, for it depends only on the functional relation connecting the four world-coordinates in the equation of the curve. Hence, if the Law of Gravitation be regarded as a generalization of the Law of Inertia, it follows necessarily that the *acceleration*, which we *interpret as a gravitational effect*, must be *independent of the mass*.

This result can be expressed in another way. It is usual to define the *inertial mass* of a body as the constant ratio which exists between an arbitrary force acting on the body and the acceleration thereby produced. (We shall not take account of the relativity correction here.) We arrive thus at a unit of mass, the magnitude of which depends on the units of force and acceleration employed. On the other hand, we can also determine the mass of bodies by measuring the forces of attraction which they experience in a gravitational

[3] In this connection it is assumed, of course, that *no external forces* are acting on the particle. Under this heading we include electromagnetic forces, but, in view of the new conceptions, not so-called gravitational forces.

field as compared with the force exerted on the above unit-mass in the same field. The number representing this ratio then determines the *gravitational mass* of the body. If, now, the acceleration in the gravitational field is actually independent of the inertial mass, then, according to Newton's Second Law of Motion, the force of gravitation must be exactly proportional to the inertial mass. Hence an inertial mass of twice the previous magnitude must be associated with twice as large a value of the gravitational force, and thus with twice the *gravitational* mass, and so on. In general, *gravitational and inertial mass* must be *identical.*

This conclusion already found definite expression in Galileo's observation that all bodies, whatever their mass, fall at the same rate in empty space. In classical mechanics it was taken for granted that the inertial mass must be proportional to the weight of a body. The correctness of this assumption was confirmed in detail by accurate measurements carried out by *Eötvös* in 1890. These measurements were based on the fact that the weight of a body represents the resultant of two forces, one of which is the attraction of the earth, and the other the centrifugal force due to the rotation of the earth. Whereas the former of these is determined by the gravitational mass of the body, the latter depends on its inertial mass. The experiments of Eötvös were carried out with extraordinary precision, and they showed that, even when the masses differed widely in magnitude and in chemical constitution, no difference whatsoever could be detected in the direction of the resultant force. The law of the identity of gravitational and inertial mass appears to be completely established by these measurements. Whereas, however, this law was regarded as self-evident prior to Einstein's theory, and physicists gave it no further thought, it was Einstein's work that first revealed the profound reason for this remarkable law.

The new Law of Gravitation is, of course, still incomplete, so long as we do not know the relation expressing the way in which the *Minkowski-world* is *curved by matter.* In order to reach a solution of this question, let us imagine a *dual system of sets of curves* on an arbitrarily curved surface. We shall distinguish these curves as x- and y-curves, and no two x-curves nor any two y-curves shall intersect (Fig. 129). We can then suppose each of the successive x-curves to be denoted by consecutive numbers, and similarly for the y-curves. We can also imagine, however, nine intermediate curves to be drawn between, say, the curves $(x = 6)$ and

($x = 7$), which likewise shall not intersect each other, and we can assign to these curves the numbers 6·1, 6·2, . . . 6·9. Similarly, we can insert nine new curves between the curves ($x = 6·1$) and ($x = 6·2$), denoting them by the numbers 6·11, 6·12, . . . 6·19. In this way we can finally assign to every point on the surface two numerical values, an x- and a y-value, and regard these in the broader sense of the word as the coordinates of the point in question. These values are called the *Gaussian coordinates* of the point. The *Cartesian* coordinate systems used in plane geometry represent, of course, only *special cases* of Gaussian systems.

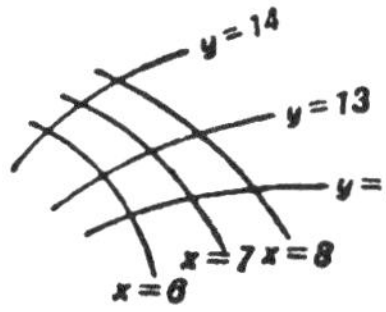

Fig. 129.

Having thus established a Gaussian coordinate system, let us consider three neighbouring points in the surface, which make up a *triangle*. In the particular case of plane coordinate geometry, the coordinates of the three points would completely determine the lengths of the sides, the angles, and the area of the triangle. In the geometry of an arbitrarily curved surface, however, this is not the case. The statement of the coordinates does *not* then suffice for the determination of the lengths of the sides, the angles, and the area of the triangle. In addition to this, we require the values of three further quantities for the region considered, which we may denote by g_{xx}, g_{xy}, g_{yy}; and, as was recognized by Gauss, these three quantities represent the *components of a tensor*. Because this tensor determines the measure relations in any particular region, it is called the *metric fundamental tensor*. Its value must be unequivocally *stated* for the place at which we supposed the small triangle to be constructed, if we wish to calculate the sides, the angles, and the area of the triangle from the coordinates of the three points. It determines really the kind of geometry,[4] and therefore it enters necessarily into all formulæ of a

[4] It is, of course, well known that the geometry of spherical surfaces is governed by other theorems than those applying to plane, Euclidean, geometry. For example, Euclid's well-known theorem, which states that the sum of the angles in a triangle amounts to two right angles, does not hold in spherical geometry. In point of fact, if we imagine a spherical triangle on the terrestrial globe formed by the northern quadrants of the meridians of 0° and 90° east longitude, and by the corresponding quadrant of the Equator, this triangle contains three right angles, of which the sum is 270°.—The theorems of plane geometry apply, moreover, not only to planes, but also to such surfaces as can be produced by the bending of a plane surface, *e.g.*, cylindrical or conical surfaces.

general theory of surfaces. Conversely, we can also determine the fundamental tensor from measurements on the above-mentioned triangle, without our requiring to know anything as to how the surface is " embedded in space " at the place in question.

Thus a general geometry of surfaces only has any sense and only becomes possible when the fundamental tensor is given for every position on the surface. Now what holds good for a general two-dimensional geometry likewise applies to any general geometry of an arbitrary number of dimensions. The fundamental tensor is symmetrical, and hence in two-dimensional geometry it has three, in three-dimensional six, and in four-dimensional geometry ten components; in general, of course, it will *vary continuously from place to place.* Every geometrical manifold may thus also be regarded as the field of its metric fundamental tensor or, in brief, as a *metric field.*

Now purely mathematical investigations show that a scalar, which is known as the *Riemann scalar* and is completely independent of the coordinate system, can be derived from the fundamental tensor. This scalar, of course, will also in general vary continuously from place to place. Let us imagine any region to be delimited within the geometrical manifold, and the Riemann scalar to be integrated over it. Then we can calculate the *variation of this integral* on the assumption that, *within* the region, the components of the fundamental tensor may be varied arbitrarily and independently of each other, conditionally upon the variations vanishing at the boundary itself. The process is entirely similar to that of the variation, in mechanics, of a motion between the initial and final positions, on the condition that the actual and the varied motion shall coincide at these positions. If, now, we calculate the variation of the integral as stated, we obtain a tensor *independent of the coordinate system,* which, in our considerations, may be called the *curvature tensor.* (In order to avoid any misunderstanding, however, it should be expressly noted that the title of " curvature tensor " is often used in another sense.) The derivation of this tensor follows along purely mathematical lines, quite independently of all physics.

Einstein's Law of Gravitation, which states that the world-line of a particle shall be a geodesic, is completed by the further postulate that, at every point in the Minkowski-world, the *curvature tensor shall be identical with the matter tensor* (§ 166), apart from a *universal propor-*

tionality factor determined by the usual physical units of measure.

Since both the geodesic and the curvature tensor are independent of the Gaussian coordinate system used, it follows that Einstein's Law of Gravitation or, in other words, the generalized Law of Inertia, also holds for *any Gaussian coordinate system whatsoever*. As long as the theory of relativity confined itself to Cartesian coordinate systems in its treatment of the Minkowski-world, it could use the displacement of the origin and the rotation of the axes in a four-dimensional coordinate system only to effect transitions between such spatial coordinate systems as are *moving uniformly with respect to each other*; for no four-dimensional coordinate transformations other than *linear* were available. The application of Gaussian coordinate systems opened up an entirely new field. A change, which in space appears as a transition between two coordinate systems *moving quite arbitrarily with regard to each other*, is always represented in a four-dimensional manifold by a transformation from one to some other Gaussian coordinate system.

The generalized Law of Inertia, when in the form of Einstein's Law of Gravitation, must therefore hold in exactly the same way for two different spatial coordinate systems, no matter how these are moving relatively to each other. On a transition from one system to the other, however, the *components* of the fundamental tensor are transformed, and it is just the alterations thus connected with the transition which were interpreted by classical mechanics as *inertial forces*, and these differ according to the choice of reference system.

From this new point of view, the battle that once raged so violently between the Ptolemaic and the Copernican systems of the universe appears quite pointless. If we imagine the earth at rest and the sky with the fixed stars rotating around it, the curvature of the Minkowski-world must yet be such that measurements of the earth's surface will show its oblateness, as *Thirring* [5] has demonstrated on the basis of Einstein's theory of gravitation.

Just as the generalized Law of Inertia is *invariant* in arbitrary four-dimensional coordinate transformations, and thus the equations referring to the components hold for any coordinate system, so also the *equations of the electromagnetic field* (§ 165) can be reduced to a form which holds for any

[5] "*Über die Wirkung rotierender ferner Massen in der Einsteinschen Gravitationstheorie*," *Physikal. Zeitschr.*, **19** (1918), p. 33.

coordinate system. The relation connecting a vector with its rotation, or a skew-symmetrical tensor with its vector divergence, is something quite independent of the coordinate system; but the equations connecting the *components* are, of course, more complicated when based on Riemannian geometry than in Euclidean geometry.

The postulate of relativity thus appears fulfilled in the most complete manner by the new theory which, with the help of Einstein's Law of Gravitation, bases the description of the Minkowski-world on Riemannian geometry. For this reason this theory, which was developed by Einstein in 1915, is known as the *general theory of relativity*, in contrast to the special theory of relativity which represents only a special case.

§ 168. Gaussian Coordinate Systems.

In order to be able to develop the mathematical bases of Einstein's theory of gravitation in an exact manner, we shall first consider *two neighbouring points on an arbitrarily curved surface*. Let the coordinates of these points relatively to an arbitrary Gaussian coordinate system be x, y and $(x + dx)$, $(y + dy)$ respectively; and let the *distance* between the two points (measured in the surface) be ds. Since we can regard an *infinitesimally small portion of the surface as plane*, we can also suppose a Cartesian coordinate system to be constructed in the small region of the surface containing the two points considered; with reference to this system, let the coordinates of the first point be ξ and η, and those of the second $(\xi + d\xi)$ and $(\eta + d\eta)$. The quantities x and y can then undoubtedly be regarded as *functions* of the two quantities ξ and η; and, if we solve the two equations expressing this connection with respect to ξ and η, we obtain two other equations

$$\xi = \varphi_1 (x, y)\,; \qquad \eta = \varphi_2 (x, y)\,. \tag{1}$$

Differentiating these, we find

$$\begin{cases} d\xi = \dfrac{\partial\varphi_1}{\partial x}\,dx + \dfrac{\partial\varphi_1}{\partial y}\,dy\,, \\[2ex] d\eta = \dfrac{\partial\varphi_2}{\partial x}\,dx + \dfrac{\partial\varphi_2}{\partial y}\,dy\,. \end{cases} \tag{2}$$

In the Cartesian coordinate system, the square of the distance between the two points is given by

$$ds^2 = d\xi^2 + d\eta^2\,, \tag{3}$$

and hence, in the Gaussian system, we have by eqns. (2)

$$(4)\quad \begin{cases} ds^2 = \left[\left(\frac{\partial\varphi_1}{\partial x}\right)^2 + \left(\frac{\partial\varphi_2}{\partial x}\right)^2\right] dx^2 + \left[\left(\frac{\partial\varphi_1}{\partial y}\right)^2\right. \\ \qquad \left. + \left(\frac{\partial\varphi_2}{\partial y}\right)^2\right] dy^2 + 2\left[\frac{\partial\varphi_1}{\partial x}\cdot\frac{\partial\varphi_1}{\partial y} + \frac{\partial\varphi_2}{\partial x}\cdot\frac{\partial\varphi_2}{\partial y}\right] dx\,dy\,. \end{cases}$$

We shall denote by x_h either of the two coordinates x or y, according as the subscript h is put equal to 1 or 2, and thus we can write eqn. (4) in the form

$$(5)\qquad ds^2 = \sum_h \sum_k g_{hk}\,dx_h\,dx_k \qquad (h = 1, 2\,;\ k = 1, 2)\,;$$

where, by eqn. (4),

$$(6)\qquad g_{hk} = \frac{\partial\varphi_1}{\partial x_h}\cdot\frac{\partial\varphi_1}{\partial x_k} + \frac{\partial\varphi_2}{\partial x_h}\cdot\frac{\partial\varphi_2}{\partial x_k}\,.$$

Hence

$$(7)\qquad g_{hk} = g_{kh}\,.$$

If, in eqns. (5) to (7), we give the subscripts h and k three, four, or more values, instead of only two, and if, correspondingly, we introduce φ_3, φ_4, *etc.*, in addition to φ_1 and φ_2, we obtain the fundamental formulæ of a *Riemannian geometry of an arbitrary number of dimensions.* In two-dimensional geometry there are four quantities g_{hk}, of which two are the same ; in three-dimensional geometry there are nine quantities, of which three pairs are alike ; and in four-dimensional geometry there are sixteen quantities, including six pairs alike, according to the scheme

$$(8)\qquad \begin{cases} g_{11} & g_{12} & g_{13} & g_{14} \\ g_{21} & g_{22} & g_{23} & g_{24} \\ g_{31} & g_{32} & g_{33} & g_{34} \\ g_{41} & g_{42} & g_{43} & g_{44}\,, \end{cases}$$

where

$$g_{12} = g_{21}, \qquad g_{13} = g_{31}, \ etc.$$

Eqn. (5) represents the *generalization of the Theorem of Pythagoras in Riemannian geometry.* We see at once from this equation that *Cartesian coordinate systems* represent special cases of Gaussian systems, for orthogonal Cartesian systems are obtained by equating the quantities g_{hk} to unity or zero according as the two subscripts are the same or different. In such systems

$$(9)\qquad g_{hk} = \begin{cases} 1\ (h = k) \\ 0\ (h \neq k) \end{cases};$$

and, in this special case, scheme (8) assumes the simple form

$$(10)\qquad \begin{cases} 1 & 0 & 0 & 0 \\ 0 & 1 & 0 & 0 \\ 0 & 0 & 1 & 0 \\ 0 & 0 & 0 & 1\,. \end{cases}$$

§ 169. Generalized Conception of Tensors.

In the previous section we considered a Gaussian coordinate system in which the coordinates of two neighbouring points on an arbitrary surface were x, y and $(x + dx)$, $(y + dy)$ respectively. We shall now suppose this system to be replaced by a second, likewise Gaussian, coordinate system, with respect to which the coordinates of the points may be denoted by x', y' and $(x' + dx')$, $(y' + dy')$ respectively. Since the new coordinates may be regarded as functions of the old ones, we have

$$(1)\qquad \begin{cases} dx' = \dfrac{\partial x'}{\partial x} dx + \dfrac{\partial x'}{\partial y} dy\,, \\[2ex] dy' = \dfrac{\partial y'}{\partial x} dx + \dfrac{\partial y'}{\partial y} dy\,. \end{cases}$$

Let us now imagine any *force*, and let its components relatively to the original coordinate system be X and Y, and those relatively to the new one be X' and Y'. Then the *work* done by this force along the element of path whose components are dx and dy, or dx' and dy' respectively, must undoubtedly be independent of the coordinate system; *i.e.*, it must be, as we say, *invariant* as regards coordinate transformations. Thus we must have

$$(2)\qquad X'\,dx' + Y'\,dy' = X\,dx + Y\,dy\,.$$

Introducing in this equation the values given for dx' and dy' by eqns. (1), we find

$$(3)\qquad \begin{cases} \left(X' \dfrac{\partial x'}{\partial x} + Y' \dfrac{\partial y'}{\partial x}\right) dx + \left(X' \dfrac{\partial x'}{\partial y} + Y' \dfrac{\partial y'}{\partial y}\right) dy \\[2ex] \qquad = X\,dx + Y\,dy\,. \end{cases}$$

If this relation is to be *generally* fulfilled, the equations

$$(4)\qquad \begin{cases} X = X' \dfrac{\partial x'}{\partial x} + Y' \dfrac{\partial y'}{\partial x}\,, \\[2ex] Y = X' \dfrac{\partial x'}{\partial y} + Y' \dfrac{\partial y'}{\partial y} \end{cases}$$

must be satisfied. Since the transition from the one coordinate system to the other is in no way privileged as compared with the reverse transition, the relation

$$(5) \qquad X' = X \frac{\partial x}{\partial x'} + Y \frac{\partial y}{\partial x'}$$

must, of course, also hold good.

Let us now pass from the two-dimensional system hitherto considered to a coordinate system of an arbitrary number of dimensions; then the *transformation formulæ* of eqns. (1) and (5) can be written in the more general form

$$(6) \qquad dx'_h = \sum_k \frac{\partial x'_h}{\partial x_k} dx_k\,,$$

and

$$(7) \qquad X'_h = \sum_k X_k \frac{\partial x_k}{\partial x'_h}.$$

On the basis of these equations we *define* a quantity as a *covariant vector* when its components obey the transformation law

$$(8) \qquad A'_h = \sum_k A_k \frac{\partial x_k}{\partial x'_h}.$$

On the other hand, we define a quantity as a *contravariant vector* when its components are transformed in accordance with the law

$$(9) \qquad A'^h = \sum_k A^k \frac{\partial x'_h}{\partial x_k}.$$

It is customary, as has been done in eqns. (8) and (9), to distinguish *covariant* vectors by writing the *index below*, and *contravariant* vectors by writing it *above*. In Euclidean geometry, of course, no distinction is made between covariant and contravariant vectors, but this is essential in Riemannian geometry.

If A_h be the components of an arbitrary covariant vector, and B^h those of an arbitrary contravariant vector, it follows at once from eqns. (8) and (9) that

$$\sum_h A'_h B'^h = \sum_k A_k B^k,$$

so that

$$(10) \qquad \sum_h A_h B^h = \text{invariant}.$$

In our further considerations we shall introduce a *simplified notation.* For this and all subsequent sections we stipulate that *everywhere that the same index occurs twice in a term, the term has to be summed for that index, although the sign of summation is dropped.*

Corresponding to the distinction between covariant and contravariant vectors, we define as a *covariant*, a *contravariant*, or a *mixed tensor of the second rank* a quantity whose components are transformed in the same way as the products formed by multiplying in pairs the components of two covariant vectors, or of two contravariant vectors, or of one covariant and one contravariant vector. Thus a covariant tensor of the second rank is transformed according to the formula

$$A'_{pr} = A_{hk} \frac{\partial x_h}{\partial x'_p} \cdot \frac{\partial x_k}{\partial x'_r} \cdot \tag{11}$$

(Following the rule given above, the double summation symbol is omitted.)

On the other hand, the transformation of a contravariant tensor of the second rank takes place in accordance with the formula

$$A'^{pr} = A^{hk} \frac{\partial x'_p}{\partial x_h} \cdot \frac{\partial x'_r}{\partial x_k} \cdot \tag{12}$$

It follows from eqns. (11) and (12) that for two arbitrary tensors, one of which is covariant and the other contravariant, the double sum

$$A_{hk} B^{hk} = \text{invariant}. \tag{13}$$

Finally, the components of a mixed tensor follow the transformation law

$$A'^{r}_{p} = A^{k}_{h} \frac{\partial x_h}{\partial x'_p} \cdot \frac{\partial x'_r}{\partial x_k} \cdot \tag{14}$$

Tensors of a rank higher than the second are defined in an analogous manner. Thus

$$A^{r}_{hkp} = B_h C_k D_p E^r \tag{15}$$

would be an example of a *mixed tensor of the fourth rank, etc.* In a four-dimensional system, a tensor of the third rank has in general 64, one of the fourth rank 256, and one of the zth rank 4^z components.

The property of invariance, as expressed in eqn. (10), enables us to derive from any mixed tensor of arbitrary rank another *tensor* of a *rank lower by two*, by equating an upper to a lower index, and then summing for this index. *E.g.*, by means of this operation we can derive a tensor of the second rank from the mixed tensor of the fourth rank defined by eqn. (15), with the help of the relation

(16) $$A_{hk} = A^{p}_{hkp} .$$

In point of fact, if for the moment we replace the summation symbol (otherwise omitted), we have

(17) $$A_{hk} = \sum_p A^{p}_{hkp} = B_h C_k \sum_p D_p E^p .$$

Since, however, the sum $D_p E^p$ is [according to eqn. (10)] invariant with respect to coordinate transformations, the quantity A_{hk} defined by eqn. (16) is actually a covariant tensor of the second rank.

This operation, by means of which we can thus derive from an arbitrary tensor[1] of rank z another tensor of rank $(z - 2)$, is known as the *contraction of a tensor*. The contraction of a mixed tensor of the second rank leads to a tensor of zero rank, *i.e.*, a *scalar*.

§ 170. The Metric Fundamental Tensor.

From § 169 eqn. (13) we are able at once to draw an important conclusion with respect to the quantities g_{hk} which, when use is made of a Gaussian coordinate system, occur in the expression for the line-element. The quantity ds^2, occurring in § 168 eqn. (5), is invariant and thus independent of the coordinate system. According to § 169 eqn. (6), the quantities dx_h and dx_k are the components of two contravariant vectors, and hence the product $dx_h \cdot dx_k$ must be a component of a contravariant tensor of the second rank. On the other hand, however, we have by § 168 eqn. (5)

(1) $$ds^2 = g_{hk}\, dx_h\, dx_k ,$$

and therefore, according to § 169 eqn. (13), the quantities g_{hk} must be the components of a *covariant tensor of the second rank*. This tensor is the *metric fundamental tensor*;[1] the geometrical manifold, in which, in general, it varies continuously from place to place, thus appears as the field of this fundamental tensor or, in brief, as the *metric field*.

Some simple calculations enable us to derive from the

[1] For any arbitrary tensor can be reduced to one whose components can be regarded as the products of the components of vectors.

[1] It is called " metric " because, in accordance with § 168 eqn. (5), it is the fundamental tensor that determines the measure relations (" metrics ").

sixteen components of the metric fundamental tensor sixteen other quantities, which are likewise of importance for subsequent considerations. We can regard the sixteen quantities in scheme (8) of § 168 as the *elements of a determinant*, and form the *minor* of each g_{hk} by striking out the column and row containing the element in question.[2] We obtain the *cofactor* of the element by multiplying the minor by $(-1)^{h+k}$, where h and k are given the values 1, 2, 3, or 4, as the case may be.[3] The *value* (g) of the determinant is obtained, in accordance with a well-known theorem, on multiplying all the elements in any row or column by the corresponding cofactors, and then adding together.[4] If we denote the cofactor of g_{hk} by G_{hk}, we have

$$g = g_{1k} \cdot G_{1k} = g_{2k} \cdot G_{2k} = g_{3k} \cdot G_{3k} = g_{4k} \cdot G_{4k} \,,$$

where, in accordance with our rule of notation, we have to take the sum for the index k. Consequently, the double sum

$$g_{hk} \, G_{hk} = 4g \,,$$

or,

$$g_{hk} \frac{G_{hk}}{g} = 4 \,. \tag{2}$$

Since the number 4 is, of course, invariant with respect to coordinate transformations, it follows from § 169 eqn. (13) that the quantities G_{hk}/g must be the components of a contravariant tensor of the second rank. This tensor is known as the *contravariant fundamental tensor*, and its components are denoted by the symbol

$$g^{hk} = \frac{G_{hk}}{g} \,. \tag{3}$$

On multiplying all the components of the covariant by all the components of the contravariant fundamental tensor, we obtain a mixed tensor of the fourth rank:

$$g_{hp}^{kr} = g_{hp} \cdot g^{kr} \,. \tag{4}$$

[2] The minor of the element g_{23} is given by

$$G_{23} = \begin{vmatrix} g_{11} & g_{12} & g_{14} \\ g_{31} & g_{32} & g_{34} \\ g_{41} & g_{42} & g_{44} \end{vmatrix} \,.$$

[3] For the cofactor G_{23} we have $h + k = 5$, and hence its sign is negative; on the other hand, G_{13} or G_{24} is positive.

[4] The values of the cofactors themselves are calculated by the same rule: *e.g.*,

$$-G_{23} = g_{11}(g_{32}g_{44} - g_{34}g_{42}) - g_{12}(g_{31}g_{44} - g_{34}g_{41}) + g_{14}(g_{31}g_{42} - g_{32}g_{41}).$$

Contraction of this tensor (by putting r equal to p, and then summing for p) leads to the so-called *mixed fundamental tensor*. Thus we have

$$(5) \qquad g_h^k = g_{hp} \cdot g^{kp} = \frac{g_{hp}\, G_{kp}}{g}\,.$$

Now when the index h is equal to the index k, the numerator of the fraction simply represents the value obtained when the elements in one row of a determinant are multiplied by the corresponding cofactors, and the products added together. This, however, gives the value (g) of the determinant, so that

$$g_h^k = 1, \text{ when } k = h.$$

Let us next consider the case in which k differs from h. We then obtain the sum $g_{hp} \cdot G_{kp}$ on multiplying the elements in the hth row by the cofactors of the elements in the kth row. The value so obtained, however, is identical with the value of the determinant formed by replacing the elements in the hth row by those in the kth row.[5] We should thus have a determinant in which the hth and k^{th} rows were identical. As may readily be seen, however, the value of a determinant in which any two rows are identical is zero. For the sign of a determinant is reversed when two of its rows are interchanged,[6] and hence, for a determinant with two identical rows, g must be equal to $(-g)$, which is, of course, only possible when g vanishes. Therefore

$$g_{hp}\, G_{kp} = 0, \text{ when } k \neq h.$$

Hence we find quite generally for the mixed fundamental tensor, for which the symbol δ_h^k is usual,

$$(6) \qquad g_h^k = \delta_h^k = \begin{cases} 1 \ (k = h) \\ 0 \ (k \neq h) \end{cases}.$$

Finally, we shall show how, with the help of the fundamental tensors, a contravariant vector can be derived from a given covariant vector, and *vice-versâ*. If A_h be an

[5] *E.g.*, the value of $g_{1p}\, G_{2p}$ is that of the determinant

$$\begin{vmatrix} g_{21} & g_{22} & g_{23} & g_{24} \\ g_{21} & g_{22} & g_{23} & g_{24} \\ g_{31} & g_{32} & g_{33} & g_{34} \\ g_{41} & g_{42} & g_{43} & g_{44} \end{vmatrix}.$$

[6] This well-known theorem follows from the previously quoted rule for the signs of the cofactors.

arbitrary covariant vector, we can form a tensor of the third rank:

$$A_h^{kp} = g^{kp} A_h .$$

On contracting this mixed tensor, *i.e.*, on putting p equal to h and summing for h, we obtain a contravariant vector:

(7) $$A^k = g^{hk} A_h .$$

Conversely, we can derive a covariant from a contravariant vector (B^k) with the aid of the relation

(8) $$B_p = g_{kp} B^k .$$

If, in particular, we replace B^k by the vector A^k as defined in eqn. (7), we find for B_p the value

(9) $$B_p = g_{kp} g^{hk} A_h = \delta_p^h A_h = A_p ;$$

i.e., the vector A_h belongs in the same way to the vector A^h as the vector A^h does to the vector A_h. Hence, instead of speaking of a covariant vector A_h and a contravariant vector A^h, we can refer simply to a single vector, whose *covariant components* are A_h, and whose *contravariant components* are A^h.

In an entirely analogous manner, we can derive a mixed tensor from a covariant tensor of the second rank (A_{hk}) by multiplying by g^{pr}, and then contracting with respect to the indices r and k:

(10) $$A_h^p = g^{pk} A_{hk} .$$

On a further multiplication by g^{si} and subsequent contraction with respect to the indices h and i, we find

(11) $$A^{ps} = g^{pk} g^{sh} A_{hk} .$$

In this way we can derive a mixed and a contravariant tensor from any covariant tensor, and thus we may also speak of the covariant, the contravariant, and the mixed components of one and the same tensor.

§ 171. Geodetic Lines.

In the last section we have formed an idea of the fundamental tensor, and we are now faced with another problem, which is likewise purely mathematical, of finding the *general equation of a geodetic line*, which is defined as being the *shortest line joining* two arbitrary points A and B within a geometrical manifold. A geodetic line must therefore satisfy the condition

(1) $$\delta \int_A^B ds = 0 ,$$

where ds denotes an element of the geodesic, and the variation is to be performed in such a way that the varied curve also leads from the point A to the point B.[1]

Let us suppose a set of $(n - 1)$-dimensional super-surfaces to be constructed in an n-dimensional manifold, the equations of these surfaces being of the form

$$\lambda = \varphi(x_1, x_2, x_3, \ldots x_n)\,. \tag{2}$$

A definite surface will then correspond to each arbitrary value of the variable parameter λ, and furthermore, since the geodesic between A and B is intersected by these surfaces, every point along the geodesic can be associated with a quite definite value of λ.

We shall now consider two neighbouring surfaces, corresponding to the values λ and $(\lambda + d\lambda)$, so that the intercept made by these two surfaces on the geodesic is given by

$$ds = w\,d\lambda\,, \tag{3}$$

where the value of the proportionality factor w follows from the equation

$$ds^2 = g_{hk}\,dx_h\,dx_k \tag{4}$$

[§ 168 eqn. (5) applied to an n-dimensional manifold]. Hence we find

$$w^2 = g_{hk}\,\frac{dx_h}{d\lambda}\frac{dx_k}{d\lambda}\,. \tag{5}$$

Let ds' be the intercept made by the two surfaces with the values λ and $(\lambda + d\lambda)$ on the *varied* line joining the points A and B. Then

$$ds' = w'\,d\lambda\,, \tag{6}$$

whilst

$$ds' - ds = \delta(ds) \tag{7}$$

and

$$w' - w = \delta w\,. \tag{8}$$

Hence

$$\delta(ds) = \delta w\,d\lambda\,; \tag{9}$$

and, since the order of integration and variation in eqn. (1) can be reversed (because the two operations are mutually independent), eqn. (1) may also be written in the form

$$\int_A^B \delta w\,d\lambda = 0\,. \tag{10}$$

[1] In view of its definition, the length of a geodesic is by no means necessarily a minimum, but only an extreme value in general.

In order to calculate the variation of w, we vary eqn. (5), thus finding

$$(11)\quad 2w\,\delta w = \frac{dx_h}{d\lambda}\frac{dx_k}{d\lambda}\,\delta g_{hk} + g_{hk}\frac{dx_h}{d\lambda}\,\delta\left(\frac{dx_k}{d\lambda}\right) + g_{hk}\frac{dx_k}{d\lambda}\,\delta\left(\frac{dx_h}{d\lambda}\right).$$

We recognize at once that the two last double summands are identical, on account of the symmetry of the fundamental tensor, and either of them may be written in the form

$$g_{hp}\frac{dx_h}{d\lambda}\,\delta\left(\frac{dx_p}{d\lambda}\right).$$

Again, inasmuch as the metric fundamental tensor is a function of the coordinates,

$$(12)\qquad \delta g_{hk} = \frac{\partial g_{hk}}{\partial x_p}\,\delta x_p\,.$$

Eqn. (11) may therefore be written as follows :—

$$(13)\qquad \delta w = \frac{1}{2w}\frac{dx_h}{d\lambda}\frac{dx_k}{d\lambda}\frac{\partial g_{hk}}{\partial x_p}\,\delta x_p + \frac{g_{hp}}{w}\frac{dx_h}{d\lambda}\,\delta\left(\frac{dx_p}{d\lambda}\right).$$

Now

$$(14)\qquad \delta\left(\frac{dx_p}{d\lambda}\right) = \frac{d\,(\delta x_p)}{d\lambda}\,,$$

and hence [2]

$$(15)\quad \frac{g_{hp}}{w}\frac{dx_h}{d\lambda}\,\delta\left(\frac{dx_p}{d\lambda}\right) = \frac{d}{d\lambda}\left(\frac{g_{hp}}{w}\frac{dx_h}{d\lambda}\,\delta x_p\right) - \frac{d}{d\lambda}\left(\frac{g_{hp}}{w}\frac{dx_h}{d\lambda}\right)\delta x_p\,.$$

If we multiply this equation by $d\lambda$ and integrate between the limits A and B, the first term on the right-hand side vanishes, because the variation of the coordinates vanishes both for A and for B, seeing that the geodesic and the varied curve coincide at their initial and final points.

Consequently, on substituting the value given by eqn. (15) in eqn. (13), and replacing δw in eqn. (10) by the value thus obtained, we find, on reversing the sign,

$$(16)\qquad \int \Theta_p\,d\lambda\,\delta x_p = 0\,,$$

where Θ_p is given by the relation

$$(17)\qquad \Theta_p = \frac{d}{d\lambda}\left(\frac{g_{hp}}{w}\frac{dx_h}{d\lambda}\right) - \frac{1}{2w}\frac{dx_h}{d\lambda}\frac{dx_k}{d\lambda}\frac{\partial g_{hk}}{\partial x_p}.$$

Inasmuch as the variation of the coordinates is quite arbitrary, however, eqn. (16) can only be satisfied when

$$(18)\qquad \Theta_p = 0\,.$$

[2] In accordance with the formula

$$A\,dB = d(AB) - B\,dA.$$

The equations obtained by giving p the values 1, 2, 3, . . . n in an n-dimensional geometry represent the equations of the geodesic.

In our previous considerations the parameter has been quite arbitrary. We can therefore stipulate that λ shall be equal to the *segment* s of the geodesic, as measured from the point A. In that case ds is equal simply to $d\lambda$, and hence w is equal to unity. Eqns. (17) and (18) then assume the simpler form

$$(19) \qquad g_{hp}\frac{d^2x_h}{ds^2}+\frac{dg_{hp}}{ds}\frac{dx_h}{ds}-\frac{1}{2}\frac{dx_h}{ds}\frac{dx_k}{ds}\frac{\partial g_{hk}}{\partial x_p}=0\,.$$

But

$$(20) \qquad \frac{dg_{hp}}{ds}=\frac{\partial g_{hp}}{\partial x_k}\frac{dx_k}{ds}\,.$$

Furthermore, we can utilize the identity

$$(21) \qquad \frac{\partial g_{hp}}{\partial x_k}\frac{dx_k}{ds}\frac{dx_h}{ds}=\frac{\partial g_{kp}}{\partial x_h}\frac{dx_h}{ds}\frac{dx_k}{ds}\,,$$

for the left-hand side of this relation does not change in value if merely other letters are substituted for the indices, and h and k are replaced by k and h respectively. Thus we can equate the second term in eqn. (19) to half the sum of the two expressions forming the identity (21). Eqn. (19) may therefore be written in the form

$$(22) \qquad g_{hp}\frac{d^2x_h}{ds^2}+\begin{bmatrix} h\,k \\ p \end{bmatrix}\frac{dx_h}{ds}\frac{dx_k}{ds}=0\,,$$

where the meaning of the expression in square brackets, which is known as *Christoffel's three-indices symbol*, is given by

$$(23) \qquad \begin{bmatrix} h\,k \\ p \end{bmatrix}=\frac{1}{2}\left(\frac{\partial g_{hp}}{\partial x_k}+\frac{\partial g_{kp}}{\partial x_h}-\frac{\partial g_{hk}}{\partial x_p}\right).$$

On account of the symmetry of the fundamental tensor, it follows at once from this definition that the three-indices symbol is also *symmetrical;* its value is not altered by an interchange of the two upper indices.

In order to simplify subsequent calculations, we shall now introduce another, likewise symmetrical, three-indices symbol, defining it by the relation

$$(24) \qquad \begin{Bmatrix} h\,k \\ r \end{Bmatrix}=g^{rp}\begin{bmatrix} h\,k \\ p \end{bmatrix}.$$

Let us then multiply eqn. (22) by $g^{r\rho}$ and sum for ρ. According to § 170 eqn. (5), the first term becomes

$$g^r_h \frac{d^2 x_h}{ds^2},$$

where, just as in eqn. (22), we have to sum for the index h. The summation for h, however, simply gives d^2x_r/ds^2, because [according to § 170 eqn. (6)] g^r_h only differs from zero when h and r are identical, and it is then equal to unity. Thus the *equation of a geodesic* finally becomes

$$(25) \qquad \frac{d^2 x_r}{ds^2} = -\begin{Bmatrix} h\,k \\ r \end{Bmatrix} \frac{dx_h}{ds}\frac{dx_k}{ds}.$$

§ 172. The Riemann Tensor.

Just as the operation of contraction enables us to derive from an arbitrary tensor a tensor of lower rank, so of course, conversely, it must also be possible to derive from an arbitrary tensor, by means of purely *mathematical operations*, a *tensor of higher rank*. For this purpose we shall start with a *scalar*, as a tensor of zero rank. Let us assume φ to be any scalar function of the coordinates, *i.e.*, to be *invariant* with respect to coordinate transformations. Let us further imagine a *geodesic* constructed quite independently of any coordinate system, and let us denote by s the length of the geodesic (again independently of any coordinate system) as measured from a definite point on it. Then it is evident that the differential coefficient $d\varphi/ds$ must likewise be a quantity independent of the coordinate system. On the other hand, if, as we have assumed, φ is a function of the coordinates, we have, of course,

$$(1) \qquad \frac{d\varphi}{ds} = \frac{\partial \varphi}{\partial x_h}\frac{dx_h}{ds}.$$

It follows from the definition of a covariant vector that, in accordance with § 169 eqn. (10) (since the dx_h are components of a contravariant vector), the four quantities

$$(2) \qquad A_h = \frac{\partial \varphi}{\partial x_h}$$

are the components of a *covariant vector.*[1]

Let us now form the second differential coefficient of φ

[1] This vector is the gradient of ϕ.

with respect to s; this must, of course, also be *invariant.* Denoting the first derivative by ψ, we have

$$(3)\qquad \frac{d^2\varphi}{ds^2}=\frac{d\psi}{ds}=\frac{\partial\psi}{\partial x_k}\frac{dx_k}{ds}.$$

But, by eqn. (1),

$$(4)\qquad \frac{\partial\psi}{\partial x_k}=\frac{\partial^2\varphi}{\partial x_h\,\partial x_k}\frac{dx_h}{ds}+\frac{\partial\varphi}{\partial x_h}\frac{\partial}{\partial x_k}\left(\frac{dx_h}{ds}\right).$$

Hence

$$(5)\qquad \frac{d^2\varphi}{ds^2}=\frac{\partial^2\varphi}{\partial x_h\,\partial x_k}\frac{dx_h}{ds}\frac{dx_k}{ds}+\frac{\partial\varphi}{\partial x_h}\frac{d^2x_h}{ds^2}.$$

Since the right-hand side of this equation contains two entirely independent summands, it is merely a matter of notation to replace the index h in the last term by the index r, and this does not involve a similar alteration in the preceding term. Inasmuch, however, as ds represents the element of a geodesic (and the dx_h are the components of this element), the equation of a geodesic [§ 171 eqn. (25)] must be applicable. Consequently, eqn. (5) becomes

$$(6)\qquad \frac{d^2\varphi}{ds^2}=\left(\frac{\partial^2\varphi}{\partial x_h\,\partial x_k}-\begin{Bmatrix}h\,k\\ r\end{Bmatrix}\frac{\partial\varphi}{\partial x_r}\right)\frac{dx_h}{ds}\frac{dx_k}{ds}.$$

The left-hand side of this equation is invariant. Hence, according to the definitions of tensors, the expression in round brackets must represent the components of a *covariant tensor*: [2] we shall denote these components by A_{hk}.[3] Taking eqn. (2) into account, we find

$$(7)\qquad A_{hk}=\frac{\partial A_h}{\partial x_k}-\begin{Bmatrix}h\,k\\ r\end{Bmatrix}A_r.$$

This tensor is called the *covariant derivative* of the vector A.

If, now, we multiply all the components of this tensor by all the components of a covariant vector B, we obtain, a covariant *tensor of the third rank.* We shall denote its components by

$$(8)\qquad C_{hkp}=A_{hp}\,B_k.$$

Hence, according to eqn. (7),

$$(9)\qquad C_{hkp}=B_k\frac{\partial A_h}{\partial x_p}-\begin{Bmatrix}h\,p\\ r\end{Bmatrix}A_r\,B_k.$$

[2] For the dx/ds are the components of a contravariant vector.

[3] The denotation both of the tensor of the second rank and of the vector by the letter A should not cause any misunderstanding. The distinction between the two is sufficiently expressed by the fact that the tensor has two indices, whereas the vector has only one.

The first term on the right-hand side of this equation suggests that we should define another tensor of the third rank in a similar manner, with the help of the equation

(10) $$D_{hkp} = B_{kp} A_h ,$$

the tensor B_{kp} being derived from the vector B_k by means of the differential operation expressed by eqn. (7). Thus we find the value of B_{kp} on substituting B for A in eqn. (7), and therefore

(11) $$D_{hkp} = A_h \frac{\partial B_k}{\partial x_p} - \left\{ \begin{matrix} k\,p \\ r \end{matrix} \right\} A_h B_r .$$

Next, we define a tensor of the second rank by means of the relation

(12) $$E_{hk} = A_h B_k ,$$

and a tensor of the third rank with the help of the formula

(13) $$E_{hkp} = C_{hkp} + D_{hkp} .$$

Then we obtain

(14) $$E_{hkp} = \frac{\partial E_{hk}}{\partial x_p} - \left\{ \begin{matrix} h\,p \\ r \end{matrix} \right\} E_{rk} - \left\{ \begin{matrix} k\,p \\ r \end{matrix} \right\} E_{hr} .$$

The tensor of the third rank, E_{hkp}, is known as the *covariant derivative* of the tensor of the second rank, E_{hk}. Admittedly, eqn. (14) has been derived on the assumption that the components E_{hk} can be represented as the products of the components of two vectors. It follows from the remarks in § 169, however, that any arbitrary tensor of the second rank can be referred to one so represented, and hence the process described by eqn. (14) enables us to derive a tensor of the third rank from any arbitrary tensor of the second rank.

Let us, therefore, replace E_{hk} in eqn. (14) by the tensor A_{hk} as defined in eqn. (7); its covariant derivative may be denoted by A_{hkp}. We shall, however, write eqn. (7) in a somewhat different form, replacing the index r by the index s; this substitution is necessary to avoid complications in the following derivation. Thus we replace E_{hk} in eqn. (14) by A_{hk}, where

(15) $$A_{hk} = \frac{\partial A_h}{\partial x_k} - \left\{ \begin{matrix} h\,k \\ s \end{matrix} \right\} A_s ,$$

thereby finding

(16) $$\left\{ \begin{aligned} A_{hkp} = {} & \frac{\partial^2 A_h}{\partial x_k \, \partial x_p} - \frac{\partial}{\partial x_p} \left\{ \begin{matrix} h\,k \\ s \end{matrix} \right\} A_s - \frac{\partial A_s}{\partial x_p} \left\{ \begin{matrix} h\,k \\ s \end{matrix} \right\} - \left\{ \begin{matrix} h\,p \\ r \end{matrix} \right\} \frac{\partial A_r}{\partial x_k} \\ & + \left\{ \begin{matrix} h\,p \\ r \end{matrix} \right\} \left\{ \begin{matrix} r\,k \\ s \end{matrix} \right\} A_s - \left\{ \begin{matrix} k\,p \\ r \end{matrix} \right\} \frac{\partial A_h}{\partial x_r} + \left\{ \begin{matrix} k\,p \\ r \end{matrix} \right\} \left\{ \begin{matrix} h\,r \\ s \end{matrix} \right\} A_s . \end{aligned} \right.$$

If now, in this equation, we interchange the indices k and p, the first term and also, because of the symmetry of the three-indices symbols, the sixth and seventh (last) terms on the right-hand side remain unaltered. The sum of the third and fourth terms is likewise unchanged. Hence, if we suppose eqn. (16) to be formed for A_{hpk}, and the resulting equation to be subtracted from eqn. (16), only those terms in which A_s occurs as a factor will remain. Thus we may put

$$(17) \qquad A_{hkp} - A_{hpk} = R^s_{hkp} A_s \,.$$

According to the definitions of tensors, the quantity $R^s{}_{hkp}$, as is expressed in that symbol itself, must be a *mixed tensor of the fourth rank*, being contravariant as regards the index s, and covariant as regards the other indices.

From eqn. (16) we find

$$(18) \quad R^s_{hkp} = \frac{\partial}{\partial x_k}\begin{Bmatrix} h\,p \\ s \end{Bmatrix} - \frac{\partial}{\partial x_p}\begin{Bmatrix} h\,k \\ s \end{Bmatrix} + \begin{Bmatrix} h\,p \\ r \end{Bmatrix}\begin{Bmatrix} r\,k \\ s \end{Bmatrix} - \begin{Bmatrix} h\,k \\ r \end{Bmatrix}\begin{Bmatrix} r\,p \\ s \end{Bmatrix}.$$

It follows from the meaning of the Christoffel's three-indices symbols [§ 171 eqns. (23) and (24)] that the tensor of the fourth rank is completely determined as soon as we know the functional relations connecting the components of the metric fundamental tensor and the coordinates. If the metric field is constant, so that the fundamental tensor is independent of the coordinates, the tensor of the fourth rank vanishes; and, indeed, since it is independent of the coordinate system, it vanishes for *all* coordinate systems. In this lies the great importance of the mixed tensor of the fourth rank, which was first derived by *Riemann*, and is therefore known as the *Riemann tensor*. The *vanishing* of the Riemann tensor is thus the *necessary condition for the constancy of the metric field.*[4]

Like any other mixed tensor, the Riemann tensor can also, of course, be contracted. On carrying out this operation, by putting the index s equal to p and summing for p, we obtain as the formula for the *contracted Riemann tensor*

$$(19) \quad R_{hk} = \frac{\partial}{\partial x_k}\begin{Bmatrix} h\,p \\ p \end{Bmatrix} - \frac{\partial}{\partial x_p}\begin{Bmatrix} h\,k \\ p \end{Bmatrix} + \begin{Bmatrix} h\,p \\ r \end{Bmatrix}\begin{Bmatrix} r\,k \\ p \end{Bmatrix} - \begin{Bmatrix} h\,k \\ r \end{Bmatrix}\begin{Bmatrix} r\,p \\ p \end{Bmatrix}.$$

Furthermore, according to § 169 eqn. (13), we can derive from the contracted Riemann tensor a scalar invariant, by means of the relation

$$(20) \qquad R = g^{hk} R_{hk} \,.$$

[4] It may be shown that this condition is also sufficient.

The quantity R is known as the *Riemann invariant*. Finally, we can obtain the *contravariant components* corresponding to the covariant components of the contracted Riemann tensor, with the aid of the equation [according to § 170 eqn. (11)]

(21) $$R^{ps} = g^{pk} g^{sh} R_{hk} .$$

§ 173. Einstein's and Newton's Laws of Gravitation.

If the Riemann invariant be integrated over an arbitrarily delimited region of the geometrical manifold, and the *variation of this integral* be then formed in such a way that the fundamental tensor is varied only within the boundaries of this region, purely mathematical considerations (which are too complicated to be reproduced here) lead to the components of a tensor

(1) $$S_{hk} = R_{hk} - \frac{1}{2} g_{hk} R .$$

It is this "*curvature tensor*" to which Einstein's theory *equates* the *matter tensor*, apart from a universal proportionality factor.[1]

In accordance with § 166, the components of the matter tensor are found by forming the products of the quantities dx_h/ds in pairs, and then multiplying by the stationary density (ϱ_0) of the proper mass. Since the quantities dx_h/ds are contravariant [according to § 169 eqns. (6) and (9)], whereas ϱ_0 is a scalar, we obtain in this way the *contravariant components of the matter tensor*

(2) $$T^{hk} = \varrho_0 \frac{dx_h}{ds} \frac{dx_k}{ds} .$$

In accordance with Einstein's theory, the relation

(3) $$T_{hk} = S_{hk}$$

must be satisfied ; *i.e.*,

(4) $$T_{hk} = R_{hk} - \frac{1}{2} g_{hk} R ,$$

or,

(5) $$\varrho_0 \frac{dx_h}{ds} \frac{dx_k}{ds} = R^{hk} - \frac{1}{2} g^{hk} R ,$$

[1] If the variation vanishes at the boundary, it may be shown that

$$\delta \int R \, d\omega = \int \left(R_{hk} - \frac{1}{2} g_{hk} R \right) \delta g^{hk} \, d\omega .$$

the contravariant components of the contracted Riemann tensor being determined by § 172 eqn. (21).

Eqn. (4) leads directly to a relation which is important for subsequent considerations. It follows from § 169 eqn. (13) that a scalar invariant, which may be denoted by T, can be formed by means of the relation

$$T = g^{hk} T_{hk} \,. \tag{6}$$

Again [by § 172 eqn. (20)], $g^{hk} R_{hk}$ is equal to R, and [by § 170 eqn. (2)] $g^{hk} g_{hk}$ is equal to four. Hence, on multiplying eqn. (4) by g^{hk} and summing for h and k, we find

$$T = R - 2R \,,$$

or,

$$R = -T \,. \tag{7}$$

Einstein's Law of Gravitation postulates that both the "*field equations*" (5) and the equation which expresses the *geodetic character of the world-line representing the motion* shall be simultaneously satisfied. Thus, by § 171 eqn. (25), the relation

$$\frac{d^2 x_r}{ds^2} = - \begin{Bmatrix} h \; k \\ r \end{Bmatrix} \frac{dx_h}{ds} \frac{dx_k}{ds} \tag{8}$$

must also exist. The two eqns. (5) and (8), which must be simultaneously satisfied, therefore form the analytical expressions of Einstein's Law of Gravitation.

In discussing these equations, we shall first consider a special case in which three greatly simplifying *assumptions* are satisfied. The *first* assumption is that the *deviations from Euclidean geometry* are *only small*. Hence, using a rectangular coordinate system, the quantities g_{hk} shall differ but little from 1 or 0, according as k is or is not equal to h. According to the definitions of cofactors and the contravariant fundamental tensor, the quantities g^{hk} also will then differ but little from 1 or 0, again according as k is or is not equal to h.[2] Our first assumption may therefore be expressed by the formula

$$\left. \begin{matrix} g_{hk} \\ g^{hk} \end{matrix} \right\} \backsimeq \delta^k_h \,, \tag{9}$$

where δ^k_h denotes the mixed fundamental tensor, whose components are always unity or zero [§ 170 eqn. (6)].

If, now, eqn. (9) be satisfied, the three-indices symbols

[2] This follows also from scheme (10) in § 168.

in curly brackets can also be reduced to a simpler form. In accordance with § 171 eqn. (24), which defines this symbol,

$$\left\{ {h\,k \atop r} \right\} = g^{rp} \left[{h\,k \atop p} \right].$$

Of the sum of the four terms on the right-hand side, three summands, however, can [by eqn. (9)] be neglected. We have only to consider the summand in which p is equal to r, and then, by eqn. (9), g^{pr} is simply equal to unity, so that, by way of approximation

$$\left\{ {h\,k \atop r} \right\} = \left[{h\,k \atop r} \right]. \tag{10}$$

In the special case considered, therefore, the curly-bracketed three-indices symbols can be replaced simply by the square-bracketed symbols.

Since the three-indices symbols contain only the partial derivatives of the g_{hk} with respect to the coordinates, it follows further from eqn. (9) that, as a result of our assumption, the *three-indices symbols* must be *small quantities*, small, at least, in the first order.

Our *second assumption* is to the effect that the *mechanical velocities* involved are *small as compared with the velocity of light*. In that case it follows from § 156 eqns. (2) and (5) that

$$\frac{dx_1}{ds}, \frac{dx_2}{ds}, \frac{dx_3}{ds} \text{ are small}; \quad \frac{dx_4}{ds} \backsimeq 1. \tag{11}$$

Moreover, according to § 166 eqn. (3),

$$T = \varrho, \tag{12}$$

and finally (inasmuch as dx_4/ds is approximately equal to unity)

$$\frac{d^2x_r}{ds^2} = -\frac{1}{c^2}\frac{d^2x_r}{dt^2}. \tag{13}$$

Finally, the *third* simplifying assumption we shall make is that the *gravitational field* shall be *static*, *i.e.*, the components g_{hk} shall be *spatially*, but *not temporally*, *variable*. The partial derivatives with respect to x_4 may thus be simply neglected, and we put

$$\frac{\partial g_{hk}}{\partial x_4} = 0. \tag{14}$$

When our three assumptions hold good, in the first place

the equation of a geodesic takes a very simple form. The expression

$$\left\{ {h\,k \atop r} \right\} \frac{dx_h}{ds}\frac{dx_k}{ds}, \tag{15}$$

it is true, represents a double sum of 16 summands (since we have to sum both for h and for k). According to eqn.(11), however, 15 of these summands can be neglected in comparison with the one term

$$\left\{ {4\,4 \atop r} \right\} \frac{dx_4}{ds}\frac{dx_4}{ds}.$$

Now it follows from eqn. (10) and the definition of the three-indices symbols [§ 171 eqn. (23)] that,

$$\left\{ {4\,4 \atop r} \right\} = \left[{4\,4 \atop r} \right] = \frac{1}{2}\left(\frac{\partial g_{4r}}{\partial x_4} + \frac{\partial g_{r4}}{\partial x_4} - \frac{\partial g_{44}}{\partial x_r} \right),$$

or since, by eqn. (14), the partial derivatives with respect to x_4 vanish,

$$\left\{ {4\,4 \atop r} \right\} = -\frac{1}{2}\frac{\partial g_{44}}{\partial x_r}. \tag{16}$$

If, further, we take into account eqns. (11) and (13), the equation of a geodesic becomes, in the special case considered, simply

$$\frac{d^2x_r}{dt^2} = -\frac{c^2}{2}\frac{\partial g_{44}}{\partial x_r}, \tag{17}$$

and, in conjunction with this equation, we have still to take eqn. (5), which, on the basis of the same assumptions, is likewise considerably simplified.

In the first place, the expression for the contracted Riemann tensor can be greatly simplified. This expression [§ 172 eqn. (19)] consists of four terms. The last two of these, which represent the products of two three-indices symbols, may be omitted, because, in accordance with our previous remarks, these symbols are small quantities, at least in the first order. Furthermore, if we put the index k equal to 4, then, by eqn. (14), the first term may also be neglected, and only the second remains. We have therefore

$$R_{h4} = -\frac{\partial}{\partial x_p}\left\{ {h\,4 \atop p} \right\}. \tag{18}$$

Hence, by eqn. (16),

$$R_{44} = \frac{1}{2}\left(\frac{\partial^2 g_{44}}{\partial x_1{}^2} + \frac{\partial^2 g_{44}}{\partial x_2{}^2} + \frac{\partial^2 g_{44}}{\partial x_3{}^2} + \frac{\partial^2 g_{44}}{\partial x_4{}^2} \right). \tag{19}$$

According to eqn. (14), the partial differential coefficient with respect to x_4 vanishes. Moreover, the sum of the second partial derivatives with respect to the spatial coordinates represents the well-known Laplace derivative, generally denoted by the symbol ∇^2 (§ 32). Hence

(20) $$R_{44} = \tfrac{1}{2}\nabla^2 g_{44}\,.$$

We shall next consider the contravariant components of the contracted Riemann tensor; according to § 170 eqn. (11),

(21) $$R^{44} = g^{4h} g^{4k} R_{hk}\,,$$

It follows, however, from eqn. (9) that, in this double sum of 16 summands, all other summands can be neglected in comparison with the one for which $h = k = 4$. Hence, since g_{44} is approximately equal to unity,

(22) $$R^{44} = R_{44}\,.$$

Moreover, by eqns. (7) and (12),

(23) $$R = -\varrho\,.$$

Let us now put $h = k = 4$ in eqn. (5). According to eqn. (11), the differential coefficients then become equal to unity, whilst ϱ_0 is equal to ϱ, and thus we obtain the relation

$$\varrho = \tfrac{1}{2}\nabla^2 g_{44} + \tfrac{1}{2}\varrho\,,$$

or,

(24) $$\varrho = \nabla^2 g_{44}\,.$$

This equation represents a partial differential equation of the second order, and it follows from § 51 eqn. (10) that its solution is given by the expression

(25) $$g_{44} = -\frac{1}{4\pi}\int\frac{\varrho\,d\tau}{r}\,,$$

where r denotes the distance of the volume-element $d\tau$ from the field-point, and ϱ the mass density in the volume-element. Finally, if we combine eqn. (25) with eqn. (17) (in which we shall replace the index r by the index h),[3] we obtain the *equation of motion in a material field* in the form

(26) $$\frac{d^2x_h}{dt^2} = \frac{c^2}{8\pi}\frac{\partial}{\partial x_h}\int\varrho\frac{d\tau}{r}\,.$$

Hitherto we have measured mass in such a way that the universal proportionality factor between the curvature tensor and the matter tensor is simply equal to unity. When, however, we measure mass density, as is usual, in gm./cm.^{3},

[3] In order that the letter r shall not occur with two different meanings in the same equation.

ϱ will be replaced in the above formulæ by the product $f\varrho$, where f denotes a universal proportionality factor. We shall introduce a new universal constant, defined by the relation

$$\kappa = \frac{f c^2}{8\pi}; \tag{27}$$

then

$$\frac{d^2 x_h}{dt^2} = \kappa \frac{\partial}{dx_h} \int \varrho \frac{d\tau}{r}. \tag{28}$$

We shall now suppose the distribution of mass, previously considered continuous, to be only approximately so; we imagine in reality a number of discrete particles to be present, and we shall describe the individually differing masses by m, and by r the individually differing distances from the field-point considered. Then we have

$$\int \frac{\varrho\, d\tau}{r} = \sum \frac{m}{r}. \tag{29}$$

Finally, we suppose a particle of mass m' to be situated at the field-point itself. On multiplying eqn. (28) by m', we obtain the relation

$$m' \frac{d^2 x_h}{dt^2} = \frac{\partial}{\partial x_h} \sum \frac{\kappa\, m\, m'}{r}. \tag{30}$$

Now the three quantities $d^2 x_h/dt^2$ are nothing else than the components of the spatial vector $d\mathbf{v}/dt$, where $\mathbf{v}$ is the velocity of the particle m'. On the other hand, in accordance with a relation derived elsewhere and frequently used, the three quantities $\partial(1/r)/dx_h$ are the components of the vector $(-\mathbf{r}/r^3)$, where $\mathbf{r}$ is the radius-vector drawn from a mass m to the field-point.[4] Eqn. (30) may therefore be written in the form

$$m' \frac{d\mathbf{v}}{dt} = -\sum \frac{\kappa\, m\, m'}{r^2} \cdot \frac{\mathbf{r}}{r}. \tag{31}$$

The left-hand side of this equation may be regarded as representing a *force* exerted on the particle m' in consequence of the presence of the other particles. Accordingly, the right-hand side may be interpreted as the resultant of the individual forces acting between the particle m' and all the other

[4] For the x_h are the coordinates of the field-point. Hence § 50 eqn. (4) applies, according to which, for a variable field-point,

$$\operatorname{grad}_a \left(\frac{1}{r}\right) = -\frac{1}{r^2}\frac{\mathbf{r}}{r},$$

where $\mathbf{r}$ denotes the radius-vector drawn from the source to the field-point.

particles. As the equation shows us, these forces must act *in the direction of the line joining the particles* concerned, and, inasmuch as they are opposite in sign to the radius-vector drawn from m to m', they will tend to draw m' towards m, *i.e.*, they must be *forces of attraction.* The magnitude of the force is, apart from a universal proportionality factor, *equal to the product of the two masses divided by the square of the distance between them.*

Thus, in consequence of the presence of the other particles, every particle moves as if a *universal force of attraction* were at work *between all masses*, governed by the law just given. This, however, is none other than the well-known *Law of Gravitation*, established by *Newton.*

§ 174. The Metrics of Space and Time in a Gravitational Field.

The derivation of Newton's from Einstein's Law of Gravitation, as a first approximation, in itself constitutes a great success for Einstein's Theory. The possibility of experimentally testing this new theory, however, is to be sought in the appearance of several phenomena which, whilst necessarily following from the more complete Law of Einstein, cannot be explained on the basis of Newton's Law.

Einstein's gravitational equations are partial *differential equations* of the second order, in which the components of the metric fundamental tensor are differentiated with respect to the coordinates. The general problem would then consist in calculating the quantities g_{hk} as functions of the world-coordinates, for a given space-time distribution of matter. The rigid solution of this problem is so complicated that it has not yet been accomplished. *Einstein*, and in a more accurate manner, *Schwarzschild*, however, have obtained the solution for the special case in which the *gravitational field is produced by a single particle at rest*, when the field must obviously possess spherical symmetry in space.

The calculations, which are too involved to be reproduced here, lead to the following solution: If we denote by M the single mass producing the field, and by κ the universal gravitational constant, and if we put, for brevity,

$$(1) \qquad \alpha = \frac{\kappa M}{c^2},$$

then, at a matter-free point with the spatial coordinates x_1,

x_2, x_3, at a distance r from the particle generating the field, we find

$$(2)\qquad \left\{\begin{aligned} g_{hk} &= \delta_h^k + \frac{2\alpha\, x_h\, x_k}{r^3}, \\ g_{h4} &= g_{4h} = 0, \\ g_{44} &= 1 - \frac{2\alpha}{r}. \end{aligned}\right\} \qquad \begin{pmatrix} h = 1, 2, 3 \\ k = 1, 2, 3 \end{pmatrix} \quad \delta_k^h = \begin{cases} 1 & (h = k) \\ 0 & (h \neq k) \end{cases}$$

Since the dimensions of the gravitational constant are gm.$^{-1}$ cm.^{3} sec.$^{-2}$, the dimensions of the quantity α defined by eqn. (1) are those of a length, and α is therefore called the *gravitational radius* of the mass. Now

$$(3)\qquad \kappa = 6{\cdot}68 \times 10^{-8}\ \text{gm.}^{-1}\ \text{cm.}^3\ \text{sec.}^{-2},$$

and thus we obtain the gravitational radius of a mass in centimetres on multiplying the number of grams by 7×10^{-29}. For example, the gravitational radius of an electron is about 6×10^{-56} cm.—an inconceivably small quantity. The earth's mass is about 6×10^{27} gm., so that its gravitational radius is approximately 4 mm. The sun is about 330,000 times as heavy as the earth, and thus its gravitational radius would be rather more than 1 km.

Eqns. (2) only hold good in space outside the mass producing the field. As eqns. (2) show us, the deviations from Euclidean metrics are thus determined by the *ratio* between the gravitational radius and the radius of the mass (assumed to be spherical). For an electron of about 10^{-13} cm. radius the ratio would be unimaginably small, *viz.*, about 10^{-43} in order of magnitude. For the earth the ratio would be about 6×10^{-10}, and for the sun approximately 2×10^{-6}. An appreciable influence of a gravitational field on the metrics of space and time is thus scarcely to be expected in atomic processes ; it is, in fact, hardly to be expected on the earth at all, but rather on the sun and in its immediate vicinity, as well as on other large stars. An experimental confirmation of Einstein's gravitational theory is thus to be sought, not in laboratory researches, but in astronomical measurements.

For the purposes of a closer investigation of the metrics in a gravitational field, let us consider quite generally two adjacent events in time and space, the corresponding world-coordinates being x, y, z, $i\ c\ t$ and $(x + dx)$, $(y + dy)$, $(z + dz)$, $i\ c\ (t + dt)$ respectively. The four-dimensional interval (ds) between the two world-points is, as we know, invariant and hence independent of the gravitational field. According to the generalized Theorem of Pythagoras [§ 168 eqn. (5)] we have, since by eqns. (2) the g_{h4} vanish,

$$(4) \qquad ds^2 = \sum_{h=1}^{h=3} \sum_{k=1}^{k=3} g_{hk}\, dx_h\, dx_k + g_{44}\, dx_4^2 ,$$

where

$$dx_4 = i\, c\, dt.$$

When, in particular, the two events are simultaneous, then ds represents the spatial interval between the two points at which the two events occur. We may distinguish between two special cases. In the first case let both points lie on the x-axis at distances r and $(r + dr)$ respectively from the particle producing the field. Then

$$(5) \qquad x_1 = r, \quad x_2 = x_3 = 0, \quad dx_2 = dx_3 = dt = 0,$$

so that we have simply

$$(6) \qquad ds^2 = g_{11}\, dx_1^2 ,$$

or,

$$dx_1 = \frac{ds}{\sqrt{g_{11}}} ,$$

or finally, by way of approximation and according to eqns. (2),

$$(7) \qquad dx_1 = \frac{ds}{\sqrt{1 + \frac{2\alpha}{r}}} = ds\left(1 - \frac{\alpha}{r}\right).$$

In this case, therefore, a *measuring-rod* lying in the x-direction, or *radially*, is apparently *shortened by the gravitational field*.

In the second special case we shall suppose the rod to be lying *tangentially* at the same point, but parallel, say, to the y-axis. Then

$$(8) \qquad x_1 = r\,, \quad x_2 = x_3 = 0\,, \quad dx_1 = dx_3 = dt = 0\,,$$

so that

$$(9) \qquad ds^2 = g_{22}\, dx_2^2.$$

But, by eqns. (2) and (8),

$$(10) \qquad g_{22} = 1 + \frac{2\alpha\, x_2^2}{r^3} = 1\,.$$

Hence the *length* of a rod lying *tangentially* in the gravitational field is *not affected* at all. A metre-rule must experience a contraction of about 6 Å.U. when raised from a horizontal into a vertical position on the earth, but this contraction is only about one-tenth of the Lorentz contraction due to the earth's motion.

When the two events represented by the two previously considered adjacent world-points occur at the same place, ds denotes the time-interval between the two events, multiplied by ic. Thus we have

$$ds^2 = g_{44}\, dx_4^{\,2}\,, \tag{11}$$

or, by eqns. (2),

$$dx_4 = \frac{ds}{\sqrt{1-\frac{2\alpha}{r}}} = ds\left(1+\frac{\alpha}{r}\right). \tag{12}$$

Hence, since dx_4 is larger than ds, *all processes* are apparently *retarded in a gravitational field.*

The alteration of the metrics of space and time in a gravitational field must obviously also have the effect that, *in a gravitational field, rays of light* are not straight but *curved*, as we might surmise without entering upon any calculations. Accurate calculations (too long to be given here) lead, in fact, to the following result :—When a ray of light passes by a gravitational mass of gravitational radius α at a distance r from it (measured from the centre of the mass), it experiences a *deflection*, of which the total magnitude [1] in circular measure is equal to $4\alpha/r$.

When the alterations of the metrics in a gravitational field are taken into account, *Kepler's Laws* also assume a *more complicated form.* Exact calculations (too involved to be reproduced here) show that the major axis of the planetary orbits must slowly rotate, in the sense in which the planet revolves. The *perihelion of the planet's orbit* must therefore *advance* slowly, and the angle of advance for each revolution is found by accurate calculations to be given in circular measure by

$$\varphi = \frac{6\pi\,\alpha}{a\,(1-\epsilon^2)}\,; \tag{13}$$

in this expression α is the gravitational radius of the sun, a the semi-major axis, and ϵ the eccentricity of the orbit.

[1] As Einstein discovered in 1911, half of this amount is due to the fact that the ray of light is of itself deflected in the gravitational field on account of its energy; this follows from the two laws of the inertia of energy and of the identity of inertial and gravitational mass.

§ 175. The Experimental Confirmation of Einstein's Theory of Gravitation.

The *perihelion motion of the planets* demanded by Einstein's theory, as opposed to the Newtonian theory, was actually discovered in 1845 by *Leverrier*, the astronomer. Leverrier observed this motion in the case of the planet nearest the sun, *Mercury*, but astronomers were unable to give a satisfactory explanation of the phenomenon, which was contradictory to Newton's theory. The perihelion motion is in the sense of the planet's motion, and, as Leverrier observed, amounts to 43 seconds of arc in a century.[1] Exactly the same value, however, was obtained by Einstein from the equation in his theory (§ 174 eqn. (13)), on inserting the known values for the major axis and the eccentricity of the orbit of Mercury, and the gravitational radius of the sun.

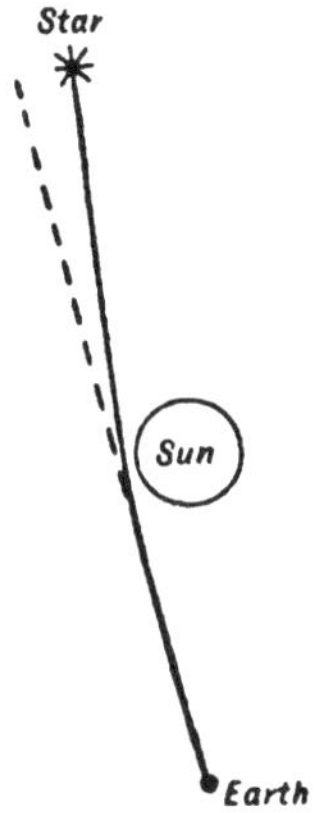

FIG. 130.

Einstein's gravitational theory has received a second striking confirmation based on the previously mentioned fact that a ray of light is deflected in passing a gravitational mass; as already stated, the total deflection amounts to $4\alpha/r$. Since, as we have already mentioned, the ratio α/r is equal to 2×10^{-6} at the surface of the sun, a *ray of light* which *grazes the sun* must experience a total deflection of 8×10^{-6} or (in the usual angular measure) of 1·7 seconds of arc.[2] If the ray passes the sun at a distance equal to γ times the sun's radius from its centre, the deflection should be $1{\cdot}7/\gamma''$. In consequence of the deflection of the ray, the star emitting it would not appear in its actual position to a terrestrial observer. Stars which appear close to the sun in the sky must therefore be *apparently displaced* by a very small amount *away from the sun*, the displacement being larger the nearer in line they are to the sun (Fig. 130).

Now the stars nearly in line with the sun are admittedly invisible as a rule, but, during a *total solar eclipse*, they can be observed and photographed. It has, in fact, been possible already to detect the so-called " Einstein effect " in several solar eclipses, the first observations having been made in

[1] This amount represents the excess of the perihelion motion actually observed over that calculated after allowing for all possible perturbations.

[2] Since 360° is equal to 2π in circular measure, a radian is equal to $360 \times 60 \times 60/2\pi''$.

1919.[3] Exact measurements of the photographs obtained of the sun and the stars surrounding it during the eclipse have confirmed the effect predicted by Einstein, not only qualitatively, but also quantitatively, with complete accuracy within small limits of error.

A third possibility of submitting Einstein's gravitational theory to an experimental test follows from the alterations of the metrics of time in a gravitational field. According to Einstein, this should lead to a *displacement towards the red of the spectral lines of stars.* Such a displacement is most difficult to detect, because of its minute amount and the simultaneous occurrence of the Doppler effect, but its existence now appears to be confirmed as a result of many years' researches [4] by *John.*[5]

§ 176. Size and Mass of the Universe.

It has long been recognized by physicists and astronomers that, in the classical theory of gravitation, Newton's Law of Gravitation *cannot be reconciled* with the concept of a *spatial infinity of the universe.* This incompatibility appears clearly from a simple consideration. Let us imagine a space whose linear dimensions are large compared with the distance between two neighbouring fixed stars, but small compared with the extent, say, of the Milky Way. We can regard the quotient of the total mass of all the stars contained in this volume and the volume itself as the *density* in that portion of the universe. We shall now suppose two spherical surfaces to be constructed about any arbitrary point as centre, the radii of these spheres being r and $(r + \Delta r)$, where r is very

[3] Half of the effect had already been predicted by Einstein in 1911, before the evolution of the general theory of relativity (*cf.* § 174 note 1). In August, 1914, it was arranged to send a German expedition to look for the predicted effect in a total eclipse observable in Southern Russia, but the outbreak of war prevented this.

[4] *Zeitschr. f. Phys.*, 21 (1924), p. 159.

[5] We may briefly mention an interesting generalization of the general theory of relativity, which, though not held by Einstein himself, was attempted by *Weyl* in 1918. Weyl's theory is based on the postulate that, in a truly general geometry free from all prejudice, not only should the problem of the transformation of direction be undetermined, as in Riemannian geometry, but also that of the *transformation of lengths.* Just as the former problem first assumes definite shape with the help of the metric fundamental tensor, so the latter problem will first be defined with the aid of a vector which, in a truly general geometry, must necessarily appear as a quantity varying from place to place. By interpreting this vector as an electromagnetic four-potential (*cf.* § 165), Weyl believes that the *electromagnetic field* of the Minkowski-world can be determined in terms of it. The author of this book has attempted a popular presentation of Weyl's theory in a paper entitled "*Die Physik als geometrische Notwendigkeit,*" *Naturwissenschaften,* 8 (1920), pp. 121–127.

large and even Δr is large as compared with the mean distance between two neighbouring stars. The two spherical surfaces will then form the boundaries of a " spherical shell " (admittedly of very large thickness). If the density in the universe be fairly uniform, inasmuch as no one point in the infinite space appears privileged as regards any other, then the total mass contained within the spherical shell will be proportional to $r^2 \Delta r$, and the force of attraction exerted at the centre by each mass will be *inversely* proportional to r^2. The total attraction exerted by a spherical shell is thus approximately proportional to the thickness of the shell, but independent of r itself. Hence, if we regard space as infinite and containing a uniform distribution of stars, a force of attraction will be exerted at every point from all directions, this force being infinite or at least indefinite, so that a stable solar system would be inconceivable.

On the other hand, if we suppose the universe to be spatially infinite, but the number of stars finite, then, on the ground of the results of the kinetic theory of gases, there could be no reason why the stars (which are comparable to molecules in that theory) should not lose themselves in the infinite space, or why there should ever be a dynamic equilibrium between them.

These difficulties of the classical theory have been surmounted by Einstein, on the assumption that, although the *universe* is *boundless*, yet it has a *finite volume*. The possibility of such an assumption follows directly from *Riemannian geometry*. Just as a circular line has finite length, but no beginning nor end, and just as the earth's surface has no beginning nor end but yet contains a definite number of square metres, so could a *curved space* (of which we can certainly form no conception) be of finite volume and yet be boundless. A geodesic drawn as a great circle on the earth's surface leads back to the starting-point, and similar considerations would apply to a three-dimensional spherical super-surface, or a so-called *spherical space*. As in the cases of a circle and a spherical surface, we have to assume a *radius of curvature* for a spherical space, and, denoting it by R, the relation (which we shall state here without giving the proof) between it and the finite volume is given by the simple expression

$$V = 2\pi^2 R^3 . \tag{1}$$

In consequence of the Principle of Relativity, the idea of space has admittedly lost its independence. On the basis of the important experimental fact that the *relative velocities of stars* are *always small as compared with the velocity of light*, however, we can, to a certain degree of approximation, split up the Minkowski-world into space and time. We can thus speak approximately of a *space*, and this Einstein regards as *spherical* in the sense already mentioned, in that it is boundless, but has a finite volume.

Again, the assumption of a spherically curved space involves the addition of a *supplementary term* in Einstein's Law of Gravitation, but this is of no importance for the derivation of Newton's Law as a special case, or of the three experimentally confirmed effects of the general theory of relativity. The *cosmological law of gravitation* equates the matter tensor, not to the curvature tensor itself, but to the difference between the curvature tensor and the product of the fundamental tensor and a universal constant; in this form also the Law of Gravitation is, of course, independent of the coordinate system. Denoting the universal, so-called *cosmological*, constant by λ, Einstein's gravitational equation in its covariant form becomes

$$R_{hk} - \frac{1}{2} g_{hk} R - \lambda g_{hk} = T_{hk}. \tag{2}$$

This equation is, of course, supplemented by the one expressing the geodetic nature of a world-line, which is not altered.

If, now, we combine these two equations with the third one, which expresses the existence of spherical metrics (in the sense previously discussed) in space split off from the Minkowski-world, calculation leads to the two relations

$$\lambda = \frac{4\pi \kappa \varrho}{c^2} = \frac{1}{R^2}, \tag{3}$$

where κ denotes the gravitational constant, ϱ the *mean mass density in the universe*, and R *the radius of curvature of the universe* [not, as in eqn. (2), the Riemann invariant].

On the other hand, denoting the *total mass of the universe* by M, we find by eqn. (1)

$$\varrho = \frac{M}{2\pi^2 R^3}, \tag{4}$$

and hence, by eqn. (3),

$$\frac{4\pi \kappa M}{2\pi^2 c^2 R^3} = \frac{1}{R^2},$$

or,

$$M = \frac{\pi c^2 R}{2\kappa}. \tag{5}$$

By means of this equation, which was derived by Einstein in 1917, the relation between the mass of the universe and its radius of curvature is given.[1] Moreover, on resolving this formula with respect to the gravitational constant, we see how the magnitude of the attraction between the smallest spheres of lead in a terrestrial laboratory depends on the size and mass of the whole universe.

[1] On the basis of astronomical observations, *de Sitter* (*Monthly Notices of the Royal Astronomical Society*: London, 1917) has estimated the mean density of matter in the universe to be about 10^{-26} gm. cm.$^{-3}$, and he has thereby obtained a value of about 20 million light-years for its radius of curvature. The distances of the furthest star clusters have been estimated at about 300,000 light-years, and those of spiral nebulæ at more than a million light-years. The length of a *geodesic* in spherically curved space would, therefore, be about a *hundred million light-years.* Since, in consequence of the curvature of space, every ray of light must return to its source after about 100 million years, it would be conceivably possible for an object in the northern celestial hemisphere to be identical with one in the southern, just as a place 10,000 km. distant to the east of any point on the equator would be identical with a place 30,000 km. to the west of that point. On the basis of de Sitter's estimation, the mass of the universe as given by eqn. (5) would be about 10^{53} grams, and the total number of all the electrons in the universe would be represented by a number of the order of magnitude of 10^{77}.

APPENDIX

SUMMARY OF THE CONTENTS OF VOLUME II

Part III

ATOMIC THEORY

Chapter XI

The Elementary Quanta

§ 87. It has been found from the movements executed by small particles, with weak electrical charges, under the dual influence of their own weight and an electric field directed vertically upwards, that the charges on such particles are exact integral multiples of an elementary quantum of electricity (e). The value of this quantum is found by observation to be $4{\cdot}77 \times 10^{-10}$ E.S.U.

§ 88. The number by which the atomic weight numbers must be divided in order to obtain the absolute masses of the atoms is termed the Loschmidt number (L). The product $L\ e$ is known from electrolytic measurements, on the basis of Faraday's fundamental law, and the value of L thus follows as $6{\cdot}06 \times 10^{23}$. Hence the mass of an atom of hydrogen is $1{\cdot}66 \times 10^{-24}$ gm.

§ 89. The specific charge of cathode ray particles is found from their electric and magnetic deflections to be about 1,800 times that of an ionized atom of hydrogen. On the assumption that the charge on each particle is one elementary quantum, the cathode rays thus consisting of electrons, the mass (m) of an electron is found to be $9{\cdot}0 \times 10^{-28}$ gm. The velocity of cathode rays is proportional to the square root of the tube potential. The negative corpuscular rays emitted by radioactive substances, or the so-called β-rays, likewise consist of electrons, and their velocities amount to from 30 to 99·8% of the velocity of light. It has been possible to demonstrate experimentally with these rays the increase in mass with the velocity demanded by the theory of relativity.

§ 90. Canal rays and anode rays consist of positive ions, whose masses are found by positive ray analysis to be always approximately integral multiples of the mass of a hydrogen atom. The positive α-rays emitted by radioactive substances have initial velocities of 5—7% of the velocity of light. The scintillation method allows of a count of the α-particles expelled by a preparation in a given time. In this way the charge on a single α-particle can be derived, and hence also its mass, since the specific charge is known from deflection experiments. It has been found that the mass of an α-particle is four times as large as that of a hydrogen atom, whilst its positive charge amounts to two elementary quanta. In hydrogen, α-particles produce long-range H-rays.

§ 91. Careful investigations on the photo-electric effect show that the equivalent potential of the liberated electrons is a linear function of the frequency of the radiation producing them. This leads to the existence of a universal constant with the dimensions of action, or of an elementary quantum of action (h); its value is found from measurements to be $6{\cdot}55 \times 10^{-27}$ erg. sec. According to a law established by Einstein, every mutual transformation between corpuscular radiation and electromagnetic wave radiation takes place in such a way that, either a single light-quantum ($h\nu$) is used to liberate a single electron or, in the reverse case, a single light-quantum is produced by the energy of a single electron.

§ 92. As applied to periodic circular motions, the quantum principle leads to the conclusion that those circular orbits are privileged, for which the angular momentum multiplied by 2π is an integral multiple of the elementary quantum of action.

Chapter XII

Theory of the Hydrogen Atom

§ 93. Observations on the passage of α-rays through matter show that an α-particle can penetrate through several thousands of atoms without experiencing any appreciable change in direction, although sometimes a single atom produces a very large deflection of the α-particle. The deflection must therefore be due to an atomic nucleus which occupies only a very small part of the total volume, and in which almost the whole mass of the atom is concentrated. According to Rutherford's fundamental conception, the nucleus must be regarded as positive and as surrounded by revolving electrons, the electrical force of attraction exerted on these being balanced by the centrifugal force.

§ 94. Bohr's fundamental hypothesis is to the effect that the angular momentum of the normal hydrogen atom multiplied by 2π shall be equal to the elementary quantum of action, but, besides the normal state, other higher quantum states shall also be possible, in which the angular momentum is an integral multiple of the normal. In any arbitrary state, the energy, which is always negative, is equal to the normal energy divided by the square of the quantum number.

§ 95. On the assumption that every transition between two different states compatible with the quantum theory is accompanied by the emission or the absorption of a light-quantum (according as the quantum number is lowered or raised), Bohr obtained for the hydrogen spectrum the formula

$$\nu = R\left(\frac{1}{s^2} - \frac{1}{n^2}\right), \qquad R = \frac{2\pi^2 e^4 m}{h^3}.$$

The frequencies of the lines thus appear representable as differences between terms.

§ 96. The spectral formula theoretically derived by Bohr agrees with that experimentally obtained by Balmer. According as the quantum number s, which corresponds to the final state for emission, is given the value 1, 2, 3, or 4 in Bohr's formula, an expression is obtained for the ultra-violet Lyman series, the optical Balmer series, the infra-red Paschen series, or the Brackett series of long wave-length.

§ 97. If the frequency in Bohr's formula for the hydrogen spectrum be multiplied by 2^2, the expression thus obtained must theoretically represent the spectrum of ionized helium. In point of fact these series, which were originally erroneously ascribed to hydrogen, have been shown by later experiments to be due to helium.

§ 98. The displacement towards the violet of the helium lines relatively to the hydrogen lines is explained, as Bohr recognized, by the simultaneous motion of the atomic nucleus. Measurements on the displacement towards the violet yield an exact determination of the specific charge of an electron.

§ 99. In Sommerfeld's development of Bohr's theory, each of the electronic orbits, regarded as ellipses, requires two quantum numbers for its definition. The principal quantum number determines the major axis of the ellipse and the subordinate number its eccentricity. The selection principle permits only of those transitions between orbits in which the subordinate number changes by unity.

§ 100. Although, as a first approximation, the energy of the electron in the hydrogen atom is independent of the subordinate number of its orbit, a dependency is found to exist within narrow limits when, in accordance with the theory of relativity, the change of mass with velocity is taken into account. In this way Sommerfeld succeeded in explaining the fine structure of the hydrogen lines ; his theory is completely confirmed by experiment.

§ 101. The resolution of spectral lines in an electric field, or the so-called Stark effect, may be regarded as a brilliant experimental confirmation of the quantum theory.

§ 102. Just as in an electric field, so also in a magnetic field there occurs a resolution of the spectral lines proportional to the field-strength ; this is known as the Zeeman effect.

Chapter XIII

Röntgen Rays

§ 103. Röntgen rays are generated by the arrest of electrons moving with high velocities.

§ 104. A space-lattice separates by diffraction rays of quite definite wave-lengths from out of an incident wave radiation. These wave-lengths depend, for given ordinal numbers, on the lattice-constant and the direction of incidence.

§ 105. The diffraction of rays by a space-lattice may also be regarded as a reflection at the lattice-planes. A set of lattice-planes singles out from an incident radiation those rays for which an integral multiple of the wave-length is equal to the difference in path between two rays which are reflected from two successive lattice-planes. On rotation of a space-lattice, all rays of the same wave-lengths are focussed at definite points lying on the circumference of a circle.

§ 106. The diffraction of Röntgen rays by crystals was demonstrated by Laue. The investigation of the structure of a crystal on the basis of the Laue photographs rendered possible an exact measurement of the wave-lengths of Röntgen rays. By means of crystals, wave-lengths of from about 100 to 13,000 X.U. can be measured.

§ 107. Duane and Hunt's equation, which follows from Einstein's law of light-quanta, states that

$$\lambda V = 1{\cdot}234 \times 10^7 \quad (\lambda \text{ in X.U., } V \text{ in volts}).$$

In complete agreement with experimental results, this equation determines, on the one hand, the excitation potential for a given wave-length, and on the other hand, the smallest possible wave-length for a given tube potential, in each case in the continuous Röntgen spectrum.

Chapter XIV

Theory of the Elements

§ 108. The foundation of Röntgen spectroscopy by Moseley is based on Barkla's discovery of characteristic radiations. As Moseley found, the elements can be arranged in a consecutively numbered series, in such a way that the square root of the frequency of a definite Röntgen line increases linearly with the number (the atomic number) of the element. The " natural sequence " of the elements thus obtained, which begins with hydrogen and ends with uranium, comprises 92 places, five of which, however, are unoccupied. The atomic number proves to be identical with the nuclear charge number.

§ 109. The natural sequence of the elements is divided by the chemically inert rare gases into seven periods. The first period contains 2 elements, the second and third periods contain 8 elements each, the fourth and fifth 18 each, the sixth contains 32 elements, and the seventh period breaks off at the sixth place with uranium. Homology among the elements allows of a distinction between eight groups, each with two sub-groups, in the periodic system.

§ 110. The phenomenon of isotopy, *viz.*, the chemical identity of substances with different radioactive properties or atomic weights, is explained by the fact that two atoms with the same nuclear charge may yet possess a different nuclear structure, and thus a different number of protons. The analysis of mixed elements was rendered practicable by positive ray analysis, which has been developed by Aston into mass-spectroscopy.

§ 111. The phenomena of radioactivity are due to a disintegration of the atomic nuclei, which, owing to the alteration in the nuclear charge number, at the same time effects a chemical transformation. An α-transformation reduces the atomic number by two, and the atomic weight by four, whereas a β-transformation raises the atomic number by one, without noticeably altering the atomic weight. The observation of the rate of formation of helium from radium, combined with the scintillation method, permits of a direct determination of the Loschmidt number.

§ 112. In radioactive equilibrium the numbers of atoms present are in inverse ratio to the disintegration constants. Equal numbers of atoms of all substances which are in equilibrium with each other disintegrate in the same time.

§ 113. At the present time three transformation series are known, commencing with uranium, actinium, and thorium respectively, and all ending with stable species of lead. Besides these series, potassium and rubidium are also known to be radioelements. The age of uranium minerals can be calculated from their content of uranium lead.

§ 114. The range of α-rays is proportional to the cube of their initial velocity. The logarithms of the disintegration constants are linear functions of the logarithms of the ranges. β-ray spectra are produced by the deflection of β-rays in magnetic fields. For γ-rays it has been possible to measure wave-lengths as small as 20 X.U., and to estimate wave-lengths as low as 5 X.U.

§ 115. Using α-rays, Rutherford has succeeded in disrupting the nuclei of B, N, F, Na, Al, and P, hydrogen nuclei being expelled from the artificially disrupted nuclei.

§ 116. In elements with an atomic weight divisible by four, the nuclei are very possibly built up solely of positive and neutralized α-particles. Elements of even atomic number are of more frequent occurrence than those of odd atomic number, and are generally of even atomic weight. Elements of odd atomic number are either pure elements, or possess two isotopes, the atomic weights of which are generally odd and generally differ by two. The linear dimensions of the nuclei prove to be of the order of magnitude of 10^{-12} cm. The comparatively enormous internal energy of the atomic nuclei is manifested, in accordance with the Law of the Inertia of Energy, in the deviation from whole numbers of the atomic weights referred to hydrogen.

Chapter XV

General Theory of Spectra and Atomic Structure

§ 117. The K-, L-, M-, and N-series of Röntgen spectra arise from transitions to a one-, two-, three-, and four-quantum final state. Since the excitation potential necessary for the generation of a definite Röntgen line is always found to be equal to the characteristic potential of the final level, a Röntgen line can obviously be produced only when an electron in the emitting atom has been previously removed from its normal orbit to beyond the limits of the atom. This hypothesis is afforded a direct experimental confirmation by the absorption edges in the continuous Röntgen spectrum.

§ 118. An explanation of the multiplicity of observed Röntgen lines demands an assumption of several modifications in the various quantum levels. According to Bohr, each level can be represented by three quantum numbers in the symbolical form n (k_1, k_2) ; n is the principal quantum number, k_1 cannot be greater than n, and k_2 is either equal to k_1, or less by one, a distinction being thereby made between normal and anomalous levels. Only those transitions between levels are apparently possible in which k_1 changes by unity, and k_2 either remains unaltered or also changes by unity. The wave-lengths of the γ-rays of a radioelement can be calculated from its magnetic β-ray spectrum, provided that the energy-levels of an isotopic element are known from Röntgen-spectroscopic measurements.

§ 119. In an excited state of an atom, an electron can describe a path of higher quantum number than the normal one. Excitation may be due to the absorption of light-quanta, or, in the case of thermal luminescence, to the transformation of translational energy. A sufficient introduction of energy may lead to the ionization of atoms, to which, *inter alia*, the phenomenon of the conductivity of flames may be referred. Besides the arc spectrum as the spectrum of the neutral atom, elements also exhibit spark spectra as the spectra of their ions. According to the spectroscopic displacement law, the spark spectra of the elements in a group of the periodic system agree in character with the arc spectra of the preceding group.

§ 120. Optical absorption spectra arise on the transition of an electron to a higher quantum orbit, either from the fundamental orbit or from a previously excited orbit. Optical emission spectra arise on the spontaneous return, in either one or more stages, from unstable states to the fundamental state. The lines in optical series converge with decreasing wave-length, decreasing intensity, and decreasing sharpness towards a limit which, in the case of emission, corresponds to the final level of the transition. The most important lines can be referred to four series of terms, which are called the *s*-, *p*-, *d*-, and *f*-terms, and characterized by the subordinate quantum numbers 1, 2, 3, and 4. The principal series, the diffuse subordinate series, the sharp subordinate series, and the Bergmann (or fundamental) series correspond, in the case of emission, to the transitions $p \rightarrow s$, $d \rightarrow p$, $s \rightarrow p$, and $f \rightarrow d$. In the arc spectra of the alkali metals all terms are double, with the exception of the *s*-terms, which are always simple. In the arc spectra of the elements of the second group, there is a system of simple terms as well as one of triple terms, and spectroscopic combinations may occur between both systems.

§ 121. The electrons revolving about an atomic nucleus may be divided up, according to the types of their normal orbits, into groups and sub-groups characterized by the values of the principal and the subordinate quantum number. The electronic groupings in the atoms of the rare gases can be derived from their atomic numbers and the scheme of Röntgen levels. The groups thus found consist of r sub-groups, each containing $2r$ electrons, where $r = 1, 2, 3, 4$. For emanation, the values to be given to r in the 1-, 2-, 3-, 4-, 5-, 6-quantum groups are 1, 2, 3, 4, 3, 2. The alkali metals contain an additional, valency, electron, and the principal quantum number of this electron is larger by unity than the highest principal quantum number normally occurring for the preceding rare gas; the subordinate number is unity, so that, in perihelion, the orbit approaches very close to the atomic nucleus. In the case of Al, Ga, In, and Tl, for which the optical fundamental term is a *p*-term, 3_2-, 4_2-, 5_2-, and 6_2-orbits respectively obviously first occur.

§ 122. According to Bohr's theory, the development in the periodic system of the outermost group of electrons is interrupted by the completion of inner electronic groups. The completion from 8 to 18 electrons takes place in the three-quantum group between Sc and Cu, in the four-quantum group between Y and Ag, and in the five-quantum group between La and Au. Further within the atom, the four-quantum group is completed from 18 to 32 electrons throughout the rare earths between Ce and Lu, so that the 72nd element, which succeeds Lu, cannot be a rare earth, but must be a homologue of zirconium, as has actually been confirmed by the discovery of hafnium. Elements belonging to series within which an inner group of electrons is being completed frequently exhibit, in consequence of the asymmetry of the inner atomic structure, the phenomena of paramagnetism, multiple valency, and the formation of coloured salts. A direct experimental confirmation of Bohr's theory of the periodic system is afforded by the course of the curves representing the relation between the square roots of the Röntgen-spectroscopic terms and the atomic numbers.

§ 123. The radiation emitted by a body whose atoms have been excited by primary radiation is known as fluorescent radiation; it is termed resonant radiation when the secondary and primary wave-lengths are identical. Such is the case with the alkali metals when the primary wave-length is one of the doublet with the greatest wave-length in the principal series. Fluorescence in gases can be extinguished by increasing the pressure; observations made in this connection have shown that the value of the mean duration of an excited atomic state is about 10^{-8} sec. The phenomena observed of sensitized fluorescence can be referred to a transference of excitational energy by molecular collisions.

§ 124. The method of electronic collisions enables the excitation and the ionization potential to be accurately determined for many elements. The values thus experimentally obtained agree perfectly with those calculated from optical spectra on the basis of Bohr's theory.

§ 125. An excited atomic state, from which a spontaneous return to the normal state appears impossible, in view of the selection principle, is termed metastable. It has been possible to produce a metastable state of helium atoms with the help of electronic collisions.

§ 126. Magnetism in bodies is to be referred to the magnetic moments associated with the individual electronic orbits; these moments may, however, be subject to a complete or partial mutual compensation within the atoms or molecules. The reality of molecular currents has been demonstrated by Einstein and de Haas's experiment.

§ 127. Whereas the formation of positive ions is presumably due to a separation of loosely bound electrons, the formation of negative ions may be ascribed to an affinity for electrons peculiar to some elements. The combination of oppositely charged ions gives rise to heteropolar molecules, but homopolar combinations between neutral atoms and between whole molecules are evidently also possible. Through the absorption of light-quanta, molecules can be brought into a state in which, because of their increased energy, they are enabled to partake in chemical reactions not possible for them in their normal state; photochemical phenomena are based on this absorption. According to Einstein's law of photochemical equivalence, the number of chemically reacting molecules should be equal to the number of light-quanta absorbed. The converse of photochemical phenomena is represented by chemico-luminescence.

§ 128. The formation of crystals is to be regarded as a special case of the formation of molecules. The most important types of crystal lattices are represented by rock-salt, calcite, zinc blende, diamond, and graphite.

§ 129. Molecular spectra may be explained on the assumption that the internal energy of molecules consists not only of the internal energy of the combined atoms, but also of an oscillational energy of the atoms and of a rotational energy of the molecules, and that all three kinds of energy can assume only quite definite values determined by quantum relations. Changes in the rotational energy alone give rise to rotation spectra lying in the far infra-red. Variations of both the rotational and the oscillational energy produce the oscillatory rotation spectra observed in the near infra-red. Finally, simultaneous alterations of all three kinds of energy lead to the formation of optical band spectra.

§ 130. When a molecule is penetrated by a light-wave of a frequency which does not differ greatly from a characteristic frequency of the molecule, a kind of resonance ensues, and this is manifested in the phenomena of dispersion (both normal and so-called anomalous).

Part IV

THEORY OF HEAT

Chapter XVI

Statistics

§ 131. The consideration of individual distributions of state leads to the concept of statistical probability; by this is understood the number of complexions through which a definite distribution can be realized. If w_i be the fraction of individuals in the state characterized by the index i, and N be their total number, the probability logarithm is given by the relation

$$\log_e W = -N \sum_i w_i \log_e w_i .$$

It follows directly from the concept of statistical probability that the transition between two distributions in the direction of increasing statistical probability is more probable than that in the reverse direction. On the other hand, since there must be an upper limit for the statistical probability, every system in which the distribution of state is constantly changing must exhibit an apparent tendency to assume the most probable distribution and, when this has been attained, to persist in it.

§ 132. Let a system possess a definite amount of a certain quantity, this amount being equal to the sum of the component amounts associated with the separate individuals. Moreover, let the component amount associated with an individual in a definite individual state be completely determined by quantities which, at any moment, have the same value for all individuals in the system, and which are therefore called parameters of the system. A distribution of which the statistical probability represents an extreme value for those variations of the distribution in which the total amount, the number of individuals, and the parameters remain unaltered, is termed canonical. It may be shown that a canonical distribution is the most probable one, every system tending to assume it. Infinitesimal canonical variations occur without any change in the statistical probability, and thus appear to be equally probable in either direction; they are apparently reversible.

§ 133. If the quantity previously considered be identified with an amount of energy (E), two important formulæ result. If dQ be an infinitesimal amount of energy introduced into the system whose energy is E, whilst the work dA is simultaneously performed by the variation of its parameters, then

$$dQ = dE + dA \,.$$

Furthermore,

$$\frac{dQ}{\Theta} = d\,(\log_e W)\,,$$

where Θ is the modulus of distribution of the energy. Thus, on the one hand the difference between the energy introduced and the work performed, and on the other hand the quotient of the energy introduced and the modulus, represent the complete differentials of two functions which depend only on the instantaneous state of the system, being completely determined by the modulus and the parameters. In accordance with a previous statement, the second function has the property of tending towards a maximum, and of remaining constant only in reversible processes.

§ 134. The probability logarithm of a composite system is equal to the sum of the probability logarithms of the component systems. It represents a maximum when all the component systems are canonically distributed and have the same modulus, and, in that case, statistical equilibrium is said to exist. When the values of the modulus are different there is a tendency towards equalization, in such a way that energy is not spontaneously transferred from a system with a smaller to one with a larger value of the modulus.

§ 135. The negative partial derivatives of the free energy with respect to the parameters may be defined as statistical forces, and, together with the parameters and the modulus of distribution, they may be regarded as quantities of state of the system. It may be shown that r equations of state must exist between these $(2r + 1)$ quantities of state, so that, of the latter, any arbitrary $(r + 1)$ can be selected as independent variables of state. A system is defined as being statistically homogeneous when it has only a single parameter, which is formed by the quotient of the volume and the number of individuals. For such a system the pressure is equal to the negative partial derivative of the free energy with respect to the volume.

§ 136. In a consideration of cyclic processes consisting of two isomodular and two adiabatic component processes, the ratio between the work performed and the energy introduced during the first isomodular component process may be regarded as the efficiency of the cyclic process. Measurements of the efficiency enable an observer to prepare an exact scale of modulus values, even when the interior of the system is completely hidden from him and he can only follow the behaviour of the system from outside.

§ 137. If the individual energy be a pure quadratic function of some (s in number) of the individual quantities of state, and if the positional coordinates of the individuals (regarded as points) be among the quantities of state upon which the individual energy does not depend, then the mean value of the individual energy is given by

$$\bar{\epsilon} = s \frac{\Theta}{2}$$

(Theorem of Equipartition). When the system is also statistically homogeneous, the simple relation

$$pV = N\Theta$$

is obtained.

§ 138. According to Rayleigh and Jeans's Principle, in a system of stationary vibrations the density of the characteristic vibrations with frequencies lying in the interval between

ν and $(\nu + d\nu)$ is equal to $4\pi\,\nu^2/v^3$ (v being the velocity of propagation of the vibrations).

§ 139. A system of stationary vibrations is in statistical equilibrium when all the component systems of different frequencies have the same value of the modulus of distribution. In a quantized vibrating system, the energy density is proportional to the fourth power of the modulus of distribution, whilst the pressure is equal to one-third of the energy density.

Chapter XVII

Thermodynamics

§ 140. The energy distributed amongst the molecules of a body is interpreted in physics as the heat content of the body. In those cases in which energy apparently disappears whilst heat is evolved, this heat can be directly measured in energy-units. Moreover, by carrying out cyclic processes and measuring their efficiencies, an exact scale of values of heat moduli can be established. The usual temperature scale represents a conventional and arbitrary scale of such values.

§ 141. The establishment of a temperature scale further permits of the creation of an arbitrary unit of quantity of heat. The energy, measured in energy-units, which is equivalent to this arbitrary unit of heat, represents the mechanical equivalent of heat.

§ 142. The theorems derived in general statistics must still hold good if, in them, modulus of distribution be made proportional to the temperature, and heat be substituted for energy. In this way the two Principal Laws of Thermodynamics are obtained: I. If an infinitesimal quantity of heat be introduced into a system, while work is simultaneously performed, the difference between heat introduced and work performed represents the complete differential of a function, the internal energy, which depends only on the instantaneous state of the system and, when r parameters exist, is therefore determined by $(r + 1)$ independent variables of state. II. The quotient of the infinitesimal quantity of heat introduced and the temperature,

$$\frac{dQ}{T} = dS\,,$$

represents the complete differential of a second function, known as entropy, which likewise depends only on the instantaneous state of the system, and is therefore also completely determined by $(r + 1)$ independent variables of state. If the ratio between modulus of heat and temperature be denoted by the universal constant k, *i.e.*, if

$$\Theta = kT,$$

then the entropy

$$S = k \log_e W$$

(Boltzmann's entropy law). Entropy must possess the same properties as the probability logarithm. Entropy appears to tend to a maximum, and to remain constant only in reversible processes. Different bodies can only be in thermodynamic equilibrium when their temperatures are the same. When their temperatures differ, there is a tendency to equalization, in such a way that heat is not of itself transferred from a colder to a warmer body.

§ 143. The law of the increase of entropy is no law in the strict sense of the word, but only a rule which is far more likely to be obeyed than disobeyed. On account of the smallness of the molecules, noticeable exceptions are rare beyond any human power of imagination, and may hence be regarded as practically excluded.

§ 144. In isothermal processes, the work performed is equal to the decrease in free energy. Again, to processes in which no external work is performed Helmholtz's equation

$$E - F = -T\frac{dF}{dT}$$

applies. The concept of free energy (F) is closely connected with that of thermodynamic potential which, for a homogeneous system, assumes the form

$$\Phi = E - TS + pV.$$

§ 145. The coefficients of expansion, pressure, and compressibility, which characterize the changes of state of a system, are interconnected by an equation from which, if two of these quantities be known, the third can be calculated. Since both internal energy and entropy can be expressed either as functions of the temperature and the volume, or as functions of the temperature and the pressure, a series of theorems on the changes of state of a homogeneous system can be derived. For example, for an isothermal increase in volume, a body absorbs or emits heat according as its pressure coefficient at the temperature in question is positive or negative. If the expansion coefficient be positive, an adiabatic increase in pressure produces a heating effect, and an adiabatic increase in tension results in a cooling effect. If the expansion coefficient be negative, these effects are reversed.

§ 146. In Thomson and Joule's experiment, gas is allowed to flow adiabatically, under a pressure head, from one vessel into another. The experimental results permit of the reduction of the scale of a thermometer to the thermodynamic temperature scale.

§ 147. A system consisting of different phases of a substance in thermodynamic equilibrium represents a saturated complex of coexisting phases. Equilibrium is only possible when all the coexisting phases are at the same pressure and the same temperature, and when unit-masses of the phases have the same thermodynamic potential. Hence, when two phases coexist, there must be a functional relation between saturation pressure and saturation temperature. In a graphical representation in a p-T-plane, a saturation curve corresponds to this relation, and in the case of three coexisting phases, a so-called triple-point.

§ 148. In a system of molecules to which may be applied the hypotheses on which the Theorem of Equipartition is based, the equation of state assumes the simple form

$$p v = R T ,$$

where v is the specific volume referred to a gram-molecule, and R is the product of the constant k and the Loschmidt number. This equation represents the ideal gas law. The hypotheses leading to this equation are satisfied when the gas molecules are imagined as executing rapid translatory motions, the distances between them, however, being so large that any eventual forces acting between the molecules can be neglected, and the molecules treated as particles. It follows from the ideal gas law that, at constant temperature, pressure and volume are inversely proportional to each other (Boyle's Law); furthermore, that at the same pressure and temperature, equal volumes of different gases must contain the same number of molecules (Avogadro's Law). The ratio of the two specific heats is equal to 5/3 for monatomic gases, whilst for polyatomic gases it must lie between 1 and 5/3.

§ 149. By taking into account the forces of attraction between the molecules, and also the lower limit to the specific volume of a gas consequent on its molecular structure, van der Waals obtained the equation of state named after him:

$$\left(p + \frac{a}{v^2}\right) (v - b) = R T ,$$

where a and b are constants. Provided that T lies below a definite critical value, the isotherms represented by this equation have three real roots for v ($v_1 < v_2 < v_3$), for a given value of p. According to van der Waals's theory, for any specific volume smaller than v_1 the substance must be liquid, whilst for any specific volume larger than v_3 it must be gaseous ; for values between v_1 and v_3 in general, however, the isotherms are to be replaced by straight lines parallel to the v-axis, corresponding to the coexistence of two states of aggregation. Hence, above the critical temperature, liquefaction of the gas is evidently impossible.

§ 150. Modern theory refers the heat content of solid bodies to molecular oscillations, the energy of which, in the sense of the quantum hypothesis, is composed of energy-elements of magnitude $h\nu$. An approximate derivation leads to Dulong and Petit's Law, which has long been known, and according to which the specific heat referred to a gram-atom, or the so-called atomic heat, is equal to three times the gas-constant R, or to about six calories. The more modern theory, however, also affords an explanation for the deviations from Dulong and Petit's Law, which were discovered at normal temperatures in the case of diamond and are observed at low temperatures for all bodies.

§ 151. If, as Debye has suggested, solid bodies are to be regarded approximately as continua filled with stationary electrical waves, statistical theory leads to the conclusion that, at very low temperatures, the specific heat is proportional to the cube of the temperature (Debye's Law), whilst, at the absolute zero, both the temperature coefficient of the free energy and the entropy vanish (Nernst's Heat Theorem). Hence, at very low temperatures the expansion coefficient must also in general vanish.

§ 152. A system of stationary electromagnetic vibrations in thermodynamic equilibrium with any bodies is known as cavity radiation. The energy density of cavity radiation is proportional to the fourth power of the temperature, as is also the emissive power of a black body (Stefan's Law), because, as a geometrical consideration shows, the emissive power is equal to one-quarter of the energy density of the radiation, multiplied by the velocity of light.

§ 153. A spectral resolution of the emissive power leads to Planck's Law of Radiation, which gives the relation between the specific emissive power and the colour and temperature. The specific emissive power possesses a maximum for a certain wave-length, the displacement of which with the temperature is determined by the relation

$$\lambda_{max} T = b$$

(Wien's Displacement Law). It appears from the quantum theory that the universal constant b is equal to $c\, h/\beta\, k$, where β (the root of a transcendental equation) represents a pure number nearly equal to 5. From the constants of Stefan's and Wien's Laws Planck was able to calculate the first exact values for h and k, and thus also for the Loschmidt number and the elementary quantum of electricity.

Part V

THEORY OF RELATIVITY

Chapter XVIII

Special Theory of Relativity

§ 154. Classical physics appeared to demand the existence of only a single reference system in which light is propagated in all directions with the same velocity. The possibility was thus presented of detecting an absolute motion of the earth by a suitable optical experiment, such as was designed by Michelson. The actual execution of this experiment, however, showed that the expected effect definitely did not exist.

§ 155. The apparent contradiction between theoretical physics and Michelson's experiment was solved by Einstein, when he showed that the prejudice of an absolute time was untenable. Einstein's Principle of Relativity maintains that the statements of time, in terms of which a physical process is described, are relative and depend on the point of view of the observer using them, whilst they are governed by the postulate that, when measured in terms of them, light shall appear to the observer to travel with the same velocity in every direction.

§ 156. According to Minkowski, Einstein's Principle of Relativity may be geometrically interpreted by stating that the square of the product of the time and the velocity of light, taken with a negative sign, and the squares of the three spatial coordinates stand in the same relation to each other as do the squares of the four coordinates in a four-dimensional geometry. The four-dimensional manifold, which appears as a linking together of space and time by the coordinates

x, y, z, and $i\,c\,t$, is known as the Minkowski-world. The motion of a particle is represented by a world-line, the directional cosines of which with respect to a definite four-dimensional system of coordinates determine the velocity of the particle relatively to the reference system used.

§ 157. The transition from one spatial coordinate system to another which is moving uniformly in the common x-direction with a velocity v relatively to the first system appears, in a four-dimensional representation, as a rotation of the coordinate axes in the x-t plane ; the tangent of the angle of rotation is equal to the imaginary quotient of the velocity v and the velocity of light. From this and from the general formulæ for the transformation of vector components follow the equations expressing a Lorentz-transformation.

$$x' = \frac{x - v\,t}{\sqrt{1 - \frac{v^2}{c^2}}}, \quad y' = y, \quad z' = z, \quad t' = \frac{-\frac{v}{c^2}\,x + t}{\sqrt{1 - \frac{v^2}{c^2}}}.$$

§ 158. In consequence of the Principle of Relativity, the idea of simultaneity loses its absolute meaning. Two events which are apparently simultaneous to one observer do not appear simultaneous to another observer moving relatively to the first one. A moving body experiences a contraction in the direction of its motion, in the ratio $\sqrt{1 - v^2/c^2} : 1$. A " moving clock " goes more slowly than one at rest.

§ 159. According to the theory of relativity, the resultant velocity obtained by compounding two similarly directed velocities is given by

$$u = \frac{v + v'}{1 + \frac{v\,v'}{c^2}}.$$

It appears from this that velocities larger than that of light are impossible.

§ 160. It follows from the relativistic formula for the composition of velocities that light-waves partake in the velocity of motion (v) of a flowing liquid to an extent given by the fraction $\alpha\,v$, where α is equal to $(1 - 1/n^2)$, n being the refractive index of the liquid (Fizeau's experiment).

§ 161. The application of a Lorentz-transformation to the equation of a light-wave leads to the Doppler effect, which consists in the displacement of the spectral lines arising from a moving source of light towards the violet or the red, according as the source is approaching the observer or receding from him. In contrast to the older theory, the theory of relativity indicates a small displacement also in the case of a transverse motion. Furthermore, the aberration of light is explained by the Principle of Relativity.

§ 162. Relativistic dynamics shows that the momental mass (m) and the stationary mass (μ) are connected by the relation $m = \mu/\sqrt{1 - v^2/c^2}$. Force and acceleration are, in general, not similarly directed, but the direction of the acceleration includes a larger angle with the direction of motion than does the direction of the force.

§ 163. It follows from the theory of relativity that all energy possesses an inertial mass, which is equal to the quotient of the quantity of energy and the square of the velocity of light.

§ 164. In a four-dimensional manifold, skew-symmetrical tensors correspond to the axial vectors of three-dimensional geometry. The vector divergence of a skew-symmetrical tensor assumes a simple form, and the components of a skew-symmetrical tensor which results from the rotation of a world-vector are also interconnected by simple relations.

§ 165. In a four-dimensional representation, Maxwell's equations may be summarized in the statement that the four-current represents the vector divergence of the electromagnetic field tensor, which in its turn represents the rotation of a world-vector. The four-current can be resolved into convection current and charge, and the field tensor into electric and magnetic field-strengths.

§ 166. The product of mass density and the expression $(1 - v^2/c^2)$ represents a scalar. On multiplying the products (taken in pairs) of the components of the world-line-element representing the motion by this scalar, the components of the matter tensor are obtained. The space-time-components of this tensor represent the density of momentum, and its purely temporal component the density of matter.

Chapter XIX

Theory of Gravitation

§ 167. The general theory of relativity bases the description of the Minkowski-world, not on Euclidean, but on the more general Riemannian geometry, for it assumes that the Minkowski-world is " curved " in consequence of the space-time distribution of matter. The assertion that, in this curved world, the motion of a particle upon which no electromagnetic forces are acting is to be described by a geodetic world-line represents both the generalization of the Law of Inertia, and also the new Law of Gravitation. From this concept follows at once the theorem of the identity of gravitational and inertial mass. In the general theory of relativity, Cartesian are replaced by the more general Gaussian coordinate systems, and thus any transition between two reference systems moving quite arbitrarily with respect to each other may be represented by a four-dimensional coordinate transformation.

§ 168. In Riemannian geometry, the Theorem of Pythagoras is replaced by the generalized formula

$$ds^2 = \sum_h \sum_k g_{hk}\, dx_h\, dx_k .$$

For a rectangular, Cartesian, coordinate system, the g_{hk} are equal to unity or zero, according as h is or is not equal to k.

§ 169. In utilizing Riemannian geometry, a distinction must be drawn between covariant and contravariant vectors, which are transformed differently on a change of coordinate system. Quantities whose components are transformed in the same way as the products formed by multiplying in pairs (or triplets, *etc.*) the components of vectors are known as covariant, contravariant, or mixed tensors. The operation of contraction permits of the derivation from any mixed tensor of another tensor of a rank lower by two.

§ 170. The quantities g_{hk} considered in § 168 are the components of a covariant tensor of the second rank, which is termed the metric fundamental tensor. If the determinant of the covariant quantities g_{hk} be formed, the cofactors divided by the value of the determinant represent the components (g^{hk}) of the contravariant fundamental tensor.

§ 171. The equation of a geodesic in Riemannian geometry follows from its definition as the shortest line connecting two points. This equation contains Christoffel's three-indices symbols, which are formed of the components of the metric fundamental tensor and its partial derivatives with respect to the coordinates.

§ 172. Mathematical considerations based on the equation of a geodesic lead to a mixed tensor of the fourth rank, known as the Riemann tensor, which depends only on the three-indices symbols and their derivatives with respect to the coordinates. Contraction of this tensor yields a tensor of the second rank (the contracted Riemann tensor R_{hk}), and from this can be obtained a scalar

$$R = \sum_h \sum_k g^{hk} R_{hk},$$

which is called the Riemann invariant.

§ 173. If the Riemann invariant be integrated over an arbitrarily delimited region of a geometrical manifold, and the variation of this integral be then formed in such a way that the fundamental tensor is varied only within the boundaries of this region, this process leads to a tensor

$$S_{hk} = R_{hk} - \tfrac{1}{2} g_{hk} R.$$

In Einstein's field equations, this tensor is equated to the matter tensor, apart from a universal proportionality factor (f), so that

$$R_{hk} - \tfrac{1}{2} g_{hk} R = f T_{hk}.$$

In the Minkowski-world, whose curvature is described by these field equations, it is postulated that the motion of a particle shall be represented by a geodetic world-line. On the special assumptions that, first, the deviations from Euclidean geometry are small; secondly, that the mechanical velocities involved are small as compared with the velocity of light; and thirdly, that the gravitational field is static, Newton's Law of Gravitation may be derived as a first approximation. The gravitational constant (κ) is found to be equal to $f c^2/8\pi$.

§ 174. In a gravitational field produced by a stationary mass, all processes are apparently retarded in the ratio $1 : (1 + \alpha/r)$, where α denotes the gravitational radius of the mass responsible for the field, this radius being obtained on multiplying the mass by κ/c^2. Rays of light must be curved in a gravitational field, and also Kepler's Laws must assume a more complicated form than in the classical theory, owing to the necessary slow advance of the perihelion of a planetary orbit.

§ 175. Einstein's gravitational theory received apparent confirmation from three experimental facts: *viz.*, the perihelion motion of the planet Mercury; the deflection of rays of light coming from stars on passing close to the sun, which has been detected and measured in total solar eclipses; and, finally, the displacement towards the red of the spectral lines of stars.

§ 176. An extension of his Law of Gravitation has led Einstein to the assumption that space is spherically curved, and thus, despite its boundlessness, that it possesses a finite volume. Einstein has obtained the equation

$$M = \frac{\pi c^2 R}{2\kappa},$$

which connects the total mass of the universe (M) with the radius of curvature of space (R) and the gravitational constant.

SYNOPSIS OF THE SYMBOLS MOST FREQUENTLY USED

Part III.

ATOMIC THEORY

- A Atomic weight.
- E Energy.
- E ($\mathbf{E}$) Electric field-strength.
- H ($\mathbf{H}$) Magnetic field-strength.
- I Molecular moment of inertia.
- L Loschmidt number.
- M Mass of a proton.
- Q Electric charge.
- R Rydberg's constant.
- T Spectral term.
- T Half-value period.
- U Angular momentum.
- V Potential.
- k Boltzmann's constant.
- m Mass of an electron.
- m_H Mass of a hydrogen atom.
- n Principal quantum number.
- p Generalized momentum.
- q Generalized coordinate.
- s Quantum number.
- t Time.
- v ($\mathbf{v}$) Velocity.
- z Nuclear charge number.

- a Radius.
- a Lattice-constant.
- c Velocity of light.
- d Lattice - plane distance.
- e Elementary quantum of electricity.
- h Elementary quantum of action.
- i Inner quantum number.
- k Subordinate quantum number.
- α, β, γ Directional cosines.
- α Ratio of velocity in one-quantum orbit in hydrogen atom to c.
- γ Specific charge.
- θ Glancing angle.
- λ Wave-length.
- λ Disintegration constant.
- ν Frequency.
- ϱ Density.
- ω Revolution number.

Part IV

THEORY OF HEAT

A Work.
B (in § 132) Individually distributed quantity.
C Molecular or atomic heat.
E Energy.
F Free energy.
J Mechanical equivalent of heat.
K Emissive power.
L Loschmidt number.
N Number of individuals or molecules.
Q Quantity of energy or heat.
R Gas constant.
S Entropy.
T Absolute temperature.
U Statistical or thermal force.
V Volume.
W Statistical probability.
Z Number of characteristic vibrations.

---

a Stefan's constant (in § 149 the first of van der Waals's constants).
b Wien's constant (in § 149 the second of van der Waals's constants).
c_p Specific heat at constant pressure.
c_v Specific heat at constant volume.
e Specific energy.
h Elementary quantum of action.
k Boltzmann's constant.
m Individual or molecular mass.
n Number of gram-molecules.
p Pressure.
q Specific quantity of heat.
s Specific entropy (or number of degrees of freedom).
u Parameter.
v Specific volume, or velocity.
w_i Fraction of number of individuals in i^{th} state.

Θ Modulus of distribution.
Φ Thermodynamic Potential.
Ψ Free energy.

α Expansion coefficient (or the pure number 1·0823).
β Pressure coefficient (or component amount of quantity, or the pure number 4·9651).
γ Absorption coefficient.
ϵ Individual or molecular energy.

ζ Density of free energy.
η Energy density.
θ Thermometric temperature.
κ Ratio of two specific heats (in § 145 compressibility coefficient).
λ Wave-length (in § 147 specific heat of reaction).
ν Frequency.
φ Thermodynamic potential of unit-mass.
ψ Mean free energy of an individual.

Part V

THEORY OF RELATIVITY

A, B, C Arbitrary vector or tensor.
E (**E**) Electric field-strength.
F Electromagnetic field-tensor.
G Cofactor in determinant of metric fundamental tensor.
H (**H**) Magnetic field-strength.
K (**K**) Force.
R Riemann tensor or scalar (in § 176 radius of curvature of space).
S Curvature tensor.
T Matter tensor.

---

b (**b**) Acceleration.
c Velocity of light.
ds World-line-element.
g Metric fundamental tensor.
m Mass.
n Refractive index.
q Minkowski-velocity.
s Four-current.
t Time.
v Velocity.

α Gravitational radius (in § 160 dragging coefficient).
β abbreviation for $1/\sqrt{1 - v^2/c^2}$.
δ Mixed fundamental tensor.
κ Gravitational constant.
μ Proper mass.
ν Frequency.
ϱ Density of mass or charge.
τ Proper time.
$d\tau$ Volume-element.
φ, ψ Angle.

THE UNIVERSAL CONSTANTS OF PHYSICS

Constant.	Symbol.	Value.	Common logarithm.
Elementary quantum of electricity .	e	$4{\cdot}774 \times 10^{-10}$ gm.$^{\frac{1}{2}}$ cm.$^{\frac{3}{2}}$ sec.$^{-1}$ ($\pm$ 0·004)	$\overline{10}{\cdot}6789$
Mass of an electron	m	$9{\cdot}00 \times 10^{-28}$ gm. ($\pm$ 0·02)	$\overline{28}{\cdot}954$
Mass of a proton	M	$1{\cdot}662 \times 10^{-24}$ gm.	$\overline{24}{\cdot}2206$
Elementary quantum of action . .	h	$6{\cdot}545 \times 10^{-27}$ erg. sec. ($\pm$ 0·012)	$\overline{27}{\cdot}8159$
Velocity of light	c	$2{\cdot}9990 \times 10^{10}$ cm. sec.$^{-1}$	10·47698
Boltzmann's constant	k	$1{\cdot}372 \times 10^{-16}$ erg. deg.$^{-1}$	$\overline{16}{\cdot}1374$
Gravitational constant	κ	$6{\cdot}68 \times 10^{-8}$ gm.$^{-1}$ cm.^{3} sec.$^{-2}$	$\overline{8}{\cdot}825$
Loschmidt number	L	$6{\cdot}062 \times 10^{23}$	23·7826
Charge on a univalent gram-ion .	Le/c	$9{\cdot}6494 \times 10^{3}$ gm.$^{\frac{1}{2}}$ cm.$^{\frac{1}{2}}$	3·98450
	M/m	1846	3·2662
Rydberg's constant	R_∞	$3{\cdot}2910 \times 10^{15}$ sec.$^{-1}$	15·51733
	R_∞/c	109,737·1 cm.$^{-1}$	5·04035
Specific charge of an electron . .	e/m	$5{\cdot}304 \times 10^{17}$ gm.$^{-\frac{1}{2}}$ cm.$^{\frac{3}{2}}$ sec.$^{-1}$	17·7246
Specific charge of a proton . .	e/M	$2{\cdot}87 \times 10^{14}$ gm.$^{-\frac{1}{2}}$ cm.$^{\frac{3}{2}}$ sec.$^{-1}$	14·458
Ratio of velocity in one-quantum orbit in hydrogen atom to c	α	$7{\cdot}29 \times 10^{-3}$	$\overline{3}{\cdot}863$
Mechanical equivalent of heat . .	J	$4{\cdot}186 \times 10^{7}$ erg. cal.$^{-1}$	7·6218
Gas constant	R	$8{\cdot}313 \times 10^{7}$ erg. deg.$^{-1}$	7·9198
,, ,, (in thermal units) .	R	1·98 cal. deg.$^{-1}$	0·297
Stefan's constant (theoretically calculated)	a	$7{\cdot}65 \times 10^{-15}$ erg. cm.$^{-3}$ deg.$^{-4}$	$\overline{15}{\cdot}884$
Wien's constant	b	0·288 cm. deg.	$\overline{1}{\cdot}459$

ADDENDA

§ 107. In 1923 it was discovered by *Compton* that the generation of secondary, continuous, Röntgen radiation by a primary radiation is accompanied by an increase in the wave-length. A quantitative interpretation of the observed effect is afforded on the assumption that the radiation is propagated in light-quanta, which can collide with free, or approximately free, electrons in such a way that both the Law of the Conservation of Energy and the Law of the Conservation of the Total Momentum are satisfied. The liberation of energy from a light-quantum must thus lead to a diminution in its frequency. In Compton's experiments, the primary radiation emitted by a molybdenum anticathode was scattered by charcoal or paraffin wax; the loosely bound valency electrons of the carbon then behave as though they were free. The secondary spectrum appears displaced in the direction of longer wave-lengths with respect to the primary spectrum. In agreement with the experimental results, the theoretical relation expressing the (positive) displacement of the wave-length is given by

$$\Delta\lambda = 2\lambda_0 \sin^2 \frac{\theta}{2}.$$

In this expression θ is the angle of scattering, and $\lambda_0 = h/mc = 0{\cdot}024$ Å is the wave-length of a light-quantum, the mass of which (calculated according to the Law of the Inertial Mass of Energy) is equal to the mass of an electron.

§ 108. According to the most recent researches, the atomic weights of the undermentioned elements should be given the following values :—

41	Cb	93·1
67	Ho	163·4
70	Yb	173·6
72	Hf	180·8
76	Os	190·8

Changes affecting the second decimal place have not been included here.

Early in 1925 the two missing homologues of manganese,

Nos. 43 and 75, were discovered by W. Noddack and J. Tacke, working in Berlin. The respective names proposed for these two elements are masurium (Ms) and rhenium (Re).

§ 113. The correct value for the half-value period of RaE is 4·98 days.—It is probable that UZ arises by a β-transformation from UX_1, and that it undergoes a further β-transformation into UII.

§ 115. In more recent researches, Rutherford and Chadwick have been able to effect the disruption also of the nuclei of the following elements (hydrogen nuclei being expelled): Ne, Mg, Si, S, Cl, A, and K. Using other methods, the results of which partially contradict those of Rutherford, Kirsch and Pettersen, in Vienna, have found that a disruption, with the liberation of H-particles, can be effected likewise of the following elements: Be, C, O, Ti, V, Cr, Fe, Cu, Se, and Zr.

§ 117. The position of the absorption band heads in the continuous Röntgen spectrum depends on the valency state of the element. Thus, in the case of both chlorine and sulphur, three different heads were observed by Lindh, according as he used compounds of 1-, 5-, or 7-valent chlorine, or of 2-, 4-, or 6-valent sulphur.

§ 118. In 1924 Wentzel closely investigated the so-called "spark lines" in the Röntgen spectra which are produced when more than one electron is removed from the interior of an atom. *Cf.* Sommerfeld's "*Atomic Structure and Spectral Lines*," 4th edn., ch. iv., § 9.

§ 120. It has been shown that the fundamental term for titanium and vanadium is an *f*-term.

Using the so-called "hot-spark" method, *Millikan* and *Bowen* have obtained the ultra-violet spectra of atoms completely deprived of their valency electrons (so-called "stripped atoms"). They have observed the spark spectra resulting from the removal of one electron from the Li atom, of two electrons from the Be atom, of three from the B atom, of four from the C atom, and of five electrons from the N atom. Furthermore, they have examined the spark spectra produced when one, two, three, four, five, and six electrons are removed respectively from the atoms of Na, Mg, Al, Si, P, and S. In these spectra of the stripped atoms, the well-known doublet relations of Röntgen spectra hold good. Observation of these spectra has shown, moreover, that, after removal of the valency electron, two electrons still persist round the nucleus in the series Li-N, and ten electrons in the series Na-S.

Close investigations on the series of manganese and iron have led to a complete disentanglement of these extremely complicated spectra.

A new method, devised by *Ornstein* at Utrecht, has rendered possible exact quantitative measurements of the relative intensities of the lines in a complex structure. It has thus been found that simple integral ratios exist between the intensities. For example, the intensity ratio for the lines of a doublet is 2 : 1, whilst the ratio for the lines of a triplet is 5 : 3 : 1, *etc.* For further details *vide* Sommerfeld's "*Atomic Structure and Spectral Lines*," 4th edn., ch. viii, § 5.

§ 122. Since, as was mentioned above, the fundamental orbit for Ti (22) and V (23) is an orbit with the subordinate quantum number 4, Bohr's scheme of the elements, as reproduced in Table XVI, certainly requires modification. The difficulties in the theory of the periodic system have been partly overcome by Stoner (*Phil. Mag.*, October, 1924).

§ 125. Lyman [*Astrophys. Journ.*, **60** (1924), p. 1] has observed in the helium spectrum a weak line which arises through the combination of the fundamental term of parhelium with the $2p$-term of orthohelium. This represents an exception from the rule, mentioned in § 125, that no combinations between ortho- and para-terms are possible.

NAME INDEX TO VOLUMES I AND II

(Roman numerals refer to the volume, and Arabic numerals to the page)

SUBJECT INDEX TO VOLUME II

(The numbers refer to pages)

PRINTED IN GREAT BRITAIN BY THE WHITEFRIARS PRESS, LTD.,
LONDON AND TONBRIDGE.

Bei Fragen zur Produktsicherheit wenden Sie sich bitte an:
If you have any questions regarding product safety,
please contact:

Walter de Gruyter GmbH
Genthiner Straße 13
10785 Berlin
productsafety@degruyterbrill.com